M000287142

The Chemistry and Physics of Coatings

Second Edition

The Chemistry and Physics of Coatings
Second Edition

Edited by

Alastair Marrion
International Coatings, Akzo Nobel, Tyne and Wear, UK

RS•C

advancing the chemical sciences

ISBN 0-85404-604-6

A catalogue record for this book is available from the British Library

Published by The Royal Society of Chemistry,
Thomas Graham House, Science Park, Milton Road, Cambridge CB4 0WF, UK

Registered Charity Number 207890

For further information see our web site at www.rsc.org

Typeset by Alden Bookset, Northampton, UK
Printed by Athenaeum Press Ltd, Gateshead, Tyne & Wear, UK

Preface to the Second Edition

Some ten years on from the first edition, and in a new millennium, a second edition is needed to take full account of developments in the industry. In most areas progress has been relentless rather than seismic, but over a span of years much change is evident. The continuing evolution of the legislative environment in the European Union, the USA, and elsewhere has driven the growth of technologies with low emissions of volatile organic compounds. Meanwhile, economic and environmental factors have encouraged innovation in the area of higher performance and longer lasting coatings. We have tried to reflect those changes in the organisation and content of the book.

As in the first edition, we describe some of the science underlying our industry for an audience of chemists who might not have a background in polymers, or too great an interest in some of our more intricate details. Since the first edition saw some use as an introductory text for new entrants to the industry, we have extended our scope to include some topics previously considered inappropriate. Notably, new chapters on inorganic coatings, and on the difficult topic of coatings formulation, that some have described as "doing chemistry without chemical reaction", have been added.

The book attempts to deal with its subject under three headings: the chemistry, and the physics of coatings as its title suggests, and the related economic and environmental issues.

Much gratitude is due to the experts who have found time to make their contributions to the book despite all the other demands on their energies. I should also like to thank Akzo Nobel for permission to undertake the task, to use the Company's resources, and to fly close to its proprietary knowledge. The authors of Chapters 2 and 10 wish to acknowledge their indebtedness to David Wigglesworth and the late Ken Baxter, whose chapters in the first edition provided a solid foundation on which to build. As to my own sections, many colleagues at International Coatings have read, advised on, and improved them. Dr Mark Nesbit of the Centre for, Economic Botany, Royal Botanic Gardens, Kew kindly advised on botanical nomenclature, and Dr Gillian Watson of the Natural History Museum, Entomology Department, provided much appreciated information about the lac insect. The errors and omissions throughout the book nevertheless remain my own. Finally, I should like to thank my wife, Christina, for her forbearance in the face of my distraction throughout the gestation of this modest work.

                                                    Alastair Marrion

Contents

Contributors

C. Cameron, *Marine and Protective Coatings Technology Centre, International Coatings Ltd, Tyne & Wear, UK*

W.A.E. Dunk, *School of Chemical Engineering and Advanced Materials, University of Newcastle upon Tyne, Newcastle, UK*

A. Guy, *Yacht Division, International Coatings Ltd, Tyne Wear, UK*

A.R. Marrion, *Marine & Protective Coatings Technology Centre, International Coatings Ltd, Tyne & Wear, UK*

A. Milne, *Occam & Morton Consultants, Newcastle-upon-Tyne, UK*

A.B. Port, *CPFilms Inc, Martinsville, Virginia, USA*

P.A. Reynolds, *Bristol Colloid Centre, University of Bristol, Bristol, UK*

J. Warnon, *European Council of Paint, Printing Ink and Artists' Colours Industry, Brussels, Belgium*

CHAPTER 1

Economics and the Environment: The Role of Coatings

A. MILNE

1.1 INTRODUCTION

The principal concerns of our confused world are still the fundamental ones of food, shelter, and warmth, aggravated by an awareness of a growing human population, its economic and material well-being, and even survival, set against the finite resources of non-renewable fossil fuels and minerals, and the unequal access to them. World population is about 6.3 billion and growing at about 2% per annum, giving a doubling period of 35 years. Global energy production for 2005 is projected to be about 12600 million tonnes oil equivalent (TOE), an average of 2 t per person, very unevenly distributed.

The Times Atlas gives the energy distribution shown in Table 1.1, within which there are even greater extremes. Norwegians, for example, are amongst the 8 t consumers, but they also export a further 16 t per head, and import an equivalent value in consumer goods.

Table 1.1 *Distribution of per capita global energy consumption*

Population (%)	*Tonnes oil equivalent*
6.3	>8
21.3	2–8
64.3	0.2–2
17.8	<0.2

As a simple comparison, a car-owner doing 15,000 miles per annum at 30 miles gallon^{-1} (Imp) is using rather more than 0.5 t of crude oil on motoring alone.

1

Table 1.2   *Global paint market (estimated for 2002)*

Region	Production (10^6 t)	Population (10^6)	Per capita (kg year^{-1})
North America	7.1	421	16.86
Western Europe	6.1	391	15.60
Asia Pacific	6.9	3536	1.95
Rest of the World	4.0	1862	2.15

There is a close relationship between energy consumption, gross domestic product and standards of living. There is a remarkable distinction, for example, between the paint and coatings consumption of the developed and underdeveloped worlds (Table 1.2). It is difficult, however, to see how the aspirations of two-thirds of the doubled world population, consuming less than 2 TOE *per capita* can be satisfied, as well as satisfying the other demands of energy and resource conservation, acid rain reduction and global warming.

What has all this to do with coatings? A great deal. Somewhat serendipitously, but from historically very sound roots, the coatings industry finds itself an inconspicuous but significant contributor to the solution to some of the problems. We are able to say, with a degree of complacency, that we have always been in the business of conservation as well as decoration. We are not, however, allowed that degree of complacency. Seen from the outside as major users of heavy and toxic metals, volatile hydrocarbons, halogenated monomers and polymers, highly reactive chemicals and carcinogens, we are perceived as part of the problem rather than as part of the solution. Part of the purpose of this book will be to see how the industry can rise to the challenge of the changing priorities and demands being put upon it. Increasingly, environmental concern is followed by legislative constraints. The original concerns of the environmental movement, for example, were with the basic extractive industries, wood, water, coal, iron and agricultural products. More recently with oil, gas, uranium, and renewables and non-renewables of all kinds.

What follows in this book is an outline of some of the ways in which the industry is responding, and can respond, to the demands being put upon it. For the more fortunate 20% of the world population, for example, with a disposable energy budget of 5–8 TOE per head, it is not difficult to make alternative choices. There is an increasing preference for, and an increasing value being put upon, clean air, clean water, natural beauty – and an increasing willingness to pay for them. No doubt these would be preferences of the other 80% also!

The most economic solution to most problems, following directly from the Second Law of Thermodynamics, is the reduction and elimination of pollution at source. Indeed, if we can express our problem in concentration terms, we can get a first approximation to the cost, as in Equation (1.1):

$$-\Delta G = RT \ln C_1/C_0 \qquad (1.1)$$

where ΔG is the Gibbs free energy change, R is the Boltzmann constant, T is the absolute temperature and C_1 and C_0 are the starting and finishing concentrations, always remembering that in the real world, efficiencies of real reactions are rarely more than 20%. That gives us a molar cost for pollutant and effluent recovery. Hence, the emphasis on total elimination at source, and the emphasis in this book of strategies for avoidance of problems. Forethought would seem to be one of the few thermodynamically efficient processes! If the objective is energy conservation, then recovering dilute liquid and gaseous pollutants very quickly becomes self-defeating in energy terms.

In the following chapters therefore, the authors respond to demands to remove solvents, by formulating coatings on liquid oligomers, use little or no solvent, as powder coatings without solvent, use liquid only at elevated temperature for a few minutes, or in the form of aqueous latices or non-aqueous dispersions. There is an emphasis on doing more with less by radical improvements to the performance of coatings. Each new technique and material creates its own problems and opportunities. Highly reactive monomers and oligomers are more likely to be hazardous almost by definition. Water, from an environmental viewpoint, would seem to be an ideal dispersant in everything except boiling point and latent heat of evaporation, while under ambient conditions it brings its own problems at low temperature and high humidity.

1.2 FUNCTIONS OF COATINGS

Why coatings? Why not simply paints? Historically, there was no distinction. Leonardo da Vinci used essentially the same materials and methods of preparation and application as the house painters of his day, and the present professional association in the UK, 'The Oil and Colour Chemists Association', is only now seeking a title more in keeping with the present role of its members. When we make the distinction between paint and coatings, we imply a distinction between functions which are largely decorative and aesthetic in the former, and have a more serious purpose in the latter. Indeed, there are several serious purposes, and they would usually be expressed in specific terms. Some of the specific purposes are: the prevention of corrosion, either actively, by the inclusion of anticorrosive pigments, or passively by providing an adhesive and impermeable barrier; providing either slip or slip resistance; impact and abrasion resistance; resistance to contamination; or hygienic properties, such as fungal, bacterial or fouling resistance. Whatever the desired end property, what they have in common can in general be expressed in economic terms and calculated in terms of:

   (i) energy savings,
  (ii) reduced downtime,
 (iii) increased lifetime,
 (iv) capital savings,
  (v) materials substitution.

The distinction between paint and coatings, even in economic terms, is more apparent than real, since even aesthetic choices can be accommodated in standard economic theory in terms of willingness to pay for some benefit, or willingness to pay for the elimination of some undesired consequence, the 'hedonistic price principle'.

Easiest to calculate are the immediate and direct benefits, the producer surplus and the consumer surplus, since they are the everyday transactions of supplier and consumer. We can illustrate (i)–(iv) above for the case of anti-fouling use, for example, since this is one of the few areas where the benefits are readily calculated in energy and monetary terms, and since all anti-foulings (to date) have used biocides which come under specific legislation to which both suppliers and users have had to accommodate. Anti-fouling coatings are thus a paradigm for many of the constraints being imposed by increasingly stringent legislation, one of which is the possibility that when other and incompatible preferences are being expressed, the solution is sub-optimum in the very area for which the coating was designed, *i.e.* maximum energy conservation at least cost.

As coatings, anti-foulings also illustrate at least two of the less expected functions which can be built into coating materials. Permanence is usually regarded as a desideratum in a coating; self-polishing anti-foulings are designed to disappear, and smoothness, usually an aesthetic consideration in a coating, is here intimately connected to the turbulent flow over the surface, again with desirable energy-saving consequences. Marine fouling and hull roughness increase the drag on vessels, and that increase can be readily translated into fuel consumption. The world fleet of some 39,000 vessels in the 1990s burned 184 million tonnes of heavy fuel oil, at a cost of $100 (US) t^{-1}, or $18.4 billion. The base line, the cost of doing nothing, would have meant that the whole fleet would burn 40% more fuel, an increase of 72 million tonnes. In practice since most vessels operated close to maximum power, the fleet would have to have increased by a similar amount. For comparison, the UK sector of the North Sea produced at maximum, just over 100 million tonnes of oil per annum.

The increase in the price of oil, in 1973, from about £5.00 t^{-1} to £50.00, and in 1979 to £100 t^{-1}, required a radical improvement in anti-fouling performance and in anti-fouling lifetime. The improvement took the form of the self-polishing tributyl tin copolymer anti-foulings. Improvements in fouling protection and in ship roughness save 4% of fuel, 7.2 million tonnes, valued at $720 million.

At the same time ship-owners demanded much longer lifetimes between drydocking. Traditionally vessels had docked on an annual basis. This was increased to 30 months and the current maximum is 5–7 years. The average improvement in 10 years (1976–1986) was from 21 to 28 months. This had a value of $820 million. Since average docking was for about 9 days, 2 days per annum extra trading was also available, worth $420 million. Capital savings are more difficult to calculate but a typical 200,000 DWT tanker would carry about 1 million tonnes per annum, so the above 7.2 million tonnes oil saving was itself worth seven tankers at $80 million each. The annual opportunity cost was some $500 million. So, for a modest improvement in anti-fouling performance, a total saving of about $2.5 billion was obtained. This is obtained also at no nett increase in price of anti-foulings. The global anti-fouling market was about 25 million litres, at $8 L^{-1}, so the $2.5 billion benefit was purchased for approximately $200 million, a benefit–cost ratio of 11.5. Both producers (the coatings manufacturers) and the consumers (the ship-owners and operators) were happy. In the language of cost–benefit analysis there was clearly a producer and consumer benefit. Resources were saved (oil and steel to make seven vessels), and those very environmental aspects on which such weight was placed, acid rain and the alleged 'greenhouse effect', benefitted to the extent of 560,000 t and 23 million tonnes, respectively. But the producer and consumer are no longer the only participants in the transaction. Tributyl tin is a 'Red List' pesticide, whose use on ships is now banned. It has measurable environmental effects both on economic crops (oysters) and non-economic species such as the dog-whelk, *Nucella lapillus*. No one, it is thought, has yet evaluated these environmental impacts, but the case serves to illustrate some of the ramifications of environmental risk–benefit analysis. For example, using the 'hedonistic price method', one would be required to put a value on the pleasure of finding dog-whelks on a particular stretch of coast, or to put a value on bio-diversity. Exactly how much pleasure, to how many people, for what period, is discussed in the literature. Alternative methods available are the 'contingent valuation method', which claims to be able to put a value on the existence of a resource whether one uses it or not, or might wish to in the future, and the 'replacement cost method', which attempts to evaluate the cost of recreating a damaged habitat or ecosystem. Consideration of human safety and health requires one to estimate a statistical value of life. This is commonly done in the transport field where the values of investment in accident prevention have to be calculated, but the methods apply equally in the fields of industrial accidents and the handling and use of hazardous materials.

This would seem to take us into territory remote from the design of coatings fit for a variety of purposes, but the process has been considered for one of the many raw materials in the coatings field, and will be so increasingly in other areas. Such considerations are clearly the preoccupations of those 20% of the population burning 8 t of oil equivalent; they are unlikely to feature strongly in the minds of those burning less than 2 t.

Item (v) in the above list of criteria, materials substitution, is by far the most important aspect of the 'added value' of coatings. There is an inverse relationship between the value and abundance of metals in the Earth's crust. Platinum, gold, silver, copper and tin are rare, precious and coinage metals, comprising no more than 71 g t^{-1} of the Earth's crust, while iron and aluminium at 5.8 and 8.8% of the Earth's crust, respectively, are by far the most abundant. Coated steel performs many of the functions which would otherwise require much more expensive metals and alloys, such as copper, bronze or stainless steel or much more expensive coating processes such as ceramic coating, electroplating with expensive metals such as nickel or chromium, vacuum metallising or plasma coating. Coatings therefore fulfil some of the ambitions of the alchemists, in transmuting some of the properties of base metals into gold. This is 'added value'. Global production of iron and steel is about 700 million tonnes per annum, or about 9 t per person per average lifetime of 70 years. Iron and aluminium occur in large deposits, in high concentration at or near the Earth's surface. The major component of the cost is the energy required to reduce them to the metallic state. They are thermodynamically unstable in our oxidising atmosphere, and tend to revert to the original oxides, as corrosion products. The materials thus substituted at great energy cost require to be conserved.

We are also highly uneven in our admiration of corrosion products. We admire the patina on copper and bronze, but have only a limited admiration for galvanised iron, and none at all for rust. There was an attempt some years ago amongst new-brutalist architects to get us to like rust, and the metallurgists complied by producing a uniformly rusting steel, Cor-Ten® steel; but we refused to love it. So one of our primary coatings objectives has been, since the Iron Age, the substitution of semi-precious metals by iron and steel. We can get some idea of how much this is worth to the users in the form of an approximate cost–benefit.

At the stage of iron ore, about 50% Fe, the cost of iron is about £13 t^{-1}. The energy requirement is 60–360 GJ t^{-1}, and at a cost of 2.2 p (kW h)$^{-1}$, or an oil price of £67 t^{-1}, the energy content of a tonne of steel is about £100, *i.e.* most of the cost is 'pure' energy. The added value of a coating is difficult to calculate, depending both on the gauge of the steel, the thickness of the coating and the end use. For some applications the uncoated steel would be of zero value even for the shortest period of time, *e.g.* a beer can uncoated inside or outside, but for a long-term use, a bridge or a ship say, calculation is possible.

An uncoated ship in seawater would suffer severe corrosion in 5 years or less. An adequately coated one might avoid serious repair for 15–25 years. A super-tanker might cost $80 million. If by coating the lifetime is tripled, then the saving is $160 million on 16,000 t of steel, or $10,000 t^{-1}. The coating enhances the value by 50 times. If the steel plate is 15 mm thick and protected on both sides with 250 μm of a polymer coating, then 120 kg (1 m^2) of fabricated steel, worth $1250, is protected for 25 years for about $2 of polymer, a benefit–cost ratio of 600 times.

In those areas where coated steel replaces, say, stainless steel, similar benefits are achievable. Indeed, in the area of chemical resistance, coated mild steel is suitable for most likely cargoes, with stainless steel used for only the most aggressive chemicals. In the case of both steel and aluminium, the energy required for remelting is a small fraction of that required for the original reduction. Coating, in both cases, ensures that almost the total original weight of metal is available for recycling. These two illustrations of direct energy saving and resource conservation represent the principal contributions of coatings to the community.

CHAPTER 2

Present and Future Coatings Legislation and the Drive to Compliance

JACQUES WARNON

2.1 ATMOSPHERIC POLLUTION AND THE COATINGS INDUSTRY

Volatile organic compounds (VOCs) are substances which contain carbon and which evaporate readily, though a few of these materials such as the oxides of carbon are not generally classed as VOCs. Most solvents used in paint, perfumes, adhesives, inks, aftershave, *etc.* are VOCs and other VOCs are present in various household chemicals, petrol, dry cleaning fluids, car exhaust fumes, cigarette and bonfire smoke, *etc.* Within Europe some 40% of all VOC emissions to the atmosphere arise from natural sources such as that invigorating smell of pine needles in some forests, emissions from animals and agriculture in general, *etc.* Some estimates of total world VOC emissions put the natural contribution at around 90%. Most of the adverse effects of VOC emissions arise because of the type of VOC, the local concentration, or the combination with other air pollution. Because of the concentration effect much of the difficulty caused by VOCs arises from anthropogenic releases.

Carbon dioxide, nitrogen oxides (NOX) and sulfur dioxide are not VOCs but are also emitted to the atmosphere and, either alone or in association with VOCs, contribute to a range of current or potential atmospheric pollution effects, the most well-known examples of which are reviewed below.

2.1.1 The Hole in the Ozone Layer

Some 15–30 miles above the Earth, in that level of the atmosphere called the stratosphere, there is a layer in which the concentration of ozone is

8

quite high. This occurs above the first low temperature trough of the atmosphere and so there is very limited mixing of gases back to the lower level of the atmosphere (the troposphere).

Some VOCs are long lived in the lower atmosphere and slowly migrate to the stratosphere, and once there can cause particular effects. One such group is the family of chlorofluorocarbons (CFCs). These are used as coolant liquids in refrigerators, deep freezers, air conditioners, *etc.* In the stratosphere these compounds are photolytically dissociated by incoming sunlight to form chlorine oxide and chlorine which reacts with the ozone to form oxygen and also regenerates a high proportion of the chlorine oxide. Thus the process continues and the ozone layer is depleted (Scheme 2.1).

$$CFCs \xrightarrow{h\nu} Cl_2$$

$$Cl_2 \xrightarrow{h\nu} Cl^\bullet + Cl^\bullet$$

$$2O_3 + 2Cl^\bullet \longrightarrow 3O_2 + Cl_2$$

$$O_2 \xrightarrow{h\nu} O^\bullet + O^\bullet$$

$$2Cl^\bullet + O_2 \longrightarrow 2ClO$$

$$ClO + O^\bullet \longrightarrow Cl^\bullet + O_2$$

Scheme 2.1

In some areas of the world, and particularly in the Antarctic, the level of depletion is quite marked and hence the expression 'hole in the ozone layer'. There are undoubtedly other effects at play as well which probably include the influence of 'solar winds' and of volcanic eruptions, and both polar stratospheric clouds and sulfur compound particles emitted by volcanoes can provide surfaces for the catalytic formation of active chlorine. Nonetheless, there is now wide acceptance that most of the chlorine comes from CFCs and that there is therefore a direct link between CFCs and stratospheric ozone depletion. The lack of ozone at this level increases the transmission of the more harmful parts of the sun's radiation and is expected to result in increased levels of skin cancer and eye cataracts. Most of the world's developed nations had agreed to phase out the production and use of the CFCs under the so-called 'Montreal Protocol'. Recent observations have shown that measures taken so far have been successful as the 'hole in the ozone layer' seems now to have become smaller than it has been in the past.

2.1.2 Summer Smogs

Meanwhile, at lower levels, in the troposphere, there is a different effect which is that of ozone creation. At this level VOC emissions react under the influence of sunlight and in the presence of nitrogen oxides to convert oxygen into ozone. This is one part of the process that generates summer hazes and smogs which are frequently seen and become quite severe in cities such as Los Angeles and Athens among many others. Essentially what is happening is that under the effect of strong sunlight VOCs photolytically dissociate and then react with lower oxides of nitrogen to yield aldehydes and higher oxides of nitrogen. Further sunlight subsequently causes the reformation of the lower oxides of nitrogen and an excess of ozone. The process only requires continuing sunshine and more VOC to continue on and on (Scheme 2.2). The rate of reaction coupled with wind speeds and potential delayed reaction due to low intensity of sunlight can mean that the ozone and the resulting smog can be formed at substantial distances from the original emission source.

$$O_3 + H_2O \xrightarrow{h\nu} 2\,^{\bullet}OH + O_2$$

$$^{\bullet}OH + RH \xrightarrow{O_2} RO_2^{\bullet} + H_2O$$

$$RO_2^{\bullet} + 2NO \xrightarrow{O_2} R^1CHO + 2NO_2 + {}^{\bullet}OH$$

$$NO_2 + O_2 \xrightarrow{h\nu} NO + O_3$$

Scheme 2.2

The resulting smog can cause eye irritation and breathing difficulties and is generally unpleasant. However, ozone is also toxic and in many areas of the world, summer time excursion levels are at or above the levels rated as harmful to crops and trees and at least irritating to animals, including us. The World Health Organisation guideline levels are 8-h average of 120 μg m^{-3} and yet higher levels are frequently recorded in the UK and Continental Europe. Many countries have legislated to reduce both VOC and nitrogen oxides emissions as part of the overall effort to reduce tropospheric ozone creation but the political issues and the chemistry are both very complex. Emissions in one state can give rise to pollution several days later in another country hundreds of miles away and with this as background The United Nations has sponsored a Protocol calling for a reduction of 30% in emissions of VOCs by the end of year 2010s. Most European countries and the USA and Canada have signed the Protocol. The accession of the European Community to the Protocol was approved by the European Council on 13 June 2003. Now the issue is how to achieve the necessary reduction in ways which are politically and economically acceptable. The problem remains and will get worse before it gets better. The stability of some VOCs and their relative decomposition rates under various reaction conditions greatly complicate the issues.

2.1.3   Other Health Effects

Apart from health effects of ozone creation, most VOCs have some toxic effects themselves and some are very toxic whilst others are carcinogens. This problem is mainly solved in most developed countries by careful controls within industrial plants and the reduced use of household coal fires and steam trains, as organic solvents are not amongst the worst offenders. These effects are classified as local issues by most nations. The VOCs from smoking tobacco continue to affect some.

2.1.4   The Greenhouse Effect

There have always been variations in the Earth's climate and substantial data show that this has been happening for centuries. However, there is growing concern that pollution may be causing a shift that will not be part of this natural variation and will not be naturally reversible.

The greatest level of concern is over the 'greenhouse effect' and its influence on global warming. World knowledge is insufficient to predict the extent to which this might happen or the total effect if it did. Nonetheless, the principle is well understood and supported and the potential problem is under close scrutiny.

The effect is that certain atmospheric gases, such as carbon dioxide, several VOCs, ozone, CFCs, and others, will transmit a high proportion of sunshine but transmit less of the long wave radiation emitted by the Earth (just like the effect of glass in a greenhouse). So there is a net increase in the temperature of the Earth. Of course this has been happening forever and some estimates are that the Earth would on average be 30 °C cooler if the 'greenhouse effect' did not exist.

However, the issue at stake is that the concentration of these gases in the atmosphere is increasing, and it is uncertain what the effect of this will be. In theory as the gas concentration increases so the energy retained by the Earth should rise and further global warming should take place. Hence, the efforts to reduce the emissions of these gases through controls, possible carbon taxes, *etc.* This becomes an issue for the paint industry partly due to the implied need to use extra energy if incineration is used to abate VOC emissions. A global agreement was reached in 2002 with the signature of the Kyoto Protocol where most of the developed countries (but not the US) have agreed to strongly reduce their emissions of greenhouse gases, mainly by reducing their use of combustible materials from fossil origin.

2.1.5   Acid Rain

Acid rain is mainly caused by sulfur dioxide and nitrogen oxides emissions returning to Earth as the respective derived acids. However, some VOCs, especially CFCs, can also contribute to this. A small proportion of emitted CFCs return to Earth as hydrochloric acid in rainfall.

The most significant effects of acid rain are damage to vegetation and to aquatic life in freshwater. Legislative focus is on reducing emissions of sulfur dioxide from power stations and nitrogen oxides from internal combustion engines and the degradation of man-made fertilisers. So these are the major concerns, they are complex issues and this is reflected in the resulting regulations.

2.2 LEGISLATIVE REACTIONS

Many legislators around the world have been tackling the issues raised by VOC emissions as part of their overall concern with air quality and whilst it is beyond the scope of this book to examine all of these actions, it is nonetheless necessary to review some regulations in order to understand the part that legislation and legislative proposals play in influencing the strategies of the coatings industry.

Through the 1970s and 1980s some major issues of air pollution have been increasingly recognised as world problems requiring global action to achieve significant benefit. This realisation has resulted in the United Nations taking an increased part in achieving Conventions which start to address these issues. Various agencies of the United Nations Organisation are involved, *e.g.* United Nations Environment Programme (UNEP), United Nations Economic Commission for Europe (UNECE), United Nations Conference on Environment and Development (UNCED), *etc.*

These bodies provide a forum for discussion and co-operation, make strong recommendations and publish detailed Convention statements and protocols. However, ultimately it is for national governments to introduce and enforce the implementing legislation, or not, as can be the case.

Possibly the best known action is that to protect the ozone layer. In response to concerns that had been growing for some years UNEP held a series of meetings that generated an 'Agreement on Substances which Deplete the Ozone Layer'. This was achieved in Montreal in September 1987 and became known as the Montreal Protocol. The Protocol was ultimately signed by over 60 governments who made commitments to reduce consumption of CFCs by 50% by 1999 and to freeze production of halons at 1992 levels. Subsequent meetings have been held to review depletion data and causative evidence which have encouraged a commitment to even stronger action. Further work will probably serve to accelerate even more both the scope and rate of these actions. Proposals already foresee a total ban on production of CFCs, halons, 1,1,1-trichloroethane and carbon tetrachloride and place the future of hydrofluorocarbons (HCFCs) formally on the agenda for subsequent meetings. All of this has impacted on the options that the paint industry has for fighting fires (halons), for formulating aerosols (CFCs), for manufacturing chlorinated rubber (carbon tetrachloride), and will probably have some bearing on the use of some low POCP (photochemical ozone creation potential) solvents (1,1,1-trichloro-ethane) and the formulation of paint strippers (methylene chloride).

Another area which has attracted United Nations attention has been tropospheric air quality. UNECE developed a forum to discuss air quality which resulted in 35 parties, mainly national governments, ratifying or accepting a Convention on Long-Range Transboundary Air Pollution which was adopted in Geneva on 13 November 1979.

This Convention has spawned four Protocols to date. The first three covered:

- Financing the Monitoring and Evaluation of Long-Range Transmissions of Air Pollutants – adopted 28 September 1984.
- Reducing Emissions of Sulfur – adopted 8 July 1985.
- Reducing Emissions of NOX – adopted 31 October 1988.

Whilst these are not directly associated with VOCs it is, as noted previously, well established that it is the combination of VOC and NOX concentrations in the atmosphere that are the major cause of ozone creation.

The fourth Protocol addresses the issue of VOC emissions directly. This is the 'Protocol to the 1979 Convention on Long-Range Transboundary Air Pollution concerning the Emissions of Volatile Organic Compounds or their Transboundary Fluxes' which was adopted 18 November 1991.

2.2.1 Gothenburg Protocol Adopted in December 1999

The VOC Protocol is not addressed directly to industry but relies on national legislation to give legal effect to its proposals. The Protocol allows for various compliance standards and most signatories have selected the option to reduce their VOC emissions by at least 30% by 1999 using 1988 emission levels as the base-line. Some countries which felt that they had already made significant reductions before 1988 negotiated for an earlier base year, *e.g.* the USA has committed to the minimum 30% reduction by 1999 using a base-line of 1984, whilst some whose emissions were still increasing in 1988 chose 1990 as their base-line, *e.g.* Italy. Some countries which have relatively weak economies have committed only to freeze their emissions, *e.g.* Greece will freeze at 1988 levels. Canada and Norway have each chosen the option of declaring only part of their country as a Tropospheric Ozone Management Area (TOMA) and their commitment to emission reduction applies only within the TOMAs. All of these options are valid compliance methods but of course focus only on the desired result not on the means of achievement nor even on the share of the burden to be borne by different VOC emitters. The Protocol itself identifies four control options:

- substitution of VOCs by the use of alternative products which are low in or do not contain VOCs;
- reduction of VOC emissions by best management practices in housekeeping, maintenance or processing;
- recycling and/or recovery for re-use of VOCs;
- destruction of potentially emitted VOCs by control techniques such as thermal or catalytic incineration or biological treatment.

The UNECE VOC Protocol classifies VOCs into three groups based on the degree to which they contribute to the formation of episodic ozone, with the aromatic hydrocarbons, which make up a high proportion of the solvents used by the paint industry, being in the most damaging category. The Protocol also provides some numerical quantification of this importance by introducing the concept of POCP and provides detailed tables of POCP values according to various test regimes. Broadly, these reflect the reactivity of particular solvents within the relevant chemical mechanisms. Unfortunately, many of the less 'reactive' solvents give rise to other problems, *e.g.* benzene and chlorinated hydrocarbons are both classified as 'least important' yet benzene is a carcinogen and chlorinated hydrocarbons can contribute to the depletion of the ozone layer.

Maybe the greatest contribution that the Protocol will make to these complex issues is that it requires the publication of a proposed compliance strategy by each signatory. Thus, substantial international attention will be focused on these issues in the future and the proposals are likely to generate new or more focused approaches.

2.2.2 The European Union

The European Community has first adopted a directive on Integrated Pollution Prevention and Control (IPPC) which is likely to become the major framework directive on industrial environmental releases with directives on such issues as the reduction of solvent emissions being effectively daughter directives to the IPPC directive. The European Commission is preparing Best Available Technique documents covering all solvent-using sectors covered by the IPPC Directive including paint and printing ink application but excluding paint manufacture.

In 1999 the European Commission adopted a directive to address solvent emissions and this directive in part derived from their commitments under the UNECE Protocol: the Solvent Emission Directive 1999/13/EC. The directive has the structure of a series of articles, which establish the general direction, coupled with annexes which define methodology for Solvent Management Plans, provide guidance on the selection of solvents, and set specific requirements on emission limit values for each of many individual sectors of solvent use. These currently include some sectors not associated with the paint industry but do specifically include paint manufacturing processes and the usage of paint by several customer groups such as car manufacturers and the industrial coating of metal, wood and plastics. Compliance may also be demonstrated by VOC reduction at the source, by using low VOC technology as high solid or water-borne technologies.

This directive deals only with emissions from contained processes and a second directive will follow and will address the issues arising from non-contained processes, *e.g.* the use of decorative paint.

In 2001 the European Union adopted two major Directives aiming at further improvement of the air quality in Europe: the National Emission Ceilings Directive and the Ozone Directive.

2.2.2.1 National Emission Ceilings. The aim of Directive 2001/81/EC of 23 October 2001 is to limit emissions of acidifying and eutrophying pollutants and ozone precursors in order to improve the protection in the European Community of the environment and human health against risks of adverse effects from acidification, soil eutrophication and ground-level ozone and to move towards the long-term objectives of not exceeding critical levels and loads and of effective protection of all people against recognised health risks from air pollution by establishing national emission ceilings, taking the years 2010 and 2020 as benchmarks.

National emission ceilings for VOC to be attained by 2010 are listed in Table 2.1. Figures have also been fixed for the other pollutants: sulfur dioxide, nitrogen oxides and ammonia.

Table 2.1 *National emission ceilings for VOC*

Country	AT	BE	DK	FI	FR	DE	GR	IE	IT	LU	NL	PT	ES	SE	GB	EC15
VOC (kT)	159	139	85	130	1050	995	261	55	1159	9	185	180	662	241	1200	6510

2.2.2.2 Ozone in Ambient Air. The purpose of Directive 2002/3/EC of 12 February 2002 is to establish long-term objectives, target values, an alert threshold and an information threshold for concentrations of ozone in ambient air in the European Community, designed to avoid, prevent or reduce harmful effects on human health and the environment as a whole. The long-term objective for the protection of human health is to not exceed 120 μg m^{-3}, calculated as a maximum daily 8-h mean within a calendar year. As target value for 2010 Member States should not exceed 120 μg m^{-3} on more than 25 days per calendar year averaged over 3 years. Similarly, long-term objectives and target values for the protection of vegetation have been fixed.

Member States have also to take appropriate steps to ensure that up-to-date information on concentrations of ozone in ambient air is routinely made available to the public. This should happen as soon as information threshold of 180 μg m^{-3} has been reached; an alert threshold has been fixed at 240 μg m^{-3}. The amount by which the threshold is exceeded is to be calculated as 1-h average to be measured or predicted for three consecutive hours.

2.2.2.3 Future Developments. Immediately after the publication of the Solvent Emissions Directive concerning industrial installations in 1999, the European Commission started working on a product-related approach to reducing VOC emissions. Decorative coatings and vehicle refinishes were among the products under scrutiny. In December 2002, the Commission published the proposal for a 'Directive on the limitation of emissions of VOCs due to the use of organic solvents in decorative paints and varnishes and vehicle refinishing products', which is fairly close to the proposals of CEPE representing the European coating industry.

2.2.2.4 Decorative Coatings. The coatings industry has actively supported the development of a harmonised product Directive for decorative coatings. This was the only way to create a level playing field for this industry and prevent competitive distortions within the European Union. Decorative coatings were not affected by the SED. In 1999, the European Commission appointed three different consultants to study the potential for reducing VOC emissions from decorative paints. The 'Decopaint' study was completed and published 2 years later at the end of 2000. At a workshop held then by the European Commission to discuss the results of the study with all stakeholders, CEPE presented a proposal for a VOC reduction programme based on careful consideration of economic viability, product quality, as well as customer acceptance.

Throughout 2001, the CEPE POG Decorative Paints carried out a campaign to gain support for our industry's position. CEPE issued press releases to trade magazines and daily papers throughout Europe, published technical articles in trade magazines in the UK, France and Italy, and produced more than 80,000 copies of a brochure that was translated into six languages. It was circulated in nearly all the European countries to national and local authorities and customers of our industry. This action programme helped CEPE win the support of other industrial sectors and organisations.

The draft of the European Commission was issued in December 2002. As a result of this energetic advocacy acceptance of about 80% of CEPE's proposals was achieved. Nevertheless, several technical and economic points require clarification and/or modification. CEPE will continue to advocate the coating industry's position at European Parliament and Council of Ministers level.

2.2.2.5 Vehicle Refinishes. The story of vehicle refinishes is slightly different, because in contrast with decorative coatings, vehicle refinishes are affected by the SED. Though the SED does not set out specifications for the product itself, it regulates VOC emissions from the activity of vehicle refinishing.

Back in 1999, in response to the SED, CEPE published Technical Guidelines setting out Best Available Techniques for the sector. These Guidelines were published in six languages, and were used to brief national authorities, and to support the formation of realistic regulations. CEPE's Guidelines were product based and were in line with the SED.

In fact, CEPE had already argued in the early 1990s, when the public consultation on the Directive started, that the rules for refinishing would be better cast explicitly in terms of product limits. Back then, this product-based approach was rejected by the Commission in favour of a Directive applicable also to other industrial sectors.

After vehicle refinishing was covered by the SED, CEPE advocated that further regulation of this sector was *not* necessary. But the Commission included vehicle refinishes in the new draft Product Directive and, in principle, CEPE supports this as it corresponds to the approach CEPE had been advocating ever since discussions started.

Nevertheless, also in the part on vehicle refinishes, a few amendments and modifications of the draft Product Directive are necessary. A major problem is that the new Directive proposal splits the vehicle refinishing sector into car and truck repair to be controlled under the new Directive, and truck/trailer painting not on a production line to remain under the SED. This is arbitrary and illogical as both activities are usually carried out by the same body shops and with the same equipment. The split of the sector has also led to various logical errors in the text of the proposed Directive and the SED. As an easy solution, CEPE advocates to remove all vehicle refinishing from the scope of the SED and to place them in the scope of the new proposed Directive.

2.2.2.6 Next Steps. The Commission proposal is being discussed in the European Council and in the European Parliament for the first reading. It is expected to be enforced by the end of 2003 or early 2004.

For decorative paints and varnishes it will impose strict VOC limits for 12 product subcategories split in specific limits for solvent-borne and for water-borne products and in two phases scheduled in 2007 and 2010.

For vehicle refinishing products same VOC limits will be defined for solvent-borne and water-borne products, which will get different specifications for the seven subcategories of products covered by the new Directive.

It is expected that the European Commission will include other product categories in product directive covering, *e.g.* adhesives, cosmetics, cleaning products or protective coatings.

2.3 PAINT INDUSTRY RESPONSE AND STRATEGY

Paint is used by so many different market segments, each with unique facilities, challenges and requirements that to prevent emissions from all the paint manufacturing processes requires the availability of a wide range of control options. Without any visible consultation amongst themselves, the paint industry negotiators in most countries have lobbied their legislatures to avoid specifying a single compliance route. The Industry's strategy is that reduced emissions are desirable but it wishes to see regulations framed around air quality standards or emission limit values expressed as mass of total emissions. This approach leaves industry free to select the control option which achieves the legislator's desire and through this route the best control option will be developed by competitive pressure.

Unfortunately, some legislators continue to drive for emission limit values based only on concentration of VOC in the exhaust gases. This can lead to approval for high mass of emission from processes which require high air flows whilst penalising processes with low mass emissions, which need only very low air flows. The concentration approach also ignores the great benefit that results to the atmospheric emissions when low VOC paints are used. No doubt the debates and lobbying will result in improved legislative focus in due course.

The UNECE VOC Protocol lists four different control options to reduce atmospheric emission and the paint industry has selected from all four options depending on the needs of particular industry segments:

(i) *Substitution of VOCs*

Great progress has been made in formulating alternative coatings which give very high standards of performance whilst containing low levels of VOC or in some instances no VOC. Research in the industry has generated several alternative approaches to low VOC coatings, some of which are particularly useful for the needs of some sectors yet quite inappropriate for others (*cf.* Chapters 7 and 8).

Radiation cured coatings contain reactive diluents which reduce the application viscosity but which are reacted into the polymer structure during the curing process thus substantially reducing the emissions to atmosphere resulting from the use of conventional solvents in 'conventional' paint systems. However, radiation cured systems require fixed energy sources and cure rates vary according to the distance between the energy sources and the substrate. Thus, whilst these systems can result in reduced VOC emissions, their use is largely limited to the coating of sheet or flat web substrates in industrial processes, and is certainly not appropriate for decorative air-drying coatings. However, new developments are promising going in the direction of higher performance radiation curing coatings including possible use in automotive OEM coatings. For most sectors of coatings usage it is possible to formulate systems of higher solids content and this is being done. It appears likely that for most sectors it will not be possible through this route to meet the regulators demands for reduced emissions in full whilst maintaining application properties and film performance character-istics. Nonetheless, in some sectors substantial progress will be made and others, for example, for the *in situ* coating of bridges in cold, damp climates, the use of high solids systems may be the compliant technology which comes closest to meeting the overall practical needs of society.

The use of powder coatings has been growing at around 2% per annum over the last few years in Europe, and whilst in many areas of use powder coatings can offer significant cost advantages and/or film performance benefits, it is clear that much of this growth has also been stimulated by concern over solvent emissions. Powder coatings are the ideal compliant coatings for many industrial processes especially the coating of metal and heat resistant plastics and boards and their rapid growth rate reflect this exceptional capability. The need to stove these materials at relatively high temperatures prevents their use in air-drying markets, and limits their use on temperature-sensitive sub-strates such as wood and plastic films. The relatively high film builds that are necessary for good film coalescence have limited their use in can coatings. However, more recent improvements in appearance, durability and colour styling capabilities have allowed increasing use as car body coatings. Some automotive production lines, *e.g.* in Germany, include use of powder clear coats and powders start also to be used in primers, sometimes as a powder slurry dispersed into water. Another promising development includes the combination of radiation curing and powder coating, allowing low temperature cure and better film properties including a smoother appearance.

Water-borne coatings are another potentially compliant technology that has grown in market share quite markedly in the last 2–3 decades and again much, but not all, of this growth has been driven by concern over solvent emissions. Some major areas of use include electrocoat technology for automotive primers and industrial uses, water-borne spray technology for coatings for two-piece cans, automotive spray surfacers and colour coats and a wide range of decorative paints. It is, though, still difficult to find a truly full gloss high durability air-drying coating and difficult to foresee a means of using water-borne coatings to repaint the exterior of ships dry docking during winter in North Europe or Korea.

Whatever the original drivers it is obvious that without these developments in coatings technologies, the emissions of VOCs to atmosphere would have been much greater.

The remaining three options listed in the UNECE VOC Protocol can also apply to the manufacture and/or use of paint:

(ii) *Reduction by best management practices*

Techniques in this area include a lot of items of good sense which can be justified on economic as well as on environmental grounds.

At their most formal they can amount to a fully speciated and quantified mass balance within a solvent management plan, with clear objectives for reducing both fugitive and contained emissions. At their most simple they focus on replacing the lid on a paint tin when it is not in use. It is really quite surprising how solvent emissions can be reduced by always putting lids on portable mixing tanks in paint factories and closing lids in paint thinning tanks at coater's factories. The design of storage tanks so that head-space fumes can be returned to road delivery tankers and rigorous preventive maintenance on all flange joints and pumps can also make substantial reductions in VOC losses. CEPE has published a guide describing the Best Available Technique (BAT) to reduce VOC emissions during coating manufacture.

(iii) *Recycling and/or recovery for re-use*

Most paint manufacturers and paint users have made progress in reducing the volume of solvent used for cleaning purposes under their application of best management practices and some will collect the remaining usage for redistillation and re-use.

There are also many users of paint in contained processes who collect the solvent evaporated during the drying stage by adsorption onto an activated carbon bed or by absorption in some form of proprietary oil for subsequent recovery and re-use. Due to the nature of the VOCs used in paint manufacture, condensation techniques are not widely used, though they do find wide acceptance as techniques in other industries.

(iv) *Control by post-use destruction techniques*

These techniques can usually only be applied to the manufacture of coatings or their use in contained conditions.

Biological treatment of VOC loaded air streams can be a very effective system for compliance with emission legislation especially when the VOC analysis is of constant composition and where the discharge is both continuous and of relatively low concentration.

However, in paint manufacture and in paint usage these ideals are met only rarely and under these circumstances the most commonly used control technique is incineration. This can take many different forms. Catalytic incineration allows combustion to take place at lower temperatures whilst straight thermal combustion involves the lowest capital cost. Regenerative and recuperative incinerators can each show benefits in capital cost/running cost compromises depending on plant sizes and the emissions challenge. It is not unusual for a recovery system such as carbon bed adsorption to be used as a first stage technique and then for the desorbed gases to be incinerated as this often reduces the need for supplemental fuel in the incinerators.

2.4 THE NEW EU CHEMICALS STRATEGY

This is the key issue for the coating industry because the final legislation resulting from the discussions with the European institutions, national governments and all parties concerned will determine the future of our industry in a very decisive way.

In February 2001 the European Commission published a White Paper on the 'Strategy for a Future Chemicals Policy' setting out its proposals for reforming the current rules governing the chemicals sector. In the White Paper the Commission identified the objectives that must be met in order to achieve sustainable development in the chemicals industry within the framework of the single market. One of its major objectives was the creation of a single regulatory system for all substances, entitled '*REACH*' for: *R*egistration, *E*valuation and *A*uthorisation of *Ch*emicals. It is giving industry full responsibility for generating data on the inherent properties of substances and for assessing the risks of their use. The REACH system will be managed by a new European Chemicals Agency and will replace four legal instruments in the European Community:

- Directive 67/548/EEC relating to the classification, packaging and labelling of dangerous substances.
- Directive 88/379/EEC relating to the classification, packaging and labelling of dangerous preparations.
- Council Regulation (EEC) 793/93 on evaluation and control of risks of existing substances.
- Directive 76/769/EEC relating to restrictions on the marketing and use of certain dangerous substances and preparations.

In a nutshell, REACH consists of:

- Registration requires industry to obtain relevant information on their substances and to use these data to manage them safely. There will be a general obligation to register substances manufactured or imported in the European Community in quantities of more than 1 tonne. Certain substances that are adequately regulated under other legislation or that present such low risks, *e.g.* polymers of low concern, will not require registration. A limited form of registration is also required for certain isolated intermediates.
- Evaluation of substances used at volume above 100 tonnes per year, to provide confidence that industry is meeting its obligations and to prevent unnecessary animal testing.
- The risks associated with uses of substances with properties of very high concern will be reviewed and, if they are adequately controlled or if the socio-economic benefits outweigh the risks, then the uses will be granted an authorisation. Substances of very high concern are defined as: substances that are category 1 and 2 carcinogens or mutagens; substances that are toxic to the reproductive system category 1 and 2; substances that are persistent, bio-accumulative and toxic or very persistent and very bio-accumulative; and substances demonstrated to be of equivalent concern, such as endocrine disruptors.
- The Restrictions procedure provides a safety net to manage risks that have not been adequately addressed by another part of the REACH system. Any substance on its own, in a preparation or in an article may be subject to Community-wide restrictions if a risk needs to be addressed.

The Agency will manage the technical and administrative aspects of the REACH system at Community level.

It is expected that the Commission will have agreed on the text during the fourth quarter of 2003. Then the proposed regulation will be passed on to the European Parliament and the Council. This procedure will take a couple of years at least, so that its enforcement is not expected before 2006.

The chemical industry and its downstream users, including the coating manufacturers, have raised various economic, technical and regulatory issues which have to be considered very carefully by the European authorities to make the new policy work.

A major concern is the cost of the REACH system. The registration will be a very costly procedure. Just to give you some figures: The estimated cost for testing one small volume chemical is 100,000–150,000 Euros. For a high volume chemical, the cost lies between 1.1 million and 1.8 million Euros. There are about 30,000 substances which will have to be tested – or else, will be taken off the market.

Our most serious concern is that substances may disappear from the market for cost reasons, and not because they are hazardous. CEPE estimates that up to 50% of the 10,000 chemicals currently used by the paint and printing ink industry may not be produced any longer if the system is not organised in a sensible way.

CEPE has made it absolutely clear to the legislators, including the European Commission, the European Parliament and the national ministries, that losing a high number of chemicals contradicts all stated objectives of the White Paper:

- The availability of a large number of raw materials is the basis for innovation in our industry. If the number of substances is reduced significantly, innovation will be hampered, not stimulated.
- Innovation leads to improved technical and environmental performance. Losing so many available chemicals will result in inferior technical product performance and a narrower product range. Coatings are applied to protect valuable resources. Inferior product performance will have negative effects on the environment.
- Excessive testing costs would lead to higher prices for raw materials and consequently for consumer products. A steep rise in prices would threaten the competitiveness of the European industry.
- The intellectual property of companies has to be protected. This issue comes up in the context of testing and a possible disclosure of product formulations. Particularly small and medium-sized enterprises active in niche markets would be threatened and could disappear from the market if they had to disclose confidential know-how.
- Companies which carry out risk assessments must be protected from so-called free-riders, meaning companies which benefit from the work of others without sharing the financial burden.

- It must be ensured that the European industry will not suffer a competitive disadvantage on a global level. Measures must be taken to prohibit the import of articles which do not fulfil the EU standards.

To reach these goals, CEPE calls for a proportionate approach and pragmatic timetabling. The European coating industry proposes that:

- The focus should be on the most critical substances and their major uses.
- Substances which are known to be harmless should be exempted from exposure tests. The same applies to substances used only in production and not sold to end consumers.
- The costly authorisation procedure should be limited to the most dangerous substances, which are POPs and CMR substances category 1 and 2.
- Risk assessment should be simple and targeted.
- Animal testing should be minimised by using and sharing existing data.
- The disclosure of formulation details should be proportionate in order to protect intellectual property rights.
- Adequate measures must be taken to protect companies from free-riders.
- Adequate legislative measures must be taken to ensure that our industry can operate on a level playing field, particularly with regard to imported articles.

2.5 CONCLUSION

With so many different options available to the industry and with different balances and compromises being of relevance in each sector, the VOC issue is quite complex, and is addressed in part by the regulators and in part by industry. The regulators are increasingly focusing on the definition of the environmental quality standards to be achieved whilst industry is developing the best means of achieving those standards.

The new EU Chemicals Strategy still under development aiming at protecting both human health and the environment will become a major challenge for the chemical manufacturers and the users of chemicals, including the coating manufacturers.

2.6 BIBLIOGRAPHY

1. CEPE, *Best Available Techniques for Coatings Manufacture*: http://www.cepe.org.
2. CEPE, *A Guide to VOC Reduction in Decorative Coatings*: http://www.cepe.org.
3. CEPE, *Technology Guidelines for Vehicle Refinishes*: http://www.cepe. org.

4. Council Decision 2003/507/EC on the accession of the European Community, to the Protocol to the 1979 Convention on Long-Range Transboundary Air Pollution to Abate Acidification, Eutrophication and Ground-Level Ozone, *Official Journal of the European Communities*, L 179/1, 17-7-2003.
5. Council Directive on the limitation of emissions of volatile organic compounds due to the use of organic solvents in certain activities and installations, *Official Journal of the European Communities*, L 85/1, 29-3-1999.
6. Directive 2001/81/EC on national emission ceilings for certain atmospheric pollutants, *Official Journal of the European Communities*, L 309/22, 27-11-2001.
7. Directive 2002/3/EC relating to ozone in ambient air, *Official Journal of the European Communities*, L 67/14, 9-3-2002.
8. Proposal for a Directive COM (2002) 750 final on the limitation of emissions of volatile organic compounds due to the use of organic solvents in decorative paints and varnishes and refinishing products and amending Directive, 1999/13/EC, 23-12-2002.
9. White Paper COM (2001) 88 final on the Strategy for a future Chemicals Policy, 16-2-2001.

CHAPTER 3

The Rheology of Coatings

P.A. REYNOLDS

3.1 INTRODUCTION

If a film is to function effectively as a barrier it must be continuous and free from defects. Of necessity, such films formed from paints require that the paint exists in the liquid state at the point of application.

One of the principal reasons is that the discrete regions of a coating, such as spray droplets for example, must coalesce to form a continuous film. Moreover, the material needs to be handleable; it must be easily transferred from its container to the surface and applied at the required film thickness. Although this process does not require the material to be liquid, conventional paints are so under normal conditions. The post-application behaviour of a coating, on going from a liquid to a solid must be such that it 'holds up' on the surface without running or dripping and that the visible surface imperfections and undulations 'flow out' to give a smooth film of the required thickness. The principal property of the paint common to all these processes, and of critical importance, is the way in which the material flows. The science of the way in which materials flow and deform is called *rheology*. We can see that paint rheology is not a simple flow property when we recall that after application the paint must hold-up, possibly on a vertical surface, yet 'flow out'.

For these two effects to occur would seem unreasonable since they appear to be mutually exclusive. However, most coatings are formulated such that the rheology adequately accommodates both requirements. In this chapter the important issues are discussed and illustrated.

Before proceeding to discuss the specific nature of the rheology of paints and its influence on the preparation and application of coatings, let us consider rheology, the science of deformation and flow of materials, in more general terms.

26

3.2 RHEOLOGY

'Viscosity' is the word most often used to convey an impression of the ease or degree of difficulty encountered when attempting to cause a material to flow. In a kitchen, for instance, a chef would probably use the word 'consistency' to convey exactly the same meaning. The word consistency has an immediate impact on both novice and expert, and in the context of its general use, rejoices in being sufficiently vague to ensure its usefulness. A contemporary definition of consistency for rheologists is 'a general term for the property of a material by which it resists permanent change of shape' whilst older definitions add '...defined by the complete flow–force relation'. A dictionary definition talks of the '...degree of density, firmness or solidity, especially of thick liquids'. Thus, we tend to use adjectives such as thick and thin to describe fluid properties. For example, we often talk of fluids being water thin.

Conversely viscosity is a precisely defined quantity which relies on rigidly defined, measurable parameters. It still, however, is associated with the ability of a material to resist a deformation or flow. In order to appreciate the rheology of paints, in general, and viscosity, in particular, it is necessary to define the measurable parameters systematically. Table 3.1 gives definitions of some common rheological terms which will be used in the following text.

3.2.1 Shear Flow

The easiest type of flow to define and control is (simple) shear flow.

This is demonstrated in Figure 3.1a where we can see that a force applied to the top surface area of a cube of material produces a deformation in that surface and gradually diminishing deformations in what can be envisaged as layers in the remainder of the material. This force, F, per unit area, A, is termed the shear stress (σ) and has been shown to create a deformation characterised by the angle γ known as shear. If this shear stress were maintained with time on the surface of a fluid then the angle γ would change with time, however the rate of change of γ would be constant, and this is known as the shear rate, denoted $\dot{\gamma}$ (the dot denotes a time derivative). Thus, essentially the shear rate is a measure of how rapidly a material is deformed, or made to flow.

The viscosity (η) of a material is defined as the ratio of the applied force per unit area to the shear rate

$$\eta = \frac{\sigma}{\dot{\gamma}} \tag{3.1}$$

The unit of shear stress is Newtons per square metre (N m^{-2}) and that of shear rate is per second (s^{-1}) (rate of change of an angle with time).

Table 3.1 *Some common rheological definitions*

Anti-thixotropic	A slow fall, on standing of a sample, of a consistency gained by shearing
Apparent viscosity	The shear stress divided by the shear rate. It is not a constant coefficient
Coefficient of viscosity	The shear stress divided by the shear rate. A constant, defining a Newtonian fluid
Consistency	A general term for the property of a material by which it resists permanent change of shape
Dilatency	An increase in volume caused by shear. (Often confused with shear thickening.)
Newtonian fluid	A fluid for which the shear stress is proportional to the shear rate
Non-Newtonian fluid	A fluid for which the proportionality between shear stress and shear rate is not constant with shear rate
Plastic	A material which flows when a yield stress is exceeded
Pseudo-plastic	Shear thinning. Often used in a context where the shear stress is linear with shear rate at high shear rates, but no yield stress can be detected
Shear	The change of angle in a deformed body
Shear rate	The change of shear per unit time
Shear thickening	The increase in viscosity with increasing shear rate
Shear thinning	The decrease in viscosity with increasing shear rate
Shear stress	The force per unit area parallel to the area
Thixotropy	A slow recovery, on standing of a sample, of a consistency lost by shearing
Viscosity	The resistance of a material to deformation. The shear stress divided by shear rate
Yield stress	The stress which must be exceeded for a non-recoverable (viscous) deformation to result

The unit $N\ m^{-2}$ can be written as Pascal (Pa) which divided by s^{-1} gives the unit of viscosity as Pa s. Table 3.2 gives examples of the viscosity of some simple liquids.

3.2.2 Newtonian Fluids

Given these three defined properties for simple shear flow a large set of material behaviours can be described.

The simplest rheological behaviour a fluid will show occurs when an applied stress is proportional to the resulting shear rate or *vice versa*.

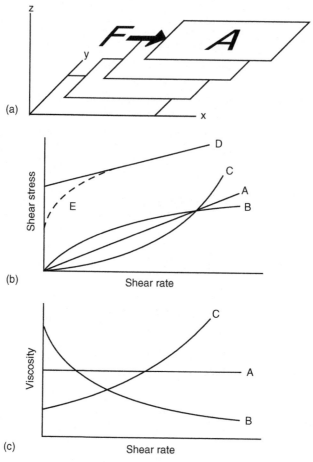

Figure 3.1 (a) *Schematic diagram of simple shear flow;* (b) *Shear stress–shear rate behaviours – A. Newtonian, B. Shear thinning, C. Shear thickening, D. Plastic, E. Pseudo-plastic;* (c) *Viscosity–shear rate behaviour – A. Newtonian, B. Shear thinning, C. Shear thickening*

The constant of proportionality between the two is the coefficient of viscosity, η, as in Table 3.2. Because viscosity is invariant with either variable in this case we say that it obeys Newton's law (postulate) and the material is Newtonian. This is shown in Figure 3.1b as a line of constant slope of shear stress with shear rate passing through the origin, and in Figure 3.1c as a line of zero slope and constant viscosity with shear rate. Many simple, low molecular weight materials, are Newtonian when liquid. Note here that viscosity changes with temperature, thus materials generally exhibit lower resistance to deformation when hot than when cold.

Table 3.2   *Examples of the viscosities of some materials*

Material	Viscosity (Pa s)	Temperature (°C)
Acetone	0.3×10^{-3}	20
Water	1.0×10^{-3}	20
Ethanol	1.2×10^{-3}	20
Propanol	2.3×10^{-3}	20
Isopropanol	2.9×10^{-3}	20
Diethylether	1.7×10^{-3}	−100
	2.8×10^{-4}	0
	1.2×10^{-4}	+100
Air	1.5×10^{-5}	20
Liquid air	1.7×10^{-4}	−192
Glycerol	1.5	20
Glucose	9.1×10^{12}	22
	2.8×10^{8}	40
	2.5×10^{1}	100

3.2.3   Non-Newtonian Fluids

For polymers, polymers in solution, particles in a continuous phase or indeed both polymers and particles in a solvent (continuous phase), it is common to find that the viscosity is no longer invariant with either shear rate or shear stress. These types of material are termed non-Newtonian. Figures 3.1b and c attempt to illustrate the general types of behaviour often observed.

The most common behaviour is shear thinning where with an increasing shear rate the material has a decreasing viscosity. Paints and coatings will generally show this type of behaviour and it will be noted that they usually contain both polymers (in solution) and pigment particles. Clearly, their absolute behaviour in terms of how much thinner they become and in what shear rate region this occurs will depend upon the exact chemical and physical nature of the materials used.

The converse of shear thinning is shear thickening, whereby the viscosity is increased by increasing the shear rate. This behaviour can be observed with materials containing a high concentration of particulates, pigments for example. The coating industry often exploits this behaviour in paint manufacture by using 'mill-bases'. Mill-bases are highly pigmented mixtures, exhibiting shear thickening, and hence when caused to flow rapidly show very high viscosity.

Aggregates of particles are in close proximity to each other hence the difficulty in moving relative to each other when in high shear flow gives rise to high viscosity. This action also leads to mutual attrition which disperses them into smaller particles.

Another class of non-Newtonian behaviour is that of plastic flow. This type of flow requires that a finite stress be applied to the material before flow is possible. That is, the material exhibits a yield stress, as demonstrated in Figure 3.1b where the curve for plastic flow intersects the shear stress axis. Very often materials, when studied at higher shear rates, appear to show a form of shear stress–shear rate behaviour which would appear to extrapolate to zero shear rate and result in a yield stress. However, when measurements are made at successively lower shear rates the curve tends towards the origin. This is called pseudo-plastic flow for obvious reasons, but is often used synonymously with shear thinning. In reality it is a class of shear behaviour.

3.2.4 Time Dependency

The remaining variable that can have a profound influence upon the rheology of a material in simple shear flow is time. It is often found that at a constant shear rate the viscosity of a material continually changes for a period of time prior to settling to a steady-state value. If, after the sample has been left to 'recover' for a significant period of time, the behaviour is seen to be repeatable that is the original steady-state value is not immediately measured, the material is showing time-dependent behaviour.

Should the viscosity decrease with time, and thereafter increase to return to its original value after a period of 'rest', the behaviour is termed thixotropic. Often confusion arises in the use of the term thixotropic since it is equated with the decrease in viscosity whereas it is the rebuilding of the 'structure' of the fluid to recover the initial high viscosity which is the thixotropic nature.

Conversely, an increase in the viscosity during shear which is lost during a period of rest is termed anti-thixotropy. This is also commonly termed rheopexy today, although the original usage of rheopexy was not synonymous with anti-thixotropy.

Thixotropy is far more commonly found than anti-thixotropy. Very often this time-dependent behaviour is tested for by observing the viscosity variation with increasing and then decreasing shear rates. This results in a thixotropic loop as shown in Figure 3.2a. Given the preceding test, true thixotropy could only be confirmed if, after a period of rest, this loop could be repeated.

In reality this thixotropic loop is a two-dimensional view of a three-dimensional surface composed of the time-dependent viscosity at constant

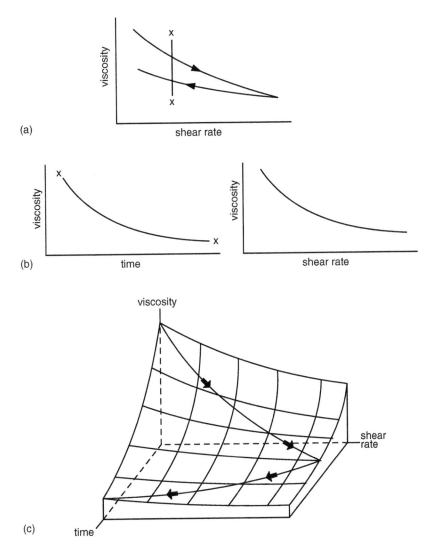

Figure 3.2   (a) *Typical thixotropic loop. Arrows indicate the direction of the shear rate sweep;* (b) *Viscosity–time at a constant shear rate and viscosity–shear rate behaviour of a thixotropic material;* (c) *Trace of the behaviour of a thixotropic material on a viscosity–shear rate–time surface*

shear rate, Figure 3.2b, and the shear rate dependent viscosity taken at a constant time after the commencement of shearing (Figure 3.2b). The thixotropic loop is seen as the trace of behaviour over the resulting surface (Figure 3.2c).

This rebuilt viscosity for thixotropy, or decayed viscosity for anti-thixotropy is often spoken of as the 'structure' of the material. The idea to be conveyed here is that the flow degrades (or builds) some micro-structural units within the material which can be regained (or lost) on standing.

3.2.5   Elasticity

Materials when subjected to a deformation may not simply respond in a viscous fashion. For instance, a spring will deform but will recover its original form when the force which produced the deformation is released. Materials may show a degree of elasticity as well as viscosity when subject to a force (per unit area or stress). This elasticity can be expressed by Hooke's law *'ut tensio sic wis'*, the stress is proportional to strain. Modern instrumentation has been developed with the intention of making ever increasingly sensitive measurements of elasticity, and apportioning elastic and viscous components to a materials rheological behaviour. Thus, terms such as viscoelasticity, linear viscoelasticity, creep compliance, forced oscillation, strain oscillation, stress oscillation, relaxation spectra and retardation spectra, will be met. This represents a specialist area of rheology which goes beyond an introductory explanation. Texts dealing with the subject are given in the bibliography. The implications of elasticity in coatings can be simply identified and described.

   Throughout the body of this chapter reference is made to structure and structuring within a paint. This structure is the result of the interactions which occur at the micro level whereby the components of a coating act so as to produce a network. This network may take different forms but may impose a mechanism that can resist sedimentation of particles due to gravity in storage or restructure rapidly to prevent sagging or running after application, for instance. Sensitive rheological measurements can interrogate these structures and be a valuable research and development tool used to design and optimise these networks. It is implicitly understood here that these structures are easily mechanically degraded so that the paint will still adopt a shear thinning profile required for spraying, for example.

   It must be emphasised that some networks which lead to very high elastic components are very definitely detrimental to coatings in service behaviour. Principally networks which are composed of entanglements of high molecular weight polymers would not allow a coating so constructed to be applied. In spray, break-up of the stream of fluid into droplets would be restricted and strings of fluid would result. In roller coating, strings would be formed giving an uneven surface coverage. Indeed even small degrees of elasticity could seriously affect a coating's ability to respond to the application and post-application processes required to give the correct coating properties. In general, elasticity within a coating which is generated from the polymers alone is to be avoided.

3.3 RHEOLOGICAL MEASUREMENTS

For determination of the absolute properties, within well-defined flow processes, simple shear flow for instance, a large number of commercial instruments are available. They are generally rotational devices in which the geometry of the device holding the sample is defined so that one part of the geometry is in rotation, whilst the other part is static, and this creates a well defined and constant shear rate throughout the sample.

It may be that the instrument sets the shear stress (actually torque) and measures the rotation rate or *vice versa*. Irrespective of which of the two quantities is set and which is measured, shear rate and shear stress are known, and thus a material's viscosity profile with shear properties can be determined. Curves similar to those shown in Figures 3.1b and c result.

3.3.1 Rheology and Coatings Reality

In rheology, we work hard to simplify the interpretation of information and in order to do this we use more and more elaborate instrumental methods. These produce well-defined flow regimes in which simple shear flow is generated. However, simple shear flow is not generally established in any real application process nor can it be considered to be established in other flow phenomena associated with coatings.

The processes designed for the application of coatings are based on other criteria than simple flow. Thus, the common methods of application, brush, spray, roller and electrodeposition, for instance were developed with consideration of criteria such as speed of application, mechanical simplicity, transfer efficiency and ease of equipment cleaning amongst a host of others. Given this mismatch we attempt to relate parts of a process to the extent of shear generated in a process.

Clearly the use of simple shear viscosity is somewhat flawed since we have to work hard at obtaining such a flow. The reality is that we can ultimately only give an estimate as to what the equivalent shear rates may be in a real process.

3.3.2 Industrial Methodologies

Within the paint industry many methods and techniques have been developed to relate a 'flow quantity' to a particular coating performance characteristic, so for a particular application an industry standard will be developed. It is beyond the scope of this chapter to review such methods but a quick consideration of some of these techniques reveals some interesting points.

One such industry standard is the Ford No. 4 efflux cup. This is essentially a machined cup of defined dimensions having a conical bottom to the cup at the centre of which is a hole of a defined size. To take a reading, the orifice is sealed with a finger and the cup is filled with material with the excess caught in an overflow trough. The finger is removed and the time for the material to stream out measured. This is a measure of the 'viscosity' having the units of seconds. It is effectively a comparative way of describing the ease with which a material will flow in a given environment.

An experienced formulator is able to assess from this information how a coating will perform, for example, in a spray application.

It will be readily appreciated that this type of measure is suitable for fluid (thin) materials, which are essentially Newtonian, thus a definition of shear rate is unnecessary. It ought to be equally apparent that as a measure for highly non-Newtonian materials the method has little use since it can give no information regarding the degree of shear thinning and how this will effect the actual application.

Another industry accepted standard instrument is the Stormer viscometer which describes the viscosity of a material by measuring the mass required to maintain a paddle immersed in the material in rotation at a constant 200 rpm. The falling weight drives the paddle through a series of gears and pulleys and the rate of 200 rpm can be quickly established with a stroboscope. The unit of viscosity is a derivative of the variables and is expressed as the Krebs Unit (KU), which is a unit specific to this instrument. The effect of measuring materials which show a time dependence is even more serious than with the Ford Cup. Thixotropic paints, for instance, will show a continuing acceleration of the paddle as the paint is reduced in viscosity by its action.

With imagination such measurements can be seen to relate to the rheology of a paint and be relevant to mixing, thus manufacturing perhaps.

As a formulation aid the 'Rotothinner' viscometer is a useful tool since it mixes the paint at the same time as making viscosity measurements. It has generally been acknowledged that for paints to be applied by spray, brush or roller they require a viscosity in the range 0.05–0.5 Pa s. Thus, this instrument allows formulation changes to be made whilst measuring the viscosity. The principle of operation is simple in that a disc shaped stirrer is immersed in a standardised container of paint and rotated at constant speed. The torque transmitted though the sample is measured *via* a spring connected to the paint container.

Those devices mentioned above are not capable of defining a shear rate clearly. The ICI Cone and Plate viscometer was designed with an acknowledgement that coatings are generally shear thinning and that brushing, spraying or roller coating of paints takes place at high shear rates, around

10,000 s^{-1}. The device uses a small diameter wide angle cone rotated at a constant speed, in contact with a plate. This accurately defines the shear rate throughout the sample and is set to give a shear rate of 10,000 s^{-1}. The viscosity is again measured through a calibrated spring torque device.

With unknown materials these estimates of paint rheology are of little value. However, under the conditions of use for which they were designed they are very often ideal tests for determining the optimum rheology for a material (of known generic form) in a particular process.

3.4 RHEOLOGICAL PROCESSES ASSOCIATED WITH COATINGS

Once in possession of a knowledge of the rheological behaviour of a material in shear flow, the task becomes one of determining which parts of the curve, shear rate–viscosity or thixotropy for instance, can be identified with various processes or phenomena. Thus, brushing, roller coating and spraying as well as shears generated in manufacture and storage can be identified with particular regions of the overall characterised behaviour.

In order to appreciate the flow or rheological characteristics associated with a paint consider how a paint proceeds from manufacture to a final dry film. Generally, a paint contains four generic components: polymers or resins, pigments, solvents and a variety of additives. For powder coatings there is no associated solvent but the thermal properties of the polymer ensure a degree of fluidity at the elevated temperatures used in curing the coating. Each component has a dramatic effect on the rheology of the material at each stage of its life.

3.4.1 Manufacture

At the paint manufacturing stage the principle action, apart from producing a homogeneous mixture, is to reduce the size of the agglomerated pigments. This involves mechanically breaking the primary particles apart, wetting their surfaces and creating a different surface such that they do not re-aggregate. Most conventional formulations rely on the binder (resin) to wet the pigment particles. As a result, considerable mechanical energy is required to 'mill' agglomerated dispersion to a 'smooth' paste, as previously mentioned often a shear thickening mill-base.

Many types of mill are commercially available to supply the energy required *via* different mechanical routes. They range from the very simple, such as the high speed disperser which is essentially a high speed rotating disc with saw-teeth at its periphery, to the more complex such as colloid mills or bead mills. The high speed disperser although simple sets up flow patterns relying on a shear thickening mill-base for its greatest efficiency. Thus, nearest the blade the high shear disperses the agglomerates as well as inducing the self-attrition mechanism, whilst the lower viscosity away from the blade ensures the mobility of the material necessary to reintroduce mill-base into the dispersing region continuously.

Bead mills, or sand mills, have chambers filled with small beads or sand and made to move rapidly by a series of rotating discs. The chamber is small compared to the volume of mill-base which is made to flow through the chamber. Here the shearing action of the beads in motion 'grinds' the pigment aggregates. A screen allows passage of the milled paint but retains the beads. Multiple passes through the mill can be made.

It turns out that rheological characteristics of a mill-base which is of optimum efficiency in one type of mill may be such that it may not be milled effectively in a different type. Mill-bases must be specially formulated for individual types of mill.

Most often the paste-like dispersions are combined by a low energy mixing method with the remainder of the formulation, and a paint of the correct 'consistency' results.

The finished paint is discharged into cans and stored.

3.4.2 In-can Stability

Since the storage time can be quite considerable the 'In-can' stability of the paint is important. The principal problem here is that the discrete, but small, pigment particles have densities far in excess of the continuous medium in which they are immersed, and hence have a high propensity to settle to the base of the can.

Such sediments can be extremely cohesive and difficult to redisperse even though the individual particles are stable to aggregation. It is easy to envisage that this settling is a low shear and slow rate process but none the less, because of the time scale involved, a problem. In order to prevent settlement the paint is often structured with additives, so that a high viscosity prevails at low shears. Structured paints often exhibit thixotropy or a yield stress value.

3.4.3 Application Process

Because paints are most frequently either highly shear thinning or thixotropic, they tend to aid the application process. High shear rates generated in these application methods destroy the inherent structure within a paint, for the duration of the application, which considerably increases the ease with which they flow in such processes. For example, brushing is considered to generate a shear rate of approximately 10^4–10^6 s^{-1}, *i.e.* high. Hence if we wanted to relate the brushing characteristics of a series of paints to their rheology, we would look at and compare their high shear viscosity.

An ICI cone and plate viscometer would be best suited for this measurement. The same argument would apply to spray coating methods, both air assisted and airless. The main problem with adjusting formulations based on high shear viscosity measurements is that this is an insensitive region of the viscosity–shear rate curve. Small variations in this region tend to be magnified many fold in the low shear region.

Another major consideration that has to be taken into account when adapting the rheology for an application process is that in order to apply a sufficiently thick coating to the substrate the high shear rate viscosities also need to be relatively high. This is a necessary condition since otherwise the stresses (gravity) acting on a thick dense paint film would cause it to run or sag.

This condition subsequently imposes a limitation on the low shear viscosities, since affecting change in the high shear region also produces change in the low shear region.

Similar limitations are imposed on industrial roller coating applications. Here large rollers, and they may be up to several metres wide, are used to lay a coating onto metal sheets or coils, for instance. The principle is that a pick-up roller rotates in a bath of the coating which ultimately, *via* several other rollers, transfers the coating at a well-defined thickness to the metal. Thus the rheology, amongst other considerations, determines the film thickness, whether the coatings will fly off the rapidly rotating rollers, and the extent to which surface imperfections in the coatings will flow out prior to the thermally induced curing. Often a Ford cup measurement is used to determine the 'viscosity' of the material for application. If the final application roller is rotating in the opposite sense to the direction of travel of the metal sheet or coil it is termed reverse roller coating. When rotating in the same sense it is termed forward roller coating. Reverse roller coating is a high shear process and minimises surface defects in the coating associated with it. Forward roller coating is a relatively lower shear rate process that sets up an inherent instability seen as ribs of coating in the direction of travel. These must flow out and level prior to curing.

Table 3.3 identifies some typical processes and estimates the equivalent simple shear rates associated with them.

3.4.4 Post-application Rheology

Several important changes often occur during and after the application of a paint. If it is atomised it will loose solvent between the spray tip of the gun

Table 3.3 *Viscosity–shear rate influence on paint properties*

Approximate range of shear rates (s^{-1})	Application	Required viscosity
10^{-6}–10^{-2}	Prevention of pigment and particulate settlement	High
10^{-2}–10^{-3}	Promotion of flow out and levelling	Low
10^{-2}–10^{-3}	Sagging and slumping control	High
10^{0}–10^{2}	Paint pick up on brush	High
10^{1}–10^{3}	Mixing, pumping and stirring	Low
10^{1}–10^{3}	Forward roller coater application	Low (but depends on required film thickness)
10^{3}–10^{6}	Brushing, reverse roller coating spraying	Low
10^{3}–10^{6}	Control of drag on brush	Low
10^{3}–10^{6}	Control of thick film build	High
10^{4}–10^{6}	Control of paint mist (fly) off rollers	High

and the substrate. This results in a lowering of its temperature and an increase in solids content, which will continue during drying. Both of these effects will act so as to raise the viscosity of the paint throughout the entire shear rate region. Once the paint is at 'rest' on the substrate surface its shear thinning nature will ensure a high viscosity in the low shear rate region. Moreover, if the paint is thixotropic it will start to restructure and increase the viscosity.

The importance of these effects is shown below. Prior to significant flow-disabling viscosities being reached, the paint droplets need to flow to coalesce and form a continuous film. Furthermore, the film needs to flow and level in order to produce a smooth, defect free, glossy finish. These are low shear processes. The time frame in which they occur is important since if the viscosity remains low for too long the paint will flow under gravity and runs or sagging and slumping of the paint will occur. The implication here of course is that a thixotropic material, where the time for restructuring is controllable, will allow significant coalescence, flow out and levelling to occur prior to runs and sagging becoming appreciable.

Whilst coatings, which are roller applied or brushed, may form continuous films during application, they still require sufficient mobility to flow out and level, again with control of the runs or sagging behaviour. Paint so applied is, therefore, also subject to the influences of shear thinning and thixotropy. Consideration must also be given to film thickness.

Many coatings, particularly those applied by roller, proceed directly from application into a high temperature cure oven, where a chemical crosslinking mechanism is initiated. Several effects are coincident. Because of the increased temperature the viscosity of the paint decreases, probably significantly. Solvent, which may of course be water, will be driven off, leaving the non-volatile polymer and pigments. The viscosity will start to rise accordingly. The crosslinking mechanism will rapidly increase the molecular weight of the polymer, which in turn will increase the viscosity very rapidly (with a third or fourth power dependence on molecular weight!). During this thermal treatment, which may be very rapid, sufficient flow must be permitted to produce the required appearance of the films.

3.4.5 Film Defects

There are other effects which can ultimately lead to poor surface appearance and although not caused by the rheology of a coating, are facilitated by it.

Flooding and floating are related in that they occur in coatings containing two or more pigments. For flooding to occur one of the pigments preferentially migrates to the surface of the coating giving it a different colour from the bulk of the paint. Floating is similar in that there is a differential concentration of pigments across the surface giving a generally mottled film appearance. This is due to the formation of Bénard cells. These are small hexagonal areas created by convection currents set up by surface tension differences developed during solvent evaporation. They circulate one of the pigments preferentially thus producing a pigment concentration difference across the surface of the paint. Clearly both processes are only possible if flow can occur in the drying coating.

Conversely orange peel, an undulation across the surface of the coating resembling the surface of an orange, is a particular defect which might be eliminated if there was adequate total film mobility during the drying or curing stage.

If the surface of a newly applied film is considered to be a series of peaks and troughs of all amplitudes and wavelengths (within reason) a good analysis of flow out and levelling is the Orchard equation. This describes the decay in amplitude with time $a_{(t)}$, from an original amplitude a_0, of a sinusoidal waveform of wavelength λ, with a film of average thickness h, and viscosity η.

$$\ln\frac{a_0}{a_{(t)}} = \frac{16}{3}\pi^4 h^3 \frac{\Gamma}{\lambda^4}\int_0^1 \frac{1}{\eta}dt \qquad (3.2)$$

This tells us that for small wavelengths surface tension, Γ, the driving force to levelling, is sufficient to collapse their amplitudes effectively to zero.

However, for large wavelengths or undulations the driving force is not adequate within the time frame of the drying or curing process in which the viscosity is continually rising. There is not enough 'flow available', so the final appearance is typically a smooth film with large wavelength undulations that are of the order of millimetres.

3.5 LOW VOC COATINGS – FLOW PROBLEMS AND SOLUTIONS

In order for coatings to become more environmentally acceptable the amount of volatile organic compounds (VOC) contained within them must be restricted, as we have seen in Chapter 2. Removal or partial removal of the organic solvents has a profound impact on the rheology of such coatings. The total removal of solvents from a conventional paint to give a zero-VOC paint may be taken as an example. The product will inevitably be highly viscous and possibly enormously shear thinning, which limits its ability to be mixed and pumped during manufacture, limits the formulator's ability to produce a sprayable coating, and limits his ability to produce acceptable surface finishes. One method of alleviating this problem is to use lower molecular weight, hence lower viscosity polymers. This solution tends to be expensive and if taken too far, can yield a highly mobile film before sufficient curing has taken place to retard sagging and slumping. Another way of achieving satisfactory spray and post-application rheology is to heat the paint until the viscosity falls into the correct region. However, a very much increased rate of curing occurs and may take place in the mixed paint prior to use, even to an extent that may prevent its application. Chemistries designed to counter this effect and retard the reaction tend to suffer from lower extents of cure or perhaps no cure at all when used in low ambient temperatures.

Amongst other low VOC technologies, solvent free or 'powder' coatings and water-based systems are most significant.

Powder coatings are paints made from solid polymers, *i.e.* polymers whose glass transition temperatures are considerably higher than ambient temperatures and whose fluidity (the inverse of viscosity) only becomes appreciable at temperatures greater than about 100 °C. When fully formulated with pigments and additives these coatings are milled into small particles of about 30 μm average size. The powders are made to flow through a gun which applies an electrostatic charge to the individual particles on their flight to the earthed substrate. The particles adhere not only to the front of the earthed article but because of the change, wrap around the rear of the substrate following the potential field. On heating the coated article in a curing oven the polymer becomes fluid, the discrete particles coalesce, and ultimately flow out and level. During this process

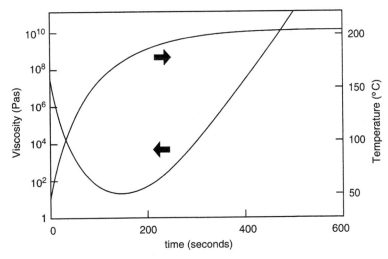

Figure 3.3   *The viscosity profile of a curing powder coating when subjected to a typical cure temperature profile*

the polymer crosslinks and hence the viscosity rises considerably. This process is shown graphically in Figure 3.3. The most significant feature of cured powder coatings in terms of their appearance is the existence of orange peel. Effectively the coatings do not remain fluid enough for long enough to allow the flow out of the longer wavelengths or in reality the larger size surface inhomogeneities. These effects can, however, be minimised by judicious formulations.

Water-borne coatings of the latex type present their own special problems that largely result from the very high molecular weight polymer existing in discrete, small particles. These particles are typically of the order of 100 nm in diameter and always less than 1 μm. In this form they can exist as dispersions of up to 50% volume fraction of polymer and still be 'water thin'. It is often observed that when the volume concentration of latex particles becomes large the viscosity increases very rapidly with small increments in concentration. They display marked shear thinning and can additionally exhibit yield stresses. When formulated with pigments and using realistic concentrations of polymers it is clear that the rheology of these paints has to be modified with various additives.

The rheology modifiers fall broadly into three classes of materials: clay-based thickeners, water-soluble polymers of high molecular weight and associative thickeners. Clay materials work by producing a structure throughout the body of the paint. This structure is sensitive to shear and as a result shows highly shear thinning behaviour and a yield stress. This will give good resistance to in-can settlement. The high molecular weight water-soluble polymers form an entangled network at low concentrations effectively thickening the continuous phase (water). By virtue of their high molecular weight they are again highly shear thinning but with no yield stress value. More often they exhibit a zero shear viscosity which, although high at concentrations generally used, does not restrict motion in gravitationally imposed stresses, for example. Degrees of flow and levelling can be influenced whilst film build is also considered. Both clays and water-soluble polymers can impart degrees of thixotropy to a paint. Associative thickeners are low molecular weight water-soluble polymers which have been modified by adding low molecular weight hydrophobic groups such as alkanes. The hydrophobes interact with each other, the hydrophobic latex surface, and other hydrophobic species such as the coalescing solvents and surfactants for instance, to build up a structure throughout the paint. In this way a shear sensitive matrix, whose rheological response can be adjusted by changing the chemistry of the associative thickener, the size and number of hydrophobes, the hydrophobicity of the polymer latex surface and the chemistry and concentration of other hydrophobic materials, is obtained. The associative thickeners do not operate by thickening the continuous phase but build individually weak mechanically sensitive contacts which are easily broken and rapidly reformed. As a result the materials do not exhibit thixotropy. Judicious use of these thickeners can enhance or suppress the high shear viscosity for film build and application, the low shear viscosity for flow, levelling, and sag control, and the medium shear viscosity for in-can stability and brush pick-up.

It is not uncommon to use mixtures of different associative thickeners or indeed mixtures of the three different classes of thickeners to produce the desired effect.

3.6 CONCLUDING REMARKS

It has been demonstrated in this chapter that the physical process of causing paints to flow, or indeed attempting to prevent flow, for a variety of reasons under certain conditions is not a simple problem. The deformation and flow properties (rheology) of coatings exhibit many and varied forms of behaviour under well controlled, laminar flow conditions. Adapting these properties to give good performance under the constraints imposed in all aspects of the coatings life, from manufacture through storage and application to post-application, is one of the naturally difficult problems for a paint formulator to consider. It is often necessary to make a series of compromises to obtain the optimum set of properties. A knowledge of the rheology of paints and how to manipulate it *via* formulation is a vital tool in the optimisation process.

3.7 BIBLIOGRAPHY

The subject of the rheology of materials is a very large area of study in its own right. It is a subject that has its roots in physics, chemistry, applied mathematics, engineering, colloid science and polymer science. Excellent texts and treatises are available in all these branches but are not for those new to the field. The most appropriate text for the general background to rheology in its most diverse form is that of H.A. Barnes, J.F. Hutton and K. Walters, *An Introduction to Rheology*, Elsevier, Amsterdam, 1989.

However, a more recent text by the same author, Howard A. Barnes, *A Handbook of Elementary Rheology*, University of Wales, Institute of Non-Newtonian Fluid Mechanics, 2000, is written for complete beginners in rheology and it is left for the reader to decide at which level their reading should commence with these two options. These are both very readable books and are recommended. An additional text is also available by Howard A. Barnes, *Viscosity*, University of Wales, Institute of Non-Newtonian Fluid Mechanics, 2002, which is a specific discourse on viscosity, including its measurement and importance in industry.

A text more specific to the paint industry is T. C. Patten, *Paint Flow and Pigment Dispersion*, Wiley Interscience, Chichester, 2nd edn, 1979. This takes the form of an undergraduate teaching textbook with worked examples and questions throughout the volume. The discussion is generally of a very high quality and it is highly regarded as a standard reference book within the paint industry. Most of the subjects covered in this chapter are dealt with in greater detail in Patten, however, application processes *per se* occupy only scant discussion when compared with the other areas. Viscoelasticity, an important area receives no treatment at all. This area of study is becoming increasingly important for coatings science.

Contained in the above book is a very detailed section dealing with specific paint and coatings industry methods for measuring rheology. Thus, many of the industry standards are included and well described.

At this level of reading it becomes necessary to understand an additional large body of non-Newtonian rheology, generally described as viscoelasticity, as alluded to in the text. This area can be interesting, challenging and often necessary to work with in order to make progress. By far the most appropriate and current text for this subject is that of J.W. Goodwin and R.W. Hughes, *Rheology for Chemists, An Introduction*, The Royal Society of Chemistry, Cambridge, 2000. This is a comprehensive well written book, which does assume some mathematical ability but has as a key feature the underlying premise that the bulk rheological properties of materials is a result of the microstructure given by the chemistry of the components of a formulation. Indeed even cursory reading of the preface demonstrates the utility of the book and its applicability to paints and coatings. This text is very highly recommended.

Beyond this treatment the field of rheology appears to diversify somewhat and is often dealt with as polymer rheology, for which I would recommend reading J.D. Ferry's *Viscoelastic Properties of Polymers*, Wiley, Chichester, 3rd edn, 1980.

CHAPTER 4

Film Formation

A.B. PORT and C. CAMERON

4.1 INTRODUCTION

Arguably the most important step in any coating process is film formation. This is the conversion of the coating from a liquid into an integral solid film after application so that the coated article can start to be usefully employed. An obvious example from most people's experience would be the painting of a domestic window frame. The first stage of film formation is completed when the window can be touched without the paint transferring to the decorator's finger. The coating is said to be 'touch dry' in this condition. Should the window be closed against its frame at this stage then it is likely that they would become stuck together under the pressure of closure.

The process of film formation must proceed further to avoid this problem and for the coating to behave more like a solid. When this condition is achieved, the coating is said to be 'hard dry' or block resistant under those particular conditions of time, temperature and pressure. For a window manufacturer wishing to stack frames in a hot warehouse, further advancement of film formation would be necessary before unwelcome adhesion or blocking could be avoided.

It is now not quite so obvious what we mean by film formation as a coating becomes 'solid'. It is clear, however, that the rate and extent of film formation are of the highest importance both in terms of the process – how quickly a coated article can be handled, overcoated, or stored, and the coating performance – what properties have developed by what time after application. Note that powder coatings must be transformed from its solid state for storage into a liquid prior to the film formation process where it is then converted to a solid film.

The rationale for film formation behaviour in coatings is best explained in terms of the polymer glass transition temperature (T_g). The glass transition temperature marks a significant change in the properties of a polymer, related to the mobility of the segments from which all polymers are constituted. As the polymer is cooled to the T_g and below, the co-operative motions responsible for both translational and rotational motions in the polymer backbone essentially become frozen, and the macroscopic flow of the material (such as the transfer of paint to the finger!) no longer occurs.

The ability of a material to flow depends on how close its temperature is to the T_g. The larger the $(T-T_g)$ – achieved by a low T_g, a high T or both– then the greater the mobility or the lower the viscosity. If the temperature at which a material is stored is equal to its T_g, so that $(T-T_g) = 0$, the mobility becomes very low indeed. To all intents and purposes the material is solid with a very high viscosity (10^7–10^{14} Pa s). As a rule of thumb, a coating becomes touch dry, with no transfer of paint to the finger, when T_g is within 20 °C of the ambient temperature.

The viscosity behaviour of liquids is often explained in terms of Eyrings' 'hole' theory, where molecular flow is achieved by a series of 'jumps' by molecules into unoccupied space, or holes. The greater the number of holes, the easier the flow. Hole size and hole concentration increase with temperature. The analogous concept of 'free volume' is familiar to most polymer scientists and is very useful for rationalising many aspects of the behaviour of polymers. The volume–temperature relationship for polymers is given in Figure 4.1.

As the temperature increases above T_g, the restrictions on polymer segment mobility become commensurately smaller as the free volume increases. The material is able to undergo liquid flow.

At temperatures equal to T_g or below, the free volume is much smaller and the material exhibits properties associated with the solid state. Very slow relaxations can still occur, and these can have a significant impact on the properties of polymers (see Chapter 5).

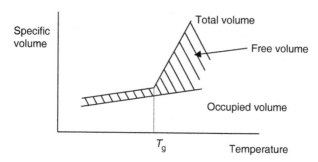

Figure 4.1 *Free volume – the 'empty space' in polymers, showing an increase at temperatures above T_g*

It is implicit therefore, in order to convert a liquid paint into a solid film at a given temperature, the T_g of the coating must increase during the film formation process to become equal to and usually exceed the prevailing temperature. There are essentially two mechanisms by which the increase in T_g can be obtained. These are solvent evaporation and chemical reaction. Note that some coatings (elastomeric systems) form films by chemical reaction from components carefully chosen to yield a product whose T_g is below the ambient temperature.

4.2   THERMOPLASTIC COATINGS

A traditional method of film formation used by the coatings industries for many years was simply to dissolve a polymer in a volatile solvent, apply it to the workpiece and allow the solvent to evaporate. Relatively few coatings technologies rely on simple solvent evaporation nowadays. The explanation for this loss in importance lies in the fact that many important polymer properties are achieved only when molecular weights (MWs) become large enough for chain entanglement to occur. The values of this entanglement MW, M_e, vary significantly for different polymers and ranges from a few thousand to in excess of 10^5 g mol^{-1}. In consequence, the viscosity of solutions of these polymers is relatively high and to maintain application viscosity, the concentration of polymer has to be low (see Chapter 7). In turn this means that many applications are necessary to build up the thick coatings necessary for some end uses. The environmental legislative pressure to reduce solvent emissions, described earlier, has meant that such low solids coatings are no longer acceptable in many industries.

Film formation from solutions of thermoplastics occurs *via* the increase in T_g with loss of solvent. Initially, loss of solvent from these systems depends only on the vapour pressure of the solvent, how quickly solvent vapour is removed from the immediate environment and the ratio of surface area to volume of the coating. If the application process involves the spraying of atomised droplets of coating, the loss of solvent during the transfer of the coating from the spray gun to the workpiece can be very substantial. As the system T_g increases, the rate of evaporation will become dependent on the rate at which the solvent can diffuse through the solvent-swollen polymer.

Solvent removal from the free surface sets up a solvent concentration gradient across the coating to act as the driving force for diffusion. The normal treatments for diffusion of small molecules through polymers are made more complex for the coating systems where the final T_g is above the film formation temperature. In these systems vitrification will occur at some stage when the nominal T_g equals the drying temperature. Rates of diffusion of solvents in the glassy state are significantly lower than in the liquid or rubbery phase which means that glassy coatings can retain significant amounts of solvent for long periods – in some cases for a number of years. For some end uses this solvent retention is unacceptable, for example, when coatings are in contact with food or drink, such as in can lacquers. Here extraction of solvents into the contents of the can would affect the taste and probably present a health hazard. In exterior durable paints, trace amounts of ethers, ketones or halocarbons can act as initiators or 'fuels' for the photo-oxidative reactions causing degradative weathering of finishes.

A further consequence of the evaporation of large volumes of solvents during film formation involving vitrification is that large tensile shrinkage stresses can be set up. These internal stresses cannot be easily relieved in thin films adhering well to rigid substrates, and will weaken the coating's resistance to applied stresses or strains as well as the strength of adhesion to its substrate. To overcome the problems of solvent retention described above, stoving at temperatures considerably in excess of the coating T_g is carried out. This will not, however, eliminate stresses from the system when coating and substrate have substantially different coefficients of thermal expansion.

There are further limitations to thermoplastic coatings over and above environmental considerations especially where thermal or chemical resistance is required. To some extent, these can be overcome using semi-crystalline polymers, but these impose their own difficulties in terms of forming stable solutions as paints. For solvent-borne coatings the more effective route to environmental compliance and higher levels of performance has been through the use of crosslinking polymers.

4.3 SOLUTIONS OF CROSSLINKING POLYMERS

4.3.1 The Crosslinking Process

Film formation from solutions of crosslinking polymer systems combines two processes. Solvent evaporation takes place as described for solutions of thermoplastic polymers. At the same time chemical reactions take place between functional groups on polymeric and/or monomeric species which result in the build-up of MW of the system. There are two main classes of crosslinking processes. These are step-growth reactions and chain-growth reactions. We will consider mainly the step-growth reactions involving two different types of functional groups which are mutually reactive, in an alternating copolymerisation. There is a wide diversity of polymer chemistries which are used to produce crosslinking systems and these are described in Chapters 6–9. The common feature, however, is that at least one of the starting components must have more than two functional groups per molecule. The functionality is very important as it determines a limit to the chemical conversion up to which the polymer system is still processable (*i.e.* flows). In a typical system, consisting of an A-functional polymer or oligomer with a B-functional crosslinker, random reactions between A and B groups will lead initially to chain extension and branching (Figure 4.2). The largest molecule in the system progressively becomes larger and larger until at a certain critical stage of the reaction its dimensions span the whole of the reaction vessel. The achievement of this state is known as gelation. Achievement of the gel state leads to the loss of solubility: the system will swell in a thermodynamically good solvent, but the crosslinks prevent the chains moving apart and dissolving. Beyond the gel point, further reaction leads to the bonding of the remaining finite molecules to the gel, increasing the crosslink density. Unreacted functional groups on the gel can also react with other unreacted groups on the gel. Eventually, all the precursor molecules may become part of the network, if the system has been formulated correctly. The concentration of chains bonded to two different network junctions determines the equilibrium mechanical properties of the crosslinked polymer network at temperatures above the T_g of the system. Many properties of crosslinked polymers and coatings are dependent upon the nature of the molecules carrying the A and B groups and the extent of reaction of the A and B groups achieved.

The gel point occurs at a particular extent of reaction dependent on the ratio of the numbers of functional groups (stoichiometry) and the effective functionality of the components. According to the theory of Flory and Stockmayer, given some assumptions, the gel point in an A + B copolymerisation is given by Equation (4.1):

$$P_{gel} = \frac{1}{\sqrt{r(f_{sA} - 1)(f_{sB} - 1)}} \tag{4.1}$$

The extent of reaction at gelation, P_{gel}, depends only on the stoichiometry r (the ratio of the number of functional groups of types A and B, $r \leq 1$) of the reaction, and the functionality of the A and B components. The $f_{sA,B}$ are the site average functionalities of the A and B functional components (Equation (4.2)):

$$f_{sA,B} = \sum_i \left(\frac{n_i f_i}{\sum_i n_i f_i}\right) f_i \tag{4.2}$$

The term in brackets is simply the fraction of A or B groups belonging to species i in the mixture relative to the total number of A or B groups. Where a mixture of A or B multifunctional components is used, it is very

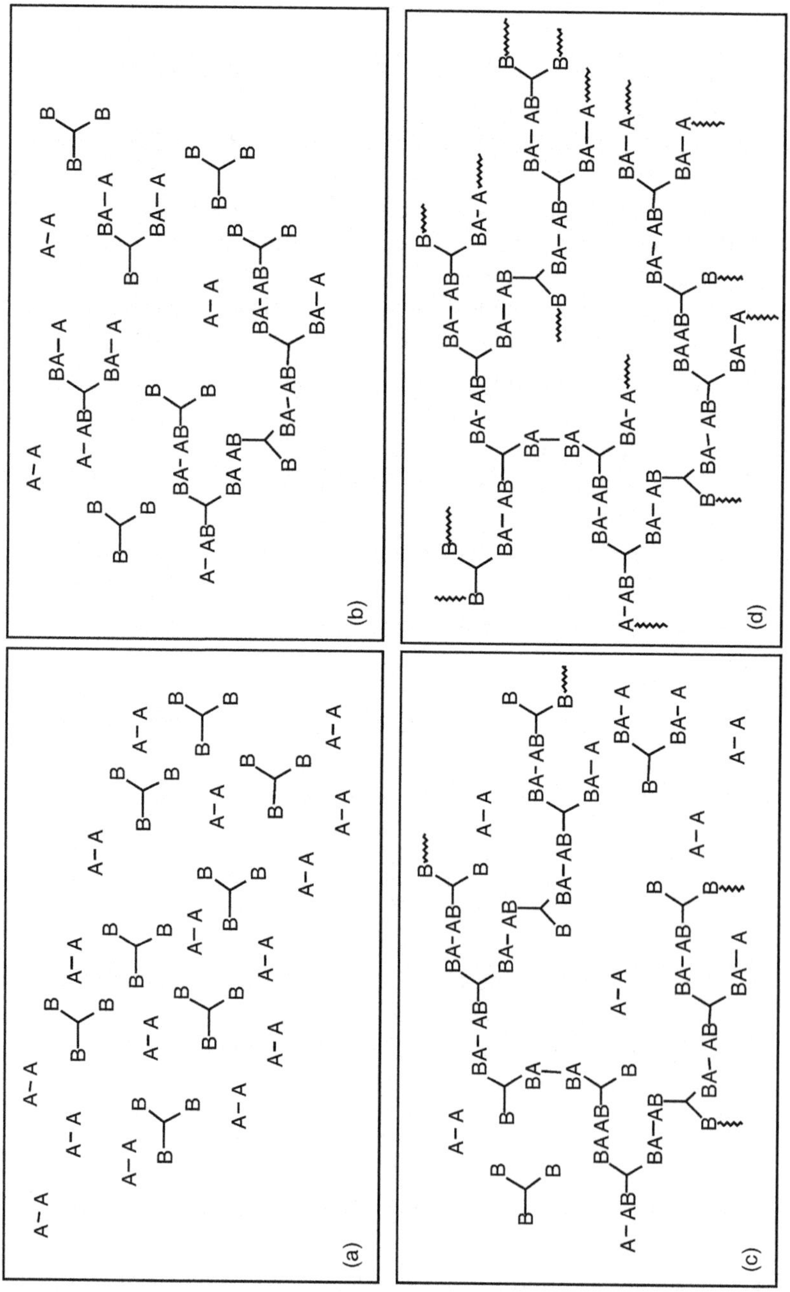

Figure 4.2 *Formation of a network of difunctional A species and trifunctional B species: (a) before reaction; (b) forming branched chains; (c) beyond the gel point; (d) when reaction is completed*

important to use the site average rather than simply the average number of functional groups per molecule. This importance arises from the random nature of the A + B reaction. Those molecules possessing greater numbers of functional groups clearly have a greater probability of being randomly selected for reaction, biasing the MW build-up towards the more highly functional species, which usually belong the larger molecules.

The onset of gelation brings about very major changes in the physical form of a crosslinking system. Creation of an infinite MW molecule causes the weight average MW (M_w) and the viscosity to diverge. Processes such as flow and levelling which depend on a coating having a finite viscosity are essentially brought to an abrupt end at gelation. Crosslinking reactions do not, however, cease at gelation. There are still a very large number of molecules and functional groups present at gelation in addition to the gel. These finite-sized species are known as the sol since they are soluble and capable of extraction by solvent. Reactions continue and the proportion of sol steadily decreases whilst that of gel increases. For stoichiometric mixtures capable of full reaction, the sol fraction will decrease to zero as the gel fraction increases to one. Off-stoichiometric mixtures or incomplete reaction will result in varying proportions of residual sol and gel. The characteristics of crosslinking systems can be summarised in Figure 4.3. The underlying cause of the fall in MW as reaction proceeds beyond the gel point is again related to the greater probability of reaction of the higher functional molecules.

There are a number of methods for identifying the gel point. One commonly used (although not universal) rheological definition of gelation

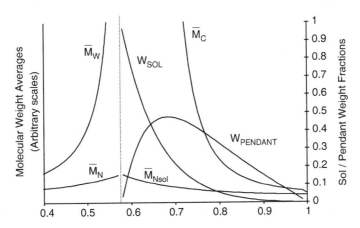

Figure 4.3 *The properties of a crosslinking system as a function conversion. Pre-gel and post-gel MWs, and weight fractions of sol and pendant chains*

is the point at which the elastic and viscous components of shear modulus are equal. More rigorous methods are known. In more practical terms gelation is marked by the rapid increase in viscosity and the appearance of a pronounced elastic solid behaviour from a previously viscous liquid. These changes provide the bases for empirical means of determining gel times or conversions such as 'snap time' and the point at which bubbles cease to rise in a reacting system.

Film formation in crosslinking systems might be thought to be complete at gelation since the viscosity has diverged to infinity at this point. This is certainly not the case. At gelation there is a large proportion of low MW sol material and a residual solvent content which depends on the kinetics of crosslinking and speed of evaporation of solvent. Typically the system T_g will be below the test temperature and the material will behave as a weak, soft or swollen rubber and may well be so fragile as to fail to resist the indentation and blocking tests described earlier. Further reaction and loss of solvent will then be necessary before defined criteria of film formation are met. Knowledge of the various rate processes is therefore required to deepen our understanding of film formation.

4.3.2 Kinetic Aspects of the Crosslinking Process

The preceding section described the changes occurring during the cross-linking process largely in terms of the extent of reaction. We must now address the rate at which these reactions occur and whether the physical state of the system influences the rate, especially when the system T_g approaches or exceeds the cure temperature. There are important reasons why the coatings technologist needs to understand and control the kinetics of the crosslinking process. Many coatings performance parameters depend critically on the extent of cure which is achieved in the various timescales available (see Chapter 5). In some industries the conditions of time and temperature for curing of crosslinking coatings are tightly specified. The reasons for this may be economic – as dictated by production line speed, size and rating of ovens, or by the sensitivity of the substrate to elevated temperatures – or simply the traditional practice of an established process. Any paint manufacturer hoping to introduce a new product in these circumstances would have to offer significant improvements in cost or performance or both to induce the customer to make major changes in his operation. Thus, it is more usual for the coatings formulator to tailor his product to an existing cure schedule.

In contrast, there are examples where curing conditions are not subject to the same control. Coating products applied to large outdoor industrial structures, such as ships, bridges or chemical plants, must achieve acceptable levels of film formation and performance whether in winter or summer and high or low humidity within production time scales which probably do not change significantly with the season. This can be very challenging.

A third aspect is that of the storage stability of the product before application. The customer or user of paint would like indefinite stability in the can, irrespective of storage conditions, as well as fast complete cure after application. For ambient temperature crosslinking systems involving the simple reaction of functional groups on polymer and curing agent, the constraints of storage and reactivity are virtually impossible to overcome – in some circumstances it is not unknown for the reaction temperature to even be lower than the storage temperature. Successful strategies have been to separate the reactive components into two or even three packs which are immediately mixed prior to use, thus negating the need for long storage times. Alternatively, a crosslinking chemistry may be employed for single pack systems which become operative only by the action of an atmospheric component such as oxygen, water, or more recently a vaporised catalyst. A number of examples of this type are described in more detail in Chapters 6 and 7.

For heat cured (stoving) systems, thermally activated crosslinking chemistries mean that one-pack systems are more common. Changes in manufacturing practice towards lower stoving temperatures and increasing use of plastics materials, whilst maintaining or increasing production line speeds, all conspire to make the job of the coatings scientist more challenging.

4.3.3 Chemical Reactivity Control

A general scheme for the crosslinking of an A-functional molecule or polymer with a B-functional polymer is given in Figure 4.4. The A and B groups diffuse together in a random process to form a cage complex

Figure 4.4 *General kinetic scheme for the random diffusion and reaction of A+B*

and either combine (with a rate constant k_r) or diffuse apart (with rate constant k_{-D}). Assuming a steady state for the complex, the overall rate constant k_{obs} for the reaction may be written as (Equation (4.3)):

$$\frac{1}{k_{obs}} = \frac{k_{-D}}{k_D}\frac{1}{k_r} + \frac{1}{k_D} \tag{4.3}$$

In crosslinking systems where molecular mobility and functional group concentrations are high ($k_D \gg k_r$), then the overall reaction rate constant of reaction may be simplified to Equation (4.4):

$$k_{obs} = \frac{k_D}{k_{-D}} k_r \tag{4.4}$$

To a first approximation, the reaction rate constant is independent of the size and mobility of reactants. In this case the temperature dependence of the rate constant is described by the Arrhenius equation (Equation (4.5)) where A is the pre-exponential factor for the reaction, E_a is the activation energy, R is the gas constant and T is the absolute temperature. In the early stages of the crosslinking reaction the Arrhenius equation:

$$\ln(k_r) = A \, \exp\left[\frac{E_a}{RT}\right] \tag{4.5}$$

provides an adequate description of the system. This is useful as many crosslinking processes are conducted under non-isothermal conditions well above T_g. When combined with knowledge or assumption of the reaction mechanism and definitions of acceptable stability and reactivity, then approximate guidelines for control and design of crosslinking kinetics can easily be obtained.

The limitation of this treatment is most apparent in isothermally cured systems. As the crosslinking reaction proceeds, the MW builds rapidly and the system T_g will increase. When T_g rises and approaches the cure temperature, then the system vitrifies. Under these conditions, diffusion processes slow down and reach the condition where $k_r \gg k_D$.

Inspection of Equation (4.3) reveals that under these conditions, the first term on the right-hand side can be neglected. Thus $k_{obs} = k_D$ so that diffusion control now operates and controls the rate of reaction.

4.3.4 Diffusion Control

The overall rate of reaction now depends on how quickly the polymer segments carrying the functional groups can move through the reaction medium to allow reaction to take place. In these circumstances, the free volume approach is a much better description of the temperature dependence of the kinetics, rather than the Arrhenius approach. Under these circumstances the reaction rate constant depends on the difference between T and T_g. Free volume theory suggests that when T_g exceeds the cure temperature by *ca.* 51 °C, the rate constant for the reaction is infinitely small.

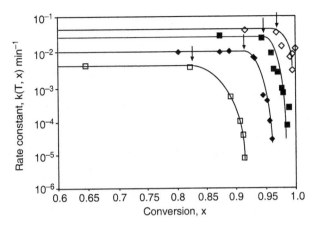

Figure 4.5 *Variation in observed rate constant, k, with epoxy conversion, x, for a simple epoxy–amine reaction*
(Reproduced with permission from K.P. Pang and J.K. Gillham, *J. Appl. Polym. Sci.*, 1990, **39**, 909)

The changeover from activation to diffusion control is demonstrated in data reported by Gillham for an epoxy–amine system (Figure 4.5). The overall rate constant for the epoxy–amine reaction was monitored as a function of epoxy conversion at a series of temperatures. It was found that the reaction rate constant was essentially independent of conversion until, over a small conversion range as the system vitrified, the rate constant decreased rapidly by up to several orders of magnitude.

The explanation for this behaviour is seen in Figure 4.6 in which the increase in T_g with conversion is typically plotted. During an isothermal

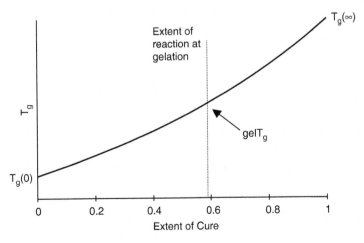

Figure 4.6 *Typical dependence of T_g on conversion*

cure the T_g rises. Early on in the reaction the T_g is usually well below the cure temperature and increases only slowly. Chemical kinetic control operates in this regime. If the cure temperature is below the ultimate T_g (the system T_g at $x = 1$, known as $T_g(\infty)$), then at some conversion below $x = 1$ the T_g will have risen to approach the cure temperature. When this occurs, the system has transformed from a viscous liquid to an amorphous solid and further reaction is controlled by diffusion. The T_g–conversion curve, particularly for epoxy–amine systems, is commonly independent of the temperature of the reaction. The T_g can therefore be taken as a measure of conversion (and also crosslink density). In addition and importantly, it implies that a specific and unique T_g corresponds to the gel point of the system. This is known as the gelT_g.

At high extents of reaction in crosslinking systems, the concentration of the residual unreacted functional groups will be low and increasingly likely to be 'fixed' on the network. Thus, the final stages of reaction may be exceedingly slow or (in practical timescales) will never occur. This is often termed the topological limit of crosslinking reactions and there is evidence to suggest that for stoichiometric rigid diepoxy–diamine systems this limit is about 90–95% conversion. If a reasonable excess of one type of functional group is present, such restrictions on the reaction do not normally occur.

Free volume theory predicts that the diffusion rate constant falls to zero when the system T_g is 51 °C above the cure temperature. In practice, however, the rule of thumb is that the reaction rate is immeasurably slow when the system T_g exceeds the cure temperature by about 25–35 °C. For coatings cured at ambient temperature, *e.g.* 25 °C, the T_g is therefore limited to 50–60 °C. If insufficient conversion has been achieved at this point, the coating will likely fail prematurely. Reactions can of course continue if the temperature is subsequently increased. In addition, by exercising choice of solvents with different volatility, the system T_g may be depressed for long enough (by solvent plasticisation) for further reaction to occur so that when the solvent does eventually leave the coating (this may take years), the final T_g may reach higher temperatures in some examples up to 50 or 60 °C above the cure temperature.

The above rule of thumb applies to crosslinking by random step-growth reactions. There are known exceptions to this vitrification limitation. The first occurs with chain-wise crosslinking. With highly heterogeneous free radical initiated crosslinking, network T_gs of 70 °C or more above the cure temperature can be achieved (note with homogeneous systems the normal rule applies). There is also an example of a step-growth system which possesses intermolecular ionic interactions which allow orientation of the reacting groups before crosslinking. In this way the $(T_g - T)$ can reach 75 °C (still only 50 °C above the cure temperature) as a result of there being very little requirement for movement prior to reaction.

Thus, it can be appreciated that the design of coating systems to achieve specified performance-related conversions under varying cure conditions is complex. It requires knowledge of conversion–performance and T_g–conversion relationships as well as reaction kinetics and solvent evaporation behaviour under different application conditions.

4.3.5 T_g–Conversion Relationships

For systems without any solvents present, very useful information can be gained from isothermal curing studies where the data are presented in terms of a time–temperature–transformation diagram as developed by Gillham and co-workers. In such diagrams (see Figure 4.7), the conditions, times, and temperature to gelation and vitrification are presented in a type of phase diagram. If the relationship between T_g and conversion is known, then time–temperature–conversion curves can be drawn.

The $T_g(\infty)$ is the maximum T_g for a particular system. It is necessary to cure within 30 °C or above this temperature to achieve it. The value of $T_g(\infty)$ is dependent on crosslink density and the structures of the component parts of the network. If the maximum achievable extent of conversion is less than 1, the $T_g(\infty)$ will not be achieved. Intuitively, we would expect that as the distance between branch points or network junctions decreases then T_g will increase. Nielsen has described an empirical relationship (Equation (4.6)) which relates the rise in T_g (ΔT_g) at full conversion to network with M_c, the number average molar mass between elastically effective network junctions.

$$\Delta T_g = \frac{3.9 \times 10^4}{M_c} \tag{4.6}$$

For highly crosslinked systems, such as epoxies and phenolics, M_c is typically 300–1000 giving ΔT_g values of 130 and 40 °C respectively.

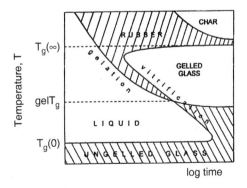

Figure 4.7   *TTT phase relationships for a thermoset polymer*

For moderately crosslinked polymers, such as those based on thermosetting polyesters, M_c is typically 1500–5000 leading to ΔT_g of 25 and 8 °C respectively. The magnitude of changes in T_g on crosslinking makes it a useful means of monitoring or following the process particularly as sensitivity is increased as full conversion is approached. Recently, theoretical treatments have successfully predicted the non-linear relationship between T_g and extent of conversion.

Figure 4.7 also shows the effect of curing at a temperature below the gel T_g: vitrification will occur *before* gelation. The system will solidify and enter the diffusion-controlled regime before any network is formed at all. This is known as B-staging, and is actively used to control some thermoset systems. Inadvertent B-staging, however, could be catastrophic for the properties of a coating.

4.4 CONSEQUENCES OF VITRIFICATION

The transformation of the liquid coating to the solid film has many consequences for the performance properties of films (Chapter 5). Three of the most important will be briefly discussed in this section.

4.4.1 Internal Stress

As a result of both solvent evaporation and chemical reaction, coatings will tend to shrink during film formation. The volume reduction due to chemical reaction alone in a typical epoxy–amine coating is commonly around 5%. While the T_g of the coating is still below the film formation temperature, there is still plenty of mobility in the system for the polymer segments and molecules to accommodate the shrinkage. Once the coating has vitrified this is no longer the case. Stresses start to develop in the coating arising from the adhesion of the film to the rigid substrate. As solvent evaporates or reaction occurs, the lateral movement of the segments and molecules to accommodate the requirement for a smaller volume cannot occur as the polymer system adheres to the substrate. The film then behaves as if it is in tension. The greater this tensile stress, the less external perturbation is required to promote unwanted effects such as delamination or cracking in the coating.

Note that environmental factors can also lead to stresses – a temperature decrease will lead to shrinkage (tensile) stresses whereas a temperature increase will lead to expansive (compressive) stresses. Similarly, swelling in water or other liquids can lead to stresses as long as the T_g remains higher than the prevailing temperature. Cyclic stresses generated by the diurnal cycle or wet/dry periods can be particularly damaging. However, such stresses are not intrinsic to the film formation process, and the term *internal stress* is reserved for this.

4.4.2 Solvent Retention

The solvent evaporation and crosslinking processes are highly interdependent. In the early stages of film formation when solvent evaporation is at its most rapid, loss of solvent leads to higher functional group concentrations and consequently faster chemical reaction. Loss of solvent, however, also leads to increased T_g, and lower species mobility. Once the system approaches vitrification, both the chemical reaction and the evaporation of solvent are slowed considerably. The result can be significant solvent retention – particularly for fast reacting systems – which can impact on the performance of the coating in a variety of ways including, reducing hardness, increasing water sensitivity (leading potentially to blistering and corrosion), and sometimes an inhomogeneous distribution of residual solvent (with many possible consequences). It has also been suggested that the existence of tensile stresses in the film actively promotes the retention of solvents in order to minimise the stress (Le Chatelier's Principle).

4.4.3 Physical Ageing

This is a phenomenon common to all vitrified materials and arises from the very restricted ability of the polymer segments to undergo any structural reorganisation in the glassy state. As a material passes into the glassy state, *e.g.* by cooling (see Figure 4.1), the rates of structural reorganisation (relaxation) are much lower than the rate of the temperature reduction. The system cannot adjust its structure in response to the lowered temperature. It is therefore in a non-equilibrium state and possesses excess volume. During storage in the glassy state, there will be a very slow reduction in volume depending on how close the storage temperature is to the T_g. This results in an increase in T_g, and thus the process is self-retarding. Physical ageing can have a significant impact on the mechanical and other properties of the coating and is discussed in more detail in Chapter 5.

4.5 SOLVENTLESS CROSSLINKING SYSTEMS

The most important class of coatings in this category are powder paints, although solvent-free liquid paints are increasing in importance. Powder coatings technology is described elsewhere in this book and, of course, the important characteristics of crosslinking systems described above apply to powder coatings. Film formation in electrostatic spray powder systems starts with a solid in particulate form on the substrate. On stoving, the individual particles soften and flow so that coalescence takes place. During this early stage in film formation, it is hoped that complete wetting of the substrate occurs, entrapped air is released, and an even film of now liquid coating covers the surface. Special additives are used to ensure this is the case. Simultaneously, as the temperature rises in the oven, the crosslinking reaction starts to occur giving rise to MW and viscosity build as described earlier. Near gelation, flow and levelling cease and the surface appearance at that time is fixed in place.

The viscosity and temperature profile during stoving might typically be as shown in Figure 4.8. The so-called 'viscosity well' describes how fluid the system becomes and for how long. It is determined by a number of factors including rate of temperature rise, crosslinking reaction kinetics, MW build characteristics prior to gelation (which depends on the functionality of the system) and extent of reaction at gelation. This is in addition to MW, viscosity and pigment volume concentration of the starting materials. The normal 'orange peel' appearance of most conventional powder coatings would indicate that there is still room for improvement in balancing the rheological and reactivity characteristics of these systems.

Powder coatings are single pack systems where it is assumed each particle contains all formulation ingredients in the right proportion. Storage prior to use places requirements both on physical and chemical stability. To avoid physical sintering, the T_g of the powder needs to be above about

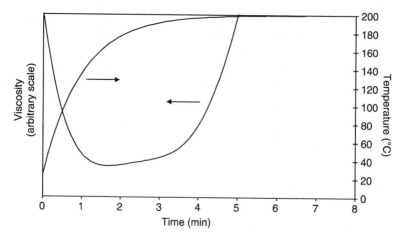

Figure 4.8 *Schematic of typical viscosity profile for a thermoset cured under non-isothermal conditions*

50 °C and depends on storage temperature and hydrostatic pressure. Chemical stability in the solid state does not usually pose too many problems since the raw materials are in a glassy state where reaction rates will be diffusion controlled and very slow. Traditional methods of manufacturing powder coatings, however, require the use of extruders to mix and disperse the ingredients into the polymer matrix. Reaction will inevitably occur in the extruder operating at typically 100–120 °C or even higher. Such pre-reaction will at best degrade the flow and levelling and hence appearance of the final coating. The problem is exacerbated by the demand for powder coatings which cure at even lower temperatures for use on temperature-sensitive substrates. Currently, the lower limit on stoving temperatures is about 120–140 °C.

Other important solventless-liquid systems include radiation cured and two-pack chemistries, which are briefly discussed in Chapter 7.

4.6   DISPERSE PHASE POLYMER SYSTEMS

The problems of using solutions of high MW thermoplastics in coatings have been successfully overcome by using disperse phase polymer technology. This approach is described in more detail in Chapter 8. The advantages of stabilised dispersions of polymer particles in an aqueous continuous phase are those of low viscosity, low VOC and the possibility of using very high MW polymers. The viscosity of a dispersed phase in a continuous liquid is more or less independent of the MW of the disperse phase polymer but depends on the volume fraction of the disperse phase. To take advantage of the properties offered by such high MW, the process of film formation in these systems is of paramount importance.

Film formation from a latex is commonly divided into three steps. Firstly, after application of these coatings to a non-porous substrate, water and organic co-solvents begin to be lost by evaporation, the solids content rises and particles approach each other more closely. Secondly, for successful film formation, the particles must coalesce and this involves overcoming the hitherto effective repulsive interparticle stabilisation forces. The driving forces for coalescence are thought to arise from capillary forces as particles get very close and/or the surface energy reduction as the sub-micron particles form a continuous film. Whatever the origin the forces are also such that substantial deformation of the spherical particles into closer packed polyhedra occurs in the latter stages of film formation. Finally, film formation requires diffusion of polymer chains across the particle interface which, for ambient systems, means that the T_g of the polymer in the interfacial regions must be below ambient. This does not necessarily mean that the particle 'bulk' T_g has to be low. Organic co-solvents or coalescing aids will plasticise the polymer for long enough to allow film formation and when lost by evaporation the coating T_g will climb to above ambient. An alternative approach is to create multilayer or gradient composition particles through controlled emulsion polymerisation techniques. Here it is possible to create a high T_g core with a low T_g shell in a particle to combine high T_g properties with low film formation temperatures.

The advantages of crosslinking the coating are difficult to achieve in ambient disperse phase systems. Such systems are beginning to appear on the market but are not yet in widespread use. For stoving systems, however, a number of crosslinking chemistries have been available for some time finding widespread application in markets such as packaging and coil coatings. Film formation and crosslinking may have to take place extremely rapidly in such end uses where stoving schedules may be less than a minute with metal temperatures rising to 250 °C or higher. Subsequent performance in hot aqueous or corrosive environments relies on the quality of film formation by coalescence and the absence of weak boundary pathways for the ingress or egress of corrosion reactants/products.

4.7 BIBLIOGRAPHY

1. J.P. Pascault, H. Sautereau, J. Verdu and R.J.J. Williams, *Thermosetting Polymers*, Marcel Dekker, New York, 2002.
2. T. Provder and M.W. Urban (eds), *Film Formation in Coatings. Mechanism, Properties, and Morphology*, ACS Symposium Series 790, ACS, Washington, DC, 2001.
3. R.F.T. Stepto (ed), *Polymer Networks*, Blackie, London, 1998.
4. Z.W. Wicks Jr., F.N. Jones and P.S. Pappas, *Organic Coatings Science and Technology*, 2nd edn, Wiley-Interscience, New York, 1999.

CHAPTER 5

Performance Properties of Coatings

A.B. PORT and C. CAMERON

5.1 INTRODUCTION

The performance of a coating is what gives it its value. The term performance covers a great variety of effects but whatever it means, and this is the subject of the present chapter, the performance of a coating film is sensitive to its chemistry, its state after film formation and its end-use. Implicit in the performance property is an estimate of the service life of the coating.

Coatings fail because of irreversible changes which occur in the film as a result of exposure to a variety of possible stresses. The changes in a coating leading to failure are almost exclusively physical and may include phenomena such as gloss loss, colour change, dirt retention, chalking, cracking, delamination, blistering, fouling and corrosion. The stresses that give rise to the failure may be large and short lived (*e.g.* impact) or small and long lived, often cyclic in nature. Moreover, the changes in the physical properties of the coating are often a consequence of underlying chemical processes occurring in the coating. In order to extend the service life of a coating, emphasis should be put on defining and understanding the mechanisms by which the controlling chemical changes occur. Without this knowledge of mechanism, it will not be possible to define or measure the controlling parameters. Unfortunately, while there are many techniques to study the 'damage' to coatings due to these chemical changes, at present there is no clear link between the chemical damage and the physical damage which results in ultimate failure. Developing this link would be a very valuable goal and only relatively recently effort has been expended on actively investigating this type of property–performance relationship. This represents a real challenge to the coatings technologist, but one in which the rewards could be significant.

Performance describes how well the coating is carrying out its function in service. It will be appreciated from other chapters that coatings are required to fulfil a broad range of roles and provide a number of effects. This in turn leads to diverse classifications for coatings performance and it would be beyond the scope of this book to attempt to describe in detail all types of performance, the coatings technologies to achieve them and the procedures used to predict or monitor those performances.

The approach that has been taken in this chapter is to concentrate on coatings performance in terms of its response to physical, mechanical, environmental and chemical stress. In particular, the maintenance of film integrity under these forms of attack is chosen, the rationale being that once the coating is cracked and/or lost from the substrate, its primary roles have been at least compromised and/or the criteria for coating failure satisfied. The retention of this performance in service will be an important factor in determining the lifetime of a coating system and so the influence of chemical and physical ageing is also addressed.

5.2   MECHANICAL PERFORMANCE

The mechanical performance of a coating describes how it responds to stresses and strains imposed on it during service. This performance is rarely the sole criterion, rather it is one of a group of properties which must be achieved and maintained. In this way the importance of mechanical properties may vary from being a key factor to being one of secondary consideration. Nevertheless, failure to resist some form of physical or mechanical abuse, which results in cracking, and any subsequent loss of coating must always be regarded as a limiting performance parameter.

The severity and frequency of the mechanical abuse clearly plays an important part in determining the relative position of mechanical performance both in the hierarchy of properties to be met and also at which stage it is considered during product development. By way of an example, coatings intended for some can and coil coating end uses have been designed around flexible and extensible polymers because of the severe deformations involved in coil forming and the manufacture of two-piece cans by the draw–redraw (DRD) process. In comparison, when mechanical performance is of secondary importance, a coating's mechanical properties may be addressed by the formulator at a relatively late stage of product development. The opportunities for such 'last minute' formulation changes are becoming fewer with the increasing constraints imposed by environmental legislation. A good example is the use of silicone elastomer technology in anti-fouling coatings, where the non-biocidal nature of the film outweighs the poor mechanical strength and abrasion resistance of the elastomer. It is now increasingly important for modifications to the formulation to avoid contributing to the VOC level. This factor obviously limits the choices available and the traditional approaches of adding fillers, extenders, plasticisers, co-film formers, *etc.* cannot be used where the penalty of higher viscosity has to be offset by additional solvent.

For this reason, and others, it becomes more important to try to target the balance of properties including mechanical performance through design of the system from the start of the coatings development. In order to do this, the key links between coating performance, material properties and coating/polymer design have to be established. Given the complexity of each of these subjects, it is not surprising that there are few quantitative performance–property–structure relationships in coatings science. The potential benefits, however, are such that this approach to coatings development is the subject of much effort in the coatings industries. If successful it should be possible to learn important lessons from each coating product development and in such a way that the knowledge gained and the emerging technology is transferable within the various coatings business sectors.

5.2.1 Performance Specification

Physical and mechanical performance requirements of coatings in service vary enormously with the different end use applications. A number of examples will serve to illustrate the diversity and complexity of the applied stresses and strains.

5.2.1.1 Marine Coatings. A general-purpose anti-corrosive and anti-abrasive paint might be required to perform equally well on a ship's side or deck or as a cargo hold coating. On the side of the ship at sea the coating experiences both diurnal temperature fluctuations and intermittent water cooling/heating. The difference in thermal expansion between paint and metal results in cyclic compressive and tensile stresses in the coating. There may also be dilational and shrinkage stresses set up in the coating as water is absorbed or desorbed. The stresses and strains involved are relatively low and are applied at low rates. In addition, there is the flexing of the ship in response to the wave motion of the sea. On berthing, as the ship is unloaded and loaded, it moves against the fenders at the dockside (Figure 5.1). Here moderate to high shear stresses are exerted on the coating at low strain rates as the ship rises and falls. Within the hold, the coating will sustain high speed impacts as a cargo falls from heights of up to 30 m. The cargo may be hard, angular particles so that local stresses and strain rates on the coating

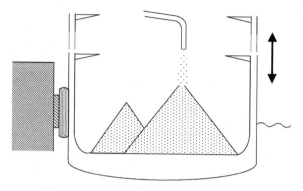

Figure 5.1 *Stresses and strains on marine coatings*

may be very high. Up on deck a complex mixture of stresses, strains and strain rates will be expected as people and machinery go about their work but with the added complication of chemical attack from liquid cargoes, lubricants, *etc.*

5.2.1.2 Can Coatings. In the DRD process for making two-piece cans, coated metal is drawn by punch and die into progressively taller, narrower can bodies. Strains in the direction of drawing may be in the order of 100% whereas strains normal to that direction are in the order of −50% (to maintain approximately constant wall thickness) (see Figure 5.2). Maximum punch speeds are a few metres per second so that strain rates will be relatively high. Metal and tooling temperatures will be above ambient due to the amount of work done on the metal during forming.

Further forming processes are carried out on the can body after drawing which may include putting expansion rings into the base, beading into the can wall, trimming the can to the correct height and providing a flange for end seaming (Figure 5.2c).

Each of these processes occurs at the high speeds commensurate with modern can making. Following this somewhat traumatic process of manufacture, the coated can body is filled, typically with foodstuff, and processed for anything up to 90 min at 130 °C for storage prior to use by the consumer. The coating must continue to provide corrosion protection during processing, transport and storage for many months. It is thus vital that these thin coatings (7–15 μm) retain their integrity and adhesion to the substrate during all of these processes.

5.2.1.3 Building Paints. In contrast to the above two examples, it would seem that a paint intended for decoration and protection of exterior

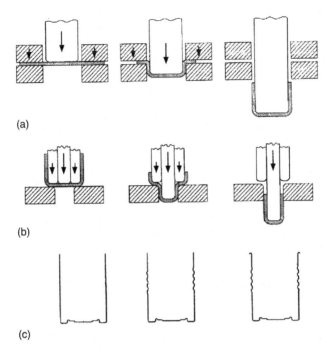

(a)

(b)

(c)

Figure 5.2 (a) *The drawing process;* (b) *the redrawing process;* (c) *further forming processes in can manufacture*

woodwork has few mechanical demands placed upon it. This substrate is, however, dimensionally unstable and expands and contracts with changes in temperature and, more importantly, with moisture ingress and egress. For a continuous substrate the stresses and strains involved are relatively small and usually occur over protracted timescales. When the wood cracks on drying, local strains may become very high. These strains must be recoverable should the wood take up water, swell, and close the cracks otherwise buckling or blistering of the coating will take place (Figure 5.3).

The retention of elasticity or plasticity under exposure to sunlight and moisture is clearly important here. Similarly, the maintenance of strong adhesion to this variable substrate is a key performance parameter.

The coatings scientist must have a clear understanding of the nature of the mechanical demands imposed in a particular end use. The critical abuses or performance criteria need to be established and characterised in terms

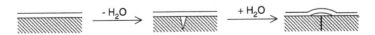

Figure 5.3 *Distortion of a building paint coating on wood*

of the types and magnitudes of stress, strain, and strain rate, temperature, and frequency or timescale. Where possible, quantification of these parameters should be undertaken although it is clear from the previous examples that this is often difficult. The analysis should be attempted at some level so that the appropriate laboratory test protocol can be devised and relevant material properties measured. In some specialised applications, a set of material properties, which have been found to relate to satisfactory performance in service, have been specified by the end-user. Here any new product under development needs to achieve the specified properties before the customer would consider the coating for his application. Examples of this type would include paints for the automotive and aviation markets. This approach occasionally causes problems when a specification is written around a particular coating technology found to be successful in an application. A new coating might well achieve satisfactory performance levels in service but fail the specification tests because the tests do not accurately reflect what is required in the application, only the properties of the previous material. In such cases the arguments for the new coating must be sound and persuasive (or the customer must want to adopt the new technology) to overcome the inertia and cost implications by which an established system resists change.

The quality of the information gained on the service environment and a coating's performance in that environment can be a major contribution to the successful development of a coating and its introduction to the marketplace. The pressures of modern materials development require that performance testing and service life prediction (SLP) must be both rapid and reliable. Recently, there has been much activity in the application of 'reliability' methodologies for SLP. This methodology attempts to make a quantitative assessment of the service life of a coating in any potential use environment and does so by developing a predictive model of how a coating property changes as a function of applied cumulative environmental stress. An excellent monograph has been published on this subject to which the reader is warmly recommended to refer. In essence there are three requirements for the reliability methodology:

- laboratory experimentation
- environmental characterisation
- model validation

The objective of laboratory experimentation is to develop a mathematical equation describing a coating degradation parameter as a function of applied environmental stress. The exposure environment must be characterised and monitored to yield the environmental stresses as a function of time in a form that is compatible with the cumulative damage model derived from laboratory experimentation. These environmental data are then used as an input into the cumulative damage model and the total damage is found by integration over the exposure period. Ideally, the degradation parameter predicted by the cumulative damage model tracks the observed change in the degradation parameter. Thereafter, the damage model can be used to predict the degradation parameter for any given set of environmental conditions.

In the absence of quality service life information, much is left to the experience of the coatings formulator and product development is guided by comparisons with previous products or those of a competitor. In either case it is usual to employ a number of mechanical performance tests and material property determinations in coatings research and development and these are described in the next section. This testing is usually carried out for two reasons – to protect the coating manufacturer from product failure claims and to demonstrate performance of the coating, sometimes against a defined specification. Sometimes, the former does not receive as much attention as it should.

5.2.2   Traditional Mechanical Performance Tests

Examination of the above three examples of coatings undergoing physical and mechanical duress suggests that the properties of hardness, flexibility, impact resistance, formability, adhesion, wear and friction would be important parameters in describing mechanical performance. Laboratory tests have been devised to provide empirical or semi-quantitative measures of these properties of coatings, usually on the substrate of interest. Some of these tests have been sufficiently widely used to be covered by national standards, *e.g.* BS3900 in the UK. There are far more tests, however, which have been devised to emulate coatings in specific applications sometimes peculiar to particular industries or even individual companies. Clearly, it would be beyond the scope of this chapter to describe all of these tests and the sections below deal with those perhaps most widely used.

5.2.2.1   Hardness. 'Pencil hardness' is quoted as the grade of pencil which either marks the coating or causes a deep scratch in the surface. This test can be somewhat subjective and operator dependent. These problems are overcome by using the more classical technique of measuring the penetration of a known indenter under a given load. Care must be taken when using this method for thin films on rigid substrates. When the depth of penetration becomes comparable with film thickness, the substrate influences the result making the coating apparently harder. Similar effects can be seen for multilayer coatings of different thicknesses. Pendulum hardness is also widely used in the coatings industry. Here a hard knife-edge mounted in a pendulum is set rocking through a given angle on a coated panel, and the number of swings counted before a certain angular decay has occurred. It should be noted that the result depends not only on hardness in terms of the depth of penetration of the knife-edge, but also on the damping properties of the coating, *i.e.* the amount of energy absorbed during cyclic deformation.

The most rapid, cheapest and most subjective tests are the thumbnail and the penknife. These are widely used in the coatings industries and despite their crude nature are still powerful in the hands of an experienced paint maker and formulator.

5.2.2.2 Flexibility. A number of tests have been devised to determine the severity of bending of a coated panel needed to cause failure in the coating as seen by cracking. The simplest of these is the T bend test where a coated panel is folded back on itself repeatedly until the coating remains intact (Figure 5.4). The most flexible coating will survive the first folding (zero-T), less flexible ones will be one-T, two-T, up to four-T.

A series of hinges with different diameters at the apex can be used to fold panels. Again the most flexible panel will survive the most severe, lowest diameter bend. This is known as a wedge bend or single diameter mandrel test. An alternative to this is the conical mandrel where a panel is folded around a single mandrel in the form of a cone with diameters ranging from 3 to 37 mm. The diameter at which folding occurs without cracking is noted; again the most flexible coating will survive the tightest bend without cracking.

As will be described later, the properties of the coating depend on temperature, rate of deformation, film thickness, age and conditioning so that care should be taken to conduct these tests consistently.

5.2.2.3 Impact Resistance. Falling weight impact testers are widely used to assess a coating's resistance to rapid loading and deformation.

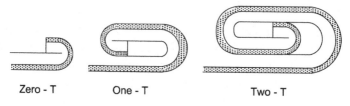

Zero - T   One - T   Two - T

Figure 5.4 *T bend tests*

Typically a flyer or indenter of known mass is dropped through a known height onto a supported coated panel. In one arrangement, indenters with one hemispherical end are used on light gauge aluminium or steel panels such that the panel is substantially deformed. Combinations of weight and height are used to deliver increasing impact energies, expressed as N m or J, until cracking and delamination are observed either on the front (forward) or back (reverse) of the panel. The panel material type and thickness should be specified since the amount of metal deformation will be affected, as will the impact energy for failure. For some end uses, heavy metal plate is used in which case deformation of the substrate is much less. Care should be taken when repeated tests at different energies are used on the same plate since the area of damage from previous impacts can extend out considerably from the point of impact. This damage may not be immediately visible to the operator but ultrasonic tests (or the penknife) will reveal polymer–substrate disruption or delamination which will compromise subsequent results if impacts are made too close to each other.

5.2.2.4 Abrasion and Wear. This is probably one of the most difficult laboratory tests to relate to in-service performance. A number of designs have been used including the measurement of coating weight loss under abrasive wheels or balls, falling sand or gravel. Many of the tests are specific to particular end-uses, and attempt to mimic the exact conditions of stresses and strains to which the coating is subjected. A discussion of these individual custom tests is beyond the scope of this book.

5.2.2.5 Formability. A number of test designs have been developed to emulate the forming, drawing and stretch forming operations carried out on coated metal. In addition to the T bend test described above, box draw tooling and the Erichsen cupping press have been used. Again it is necessary to apply representative tool/substrate temperatures and deformation rates as well as lubrication to obtain data relevant to particular operations.

5.2.2.6 Friction. A simple apparatus to measure static and dynamic coefficients of friction under load involves pulling a weighted sledge across a coated panel. Where higher pressures are applied to coatings, such as occur during the movement of coated metal into the die in the DRD process described above, hard polished balls may be used. Choice and weight of lubricant film on the coating and condition of load bearing surfaces need careful attention to avoid spurious results.

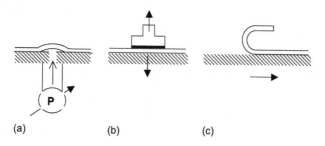

Figure 5.5 (a) *The blister test;* (b) *the bolt-pull test;* (c) *the peel test, for adhesion*

5.2.2.7 Adhesion. This critically important performance property of coatings remains an area of controversy. The penknife and its slightly more sophisticated multi-bladed version known as the cross hatch tester remain the most widely used methods of assessing coating adhesion. There are few, if any, methods for quantifying the adhesive strength of thin coating films to their substrates. Some success in terms of method and analysis has been achieved in the blister test developed by Briscoe and others. Here air or fluid pressure is applied behind the coating to grow a blister (Figure 5.5a). The so-called bolt-pull test, where a bolt (sometimes known as a 'dolly') or similar is bonded to the coating and the load to cause coating–substrate debonding is measured (Figure 5.5b), is irreproducible and analytically flawed. It does, however, give information on the locus of failure – either at the interface or cohesive within a layer of paint. Similarly, lap shear testing is problematical and irreproducible. The peel test (Figure 5.5c) is unsuitable for most glassy coating systems.

It is probably fair to say that the extremes of good and poor adhesion present no problems to the coatings scientist. At one extreme any attempts to remove the coating result primarily in cohesive failure of the coating itself, whilst at the other coating adhesion failure is only too evident. When adhesion is marginal, a sensitive quantitative method of measuring the strengths of adhesion at the interface with its substrate (metal or another coating) is still required. Without such a method, polymer design and coating formulation for improved adhesion become very difficult especially in view of the complexity and importance of surface preparation, application conditions and environmental variability.

5.2.3   Material Properties of Coatings

In addition to the traditional paint tests, increasing effort has been put into determining more fundamental materials properties of coatings. This normally requires the preparation of free films of the coatings or, in some cases, 'bulk' specimens a few millimetres in thickness. Free films are conveniently prepared by applying the coating by an appropriate method to low energy surfaces such as fluoropolymer sheet, fluoropolymer coated metal or glass, or substrates treated with release agents. Bulk specimens will usually require tooling for casting or compression moulding. This is relatively straightforward for zero VOC systems which evolve no volatiles during cure or film formation. Where solvent, water or other volatile products are generated, they must be allowed to escape, otherwise voided specimens will result.

Once the difficulties of specimen preparation are overcome, a powerful range of materials property measurement and characterisation techniques is accessible. These techniques can give valuable quantitative information on the stress–strain behaviour, dynamic and thermo-mechanical properties, and the deformation, yield and fracture response of the coatings. Indeed, the application of fracture mechanics to coatings is increasingly important for understanding performance–property relationships. Time-dependent or viscoelastic properties can be measured and can be used to predict behaviour at extremes of time and temperature not conveniently available in the laboratory. Much of the data produced in these experiments can be directly related to the large body of published information on polymeric materials properties. The theory of polymer viscoelasticity can be used to analyse the data to characterise the structure of the coatings and to suggest the micro- and macromolecular origins of the macroscopic response of the material to stress and strain. This continues to be an area of much activity and development – from both a theoretical and a practical point of view.

The samples can also be used in other experiments for the characterisation of structure and morphology of the materials and so aid the development of structure–property relationships. Examples of these techniques would include scanning and transmission electron microscopy for examination of morphology and fracture surfaces, scattering techniques for identification of fine structure and ordering, and solvent swelling methods for network characterisation.

5.2.3.1 Stress–Strain Properties. In the simplest and most commonly used stress–strain test, a specimen is deformed at constant rate and the resultant stress measured by some form of load cell (Figure 5.6a). Results are plotted as stress (load/original cross-sectional area of sample) against strain (extension/original length of specimen). Polymers and, hence, organic coatings exhibit a wide range of stress–strain behaviour from hard and brittle (A) through tough, ductile (B), to soft elastometric (C) (Figure 5.6b).

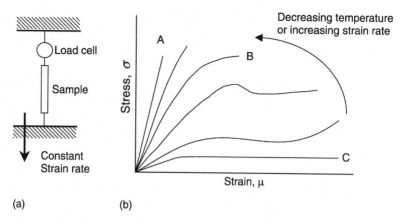

Figure 5.6 (a) *Stress–strain measurements;* (b) *typical representation of stress–strain results*

The form of the stress–strain curve is influenced not only by the structure of the polymer but also strongly by temperature and strain rate (as indicated in Figure 5.6b). This dependence is strongest when the coating T_g and test temperature are reasonably close, say ± 30–$40\,^\circ$C. Residual solvent, water or other plasticising species can cause significant shifts in T_g and hence the tensile stress–strain curves. Failure stresses and strains are also particularly difficult to measure for samples exhibiting brittle fracture. For these materials, failure is governed more by the flaws present and the material's ability to resist propagation of these flaws and cracks under stress.

In conclusion, although stress–strain tests are relatively quick and easy to do, the interpretation of the results is more difficult and needs to be done carefully. However, if the conditions of stress, strain, strain rate and temperature for an end use are known, then the generalised response of the coating under those conditions can be easily obtained through these tests. Similarly, if the desired response appropriate to the conditions of abuse is known, appropriate classes of polymer technology can be selected from experience or published data.

5.2.3.2 Rubber Elasticity Theory. Many crosslinked polymers above their glass transition temperature exhibit elastomeric or rubbery elastic behaviour. Such behaviour is normally characterised by systems which consist of long flexible polymeric chains and which are lightly crosslinked. However, coatings with a moderate to even high crosslink density are commonly characterised above their T_gs by means of the formalism. Stress–strain measurements in this rubbery state can be treated by the classical theory of rubber elasticity to yield information on the underlying network structure. Published data on so-called ideal networks can also be used to give an insight into the influence of network structure on physical and mechanical properties. In this context, the distance between branch points, functionality of branch points, network imperfections, and, by inference, the extent of cure all have an influence on stress–strain properties in the rubbery state. As will be described below, these effects may still operate in the glassy region when large strain properties of coatings are considered.

In its simplest treatment, measurements of shear (G) or tensile modulus (E) can be obtained from the slope of the initial stress–strain curve or from dynamic mechanical data. For an ideal rubber, $E = 3G$. The modulus is related to the network crosslink density as shown in Equation (5.1):

$$G \cong \frac{\rho RT}{\overline{M}_C} \qquad (5.1)$$

where R is the gas constant, T is the absolute temperature, ρ is the density and $\overline{M}_C$ is the number average molecular mass between crosslink junction points. More rigorously, equilibrium stress (f) should be plotted against $(\alpha - \alpha^{-2})$, where α is the extension ratio of length/original length. Such plots usually give good straight lines from which v, the number of moles of elastically effective network chains per unit volume, can be obtained (Equation (5.2)):

$$\frac{f}{(\alpha - \alpha^{-2})} = vRT = G = \frac{\rho RT}{\overline{M}_C} \qquad (5.2)$$

5.2.3.3 Dynamic Mechanical Properties. Dynamic mechanical testing measures the response of a polymer to a sinusoidal or other cyclic stress (or strain). For an ideal elastic material, stress and strain would be completely in phase, whereas an ideally viscous response has stress and strain 90 ° out of phase. Real polymers are viscoelastic, in other words possess both elastic and viscous components and strain lags stress by the phase angle δ. There are many designs of equipment used to generate dynamic mechanical data. The most useful of them are able to vary both the applied frequency of the cyclic stress and the temperature of the specimen independently. Depending on the geometry of the instrument, the stress may be applied in tension, shear or oscillatory torsion between parallel plates. In general, the results are expressed as a complex modulus, for example, a complex Young's Modulus, E^* given by Equation (5.3):

$$E^* = E' + iE'' \qquad (5.3)$$

where i $= \sqrt{-1}$, E' is the phase or storage or real part of the modulus and E'' is the out of phase or loss or imaginary part of the modulus. The ratio of E''/E' gives the ratio of energy lost as heat to that stored during one cycle of deformation and equates to the tangent of the loss angle, δ, in Equation (5.4):

$$\tan \delta = \frac{E''}{E'} \qquad (5.4)$$

In the last chapter it was stated that the gel point is sometimes characterised when $E' = E''$ (*i.e.* $\tan \delta = 1$). This is not a general criterion. More rigorously, the gel point occurs when $\tan \delta$ is independent of the frequency of measurement. There are numerous examples in the literature where this has been unambiguously established.

Dynamic mechanical testing is a powerful technique for characterising polymers and coatings. It is one of the most sensitive techniques for determining primary and secondary thermal transitions and for probing the structure of multiphase systems such as semi-crystalline polymers, polymer blends and toughened systems. In phase-separated systems, detailed analysis can often give information on the mechanism of phase separation. In addition to this, dynamic mechanical testing has been used in coatings for determining extent of cure (from T_g and the shape and intensity of the $\tan \delta$ peak at T_g), crosslink density (from the value of rubbery modulus), and determining the optimum stoichiometry of polymer and crosslinker (from maximum T_g) (Figure 5.7).

5.2.3.4 Time-Dependent Properties: Creep and Stress Relaxation Techniques. Creep is the time-dependent deformation of a material under

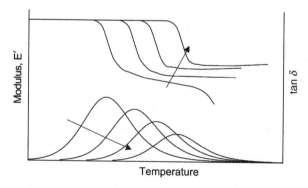

Figure 5.7 *Typical dynamic mechanical data (arrows indicate increasing extent of cure or optimisation of reaction stoichiometry)*

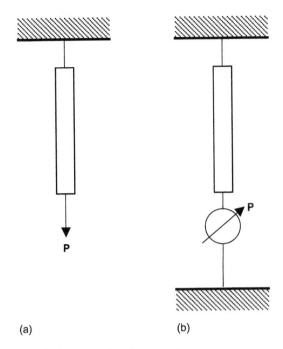

(a)   (b)

Figure 5.8   (a) *Creep and* (b) *stress relaxation experiments*

constant stress (Figure 5.8a), whereas the decay of stress at constant
elongation is termed stress relaxation (Figure 5.8b).

Data from these experiments are typically expressed as creep compliance
curves (Figure 5.9a), where compliance is the inverse of modulus and
relaxation modulus–time curves (Figure 5.9b).

At short times, the material behaves as a hard elastic solid and rates of
creep/strain rate are low (region I in Figure 5.9b). At longer times, rates of
creep increase rapidly (region II) as the material gradually alters its response

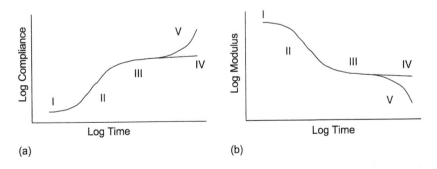

Figure 5.9   (a) *Creep compliance and* (b) *relaxation curves*

from hard to soft elastic behaviour (region III). At still longer times, a
non-crosslinked polymer will behave as though it were a viscous liquid
(region V) and large-scale irrecoverable deformation takes place. When
the polymer is crosslinked, rates of creep/strain rate remain very low
(region IV). The stabilisation of creep rates for non-crosslinked polymers
in region III is due to polymer chain entanglements acting as physical
crosslinks and so the length of this plateau is highly MW dependent.
For many polymers, the timescales necessary to observe these different
regions are inaccessible at either extreme, and data are only conveniently
measured between 1 and 10^5 s. To overcome this problem, the apparent
similarity between modulus–time and modulus–temperature curves is
exploited in the time–temperature superposition principle. Stress relaxation
data are collected over convenient timescales at *different* temperatures and
the data are then shifted along the log time axis relative to a particular
temperature to yield a master curve (Figure 5.10).

   This shift operation can either be done 'by eye' or by using the shift
factor as predicted by the WLF equation (Equation (5.5)):

$$\log a_T = \frac{C_1(T - T_{ref})}{C_2 + (T - T_{ref})} \qquad (5.5)$$

This effectively describes a material's response over many decades of time
or frequency.

   These techniques have been used to address a number of performance
issues for coatings, for example, the rapid assessment of block resistance
and cold flow of architectural paints under pressure and the physical
stability of powder paints at high storage temperatures. For abrasion

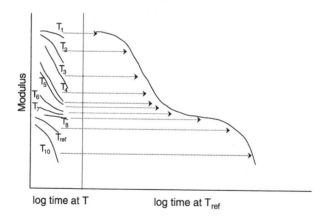

Figure 5.10   *Schematic demonstrating build up modulus curve at T_{ref} from a series of
measurements at different T by the shift method*

resistant coatings, the response to high speed impact can be gauged from the position on the master curve (relative to service temperature) corresponding to the frequency of the test impact, *i.e.* it is a hard brittle glass under these conditions, a viscoelastic solid, or a rubber.

5.2.3.5 Fracture Properties. For glassy polymers under tensile stress, ultimate strengths and elongations are governed largely by the propagation of existing flaws and cracks. Many important classes of coatings are glassy materials and the majority are thermosets. Furthermore, close examination of the mechanical abuse scenarios reveals that the majority involves (at least in some part) the generation of tensile stresses in the coatings. Thus, the fracture behaviour of thermoset glasses is an area of much effort and activity in coatings science. Whereas the science of fracture mechanics of polymers, in general, is well developed (see, for example, Kinloch and Young's text), it is less well understood for glassy thermoset polymers. When these materials are used as thin films on relatively massive and stiff substrates, little quantitative information is known.

Two material properties quantify the stability of a polymer against the initiation and propagation of cracks. These are the fracture energy, G_{1c}, defined as the critical strain energy release rate (or the energy required to form unit area of crack) and fracture toughness, K_{1c}, or critical stress intensity factor. K_{1c} relates the magnitude of stress intensity local to a crack in terms of loading and geometry. These properties can be measured for coatings systems by a number of techniques. For bulk specimens, the compact tension test geometry is favoured (Figure 5.11a), for thin films single edge notched testing (Figure 5.11b) is often carried out. Placing a notch in the sample removes the influence of the crack initiation process. Fracture energy and fracture toughness are material constants and are influenced by temperature, rate of testing and sample geometry. They are

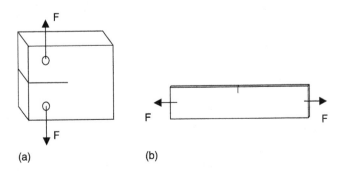

Figure 5.11   (a) *Compact tension specimen geometry;* (b) *single edge notch geometry*

fundamental engineering properties of materials and their measurement for coatings allows comparison with published data for similar classes of polymer.

Published data also provide guidelines for the design of polymer and coating systems for toughness and formability. Toughness in glassy polymers is associated with the ability to undergo extensive yielding and plastic flow ahead of a flaw or crack tip. In bulk specimens, the material itself acts as a constraint to the formation of the plastic zone and relief at the edges of the sample is negligible. For thin films, edge effects are dominant allowing a larger plastic zone to develop. These two extremes of crack geometry are known as plane strain and plane stress conditions and account for the difference in toughness of glassy materials depending on sample size. Plane stress fracture toughness values are typically 2–3 times higher than those in plane strain. The problem remains as to how to treat thin coatings on large, stiff substrates since strategies for toughening of thermoset glasses are influenced by whether plane stress or plane strain conditions apply. At present, neither the theory nor the practical techniques have been developed for determining the fracture behaviour of coated metals.

The major influence on toughness of glassy thermoset coatings is the degree of conversion (or extent of crosslinking reaction). Whenever conditions of time, temperature, stoichiometry, *etc.* lead to glassy materials with low degrees of conversion, the resulting mechanical properties (especially under tensile stresses) can be dramatically weak and fragile. In such cases the failure by cracking of coatings can occur due to the relatively weak stresses associated with polymerisation shrinkage or thermal expansion mismatches.

The greatest single improvement in toughness for such systems can be made by ensuring that the crosslinking reaction goes to high levels of conversion – typically 80–90%. There is convincing evidence that pushing the reaction nearer to completion (>95% for diepoxy–diamine-based coatings) causes the toughness to decrease slightly but this decline is modest compared to the gains to be made by maximising conversion.

Once a sufficiently high level of conversion can be assumed for a coating system, we can use the body of published information on the influence of polymer structure and properties on fracture behaviour to optimise the mechanical performance of coatings on their substrates. From fracture mechanics studies of thermosets, we would expect to see extensive plastic flow and deformation in the plastic zone ahead of the crack tip. Any structural features or conditions or properties which restrict this yielding and plastic flow will restrict the size of the zone (and hence the amount of energy absorbed there) and will render the material more susceptible to brittle failure. The following predictions can then be made which will lead to *decreased toughness*:

(i)   increased modulus, E and yield stress, σ_y;
(ii) increased crosslink density, decreased M_C;
(iii) decreased free volume, *e.g.* from physical ageing *q.v.*;
(iv) decreased temperature and increased rate of testing;
(v)   increased network junction functionality;
(vi) increased proportion of network defects or elastically ineffective network chains and low extents of cure.

In terms of network design and chain architecture, there has been some progress demonstrating that even for relatively high crosslink density glassy thermosets we can use some of the insights from rubber elasticity theory. There is a good correlation between the yield properties and the fracture properties and the network density for different junction functionalities. Indeed, when a junction functionality normalisation factor is applied (the well-known 'front' factor from rubber elasticity) to epoxy–amine materials, all the data lie on a single mastercurve, demonstrating the importance of the network structure. This is convincingly demonstrated in Figure 5.12.

Modification of the chain flexibility results in a new correlation (by moving from an aromatic epoxy networks to aliphatic epoxy networks), but the importance of crosslink density is still obviously apparent.

Local main chain relaxation mechanisms (as seen in secondary or β transitions by dynamic mechanical analysis (DMA)) can be important in

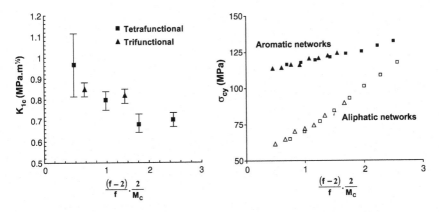

Figure 5.12 *Fracture properties and compressive yield stress of a series of epoxy–amine networks*
(Reprinted with permission of E. Crawford and A.J. Lesser, *J. Polym. Sci. B*, 1998, **36**, 1371)

enhancing toughness. It would appear that β transitions originating in main chain motions are effective in increasing toughness while those originating in side chain motions are not. Detailed correlation, however, between polymer structure and yield, deformation, and fracture is not well developed.

Real coatings will often contain large amounts of pigment, fillers and other additives which add further levels of complexity to the largely polymer-based picture above. The addition of high modulus solids to polymers will increase the modulus and yield stress of the coating but this need not always serve to weaken it. Moderate levels of well-dispersed fine pigment particles can act as reinforcement in terms of increased strength and improved toughness. Poor dispersion and wetting of pigments and fillers generally detract from mechanical performance due to increased number of flaws and voids present. Similarly, application and film formation processes which leave flaws in the coating will produce the same effect. For example, it is possible to reduce adhesion by switching from a slow evaporating solvent to a faster solvent. In this case the molecules have less time for penetration into the substrate and for orientation into a conformation in which the maximum interaction between the functional groups of the molecules and the substrate can interact to maximum extent.

5.2.3.6 Internal Stresses. Application conditions and film formation processes can also lead to the establishment of significant internal stresses within coatings. The commonest causes are from solvent or water or other volatile release and from differential thermal expansion between coating and substrate. In the worst cases, the stresses are sufficient to cause cracking of the coating and loss of adhesion. Their presence, however, will inevitably detract from impact, formability, flexibility and adhesion properties to some extent.

5.3 AGEING PROCESSES AND THE RETENTION OF PROPERTIES

The first part of this chapter dealt largely with the mechanical performance of coatings from the standpoint of design and development, to meet the stresses and strains imposed on them, in various end use applications. Whilst in many cases the critical mechanical abuse takes place soon after the coating is applied, there are many other examples where such abuse is experienced throughout the coating's service life or at some time afterwards. In such cases it is important to know the nature and the effect of the chemical and physical changes occurring in the coating as a result of exposure to its environment. It is then feasible to attempt to predict performance from suitably designed laboratory tests or from real time ageing or exposure experiments. As described in the first part of this chapter, it is necessary both to characterise the exposure environment accurately and to ensure that laboratory test protocol emulates that environment realistically. Similarly, it is necessary to be vigilant in checking the correlation between predictions from accelerated testing and real time experiments carried out by the coating laboratories, and also that these tests reflect performance in actual service.

This becomes increasingly important and more difficult as the design life of the coating lengthens. For example, some coating systems for architectural steel cladding are required and specified to retain their appearance and corrosion protection performance for the design life of the building, say 15–25 years. Clearly, the paint manufacturer cannot wait 25 years before offering a new product for this market. Furthermore, the cost implications of premature failure can be very significant; thus, the importance of a sound understanding and experimental approach is evident.

5.3.1 Exposure Environments

5.3.1.1 Weathering. Perhaps the most important category of exposure environment and its effect on paint performance is 'simple' outdoor weathering. It will be common to most people's experience that domestic outdoor paintwork especially on wood does not retain its appearance or film integrity forever. The paint's gloss level decreases, colours fade or change, and cracking, peeling and blistering may occur particularly in direct sunshine. On closer examination neither the exposure environment nor the chemical and physical processes going on are 'simple'. The paint is subjected to many environmental stresses, the most important of which are heat, oxygen, sunlight, water (liquid and vapour), mechanical and thermal stresses, atmospheric pollutants and various cleaning chemicals. It will contain decreasing levels of residual solvent or water or both retained after application, as well as pigments, fillers, additives or trace metal impurities incorporated during manufacture – all of which might have some photochemical activity. Furthermore, the paint itself will probably have experienced high temperatures during manufacture and, if it is a thermosetting chemistry, during curing or stoving (as used in prepainted metal frames, *etc.*). From this description, the thermal and photo-oxidative degradation processes of polymers used in coatings are going to be an important consideration in the design of exterior durable systems. The basic oxidation mechanism of organic polymers outlined in Scheme 5.1 has common features whether the source of initiation is thermal, photochemical,

PH	$\longrightarrow$	$P^{\cdot}$	Initiation
$P^{\cdot} + O_2$	$\longrightarrow$	$POO^{\cdot}$	Propagation
$POO^{\cdot} + PH$	$\longrightarrow$	$POOH + P^{\cdot}$	
POOH	$\longrightarrow$	$PO^{\cdot} + HO^{\cdot}$	
$PO^{\cdot} + PH$	$\longrightarrow$	$POH + P^{\cdot}$	
$HO^{\cdot} + PH$	$\longrightarrow$	$H_2O + P^{\cdot}$	
$2P^{\cdot}$	$\longrightarrow$	P-P	Termination
$2POO^{\cdot}$	$\longrightarrow$	$POOP + O_2$	

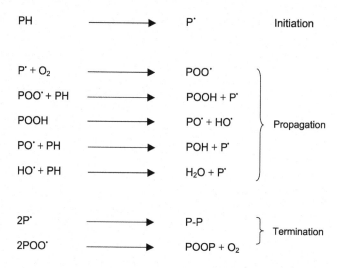

Scheme 5.1 *Basic oxidation processes (P = polymer)*

photophysical or high energy radiation. At its simplest, it is a branched chain reaction, exemplified by Scheme 5.1.

An exact understanding of the sources of initiation in the oxidation scheme is not available. However, it seems likely that they would include one or more of the following: thermal processing and curing and/or intense mechanical shear giving rise to polymer radicals which go on to form groups or species which are themselves unstable to heat or sunlight (chromophores). Chemical impurities introduced during manufacture and incorporated either into polymer or the formulation are unstable to heat or sunlight. Whatever the origin, once initiated the oxidation proceeds largely through hydrogen abstraction in chain or branching reactions. Polymer structures that contain easily abstractable hydrogen will oxidise more rapidly and so these should be avoided for exterior durable systems. Tertiary and allylic hydrogens are particularly susceptible to abstraction by free radicals (see Scheme 5.2 and Table 5.1 – note the similarity to the oxidative curing mechanism of alkyds discussed in Chapter 7).

In general, the weaker the C–H bond dissociation energy, the faster the rate of H abstraction by radical species.

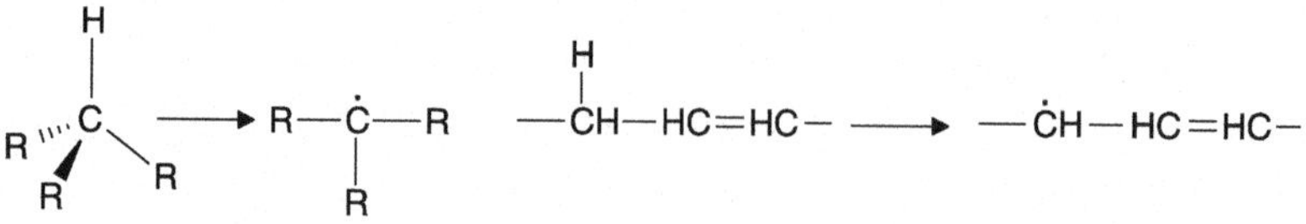

Scheme 5.2 *Easy abstraction of H from methine or allylic groups*

Table 5.1 *R–H bond dissociation energies to yield radical R*

R	D (kcal mol^{-1})	R	D (kcal mol^{-1})
Ph	111	Me$_2$CH	96
CH$_2$=CH	106	Me$_3$C	93
Me	105	CH$_2$=CHCH$_2$	86
Me$_3$CCH$_2$	100	PhCH$_2$	85
Et	100	Cyclohexadiene	73

Not all tertiary hydrogens are weak. Those attached to bicyclic structures can be anomalously strong – a good example would be adamantane (**1**). This can be attributed to the inability of the radical centre to adopt a planar sp^2 conformation.

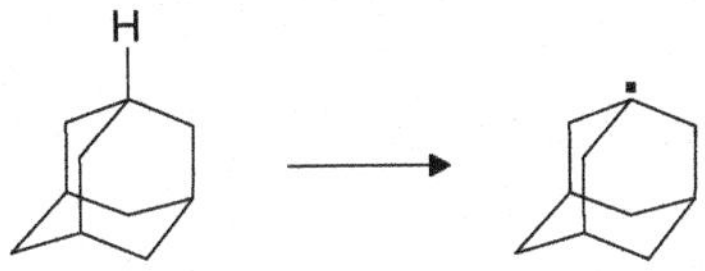

(**1**)

Most coatings contain heteroatoms and other polar functional groups, sometimes conjugated to unsaturated groups. Polar effects complicate the situation. However, structures that strongly absorb the short wavelength, high energy portion of solar radiation should be avoided. In such cases the energy can be sufficient to cause photochemical reactions leading to chain scission, radical formation and the formation of other chromophores. Epoxy resins based on bisphenol A, for example, deteriorate fairly rapidly on exposure to sunlight due to chain scission and phenoxy radical formation.

The overall effect of the complex reactions will be predominantly chain scission or crosslinking depending on the chemical structure. Chain scission usually results in a gradual softening of the coating with a lowering of T_g and as crosslink density decreases embrittlement and eventual solubilisation of thermosets will take place. This is typical of epoxies. Not all coatings give a decrease in T_g as they degrade. Table 5.2 indicates the general trend for several coating types.

The photolysis of side chain ester bonds in acrylic coatings leads to a loss of the plasticising side chains and hence an increase in T_g. Melamine coatings are commonly formulated with a considerable excess of cross-linker, and the degradation reaction is often compensated for by further self-condensation leading to only small changes in crosslink density.

Table 5.2   *Effect of photodegradation on T_g and crosslink density*

Photodegradation of	T_g	Crosslink density
Acrylic–urethane	Increases	Decreases
Acrylic–melamine	Increases	Little change
Epoxy–amine	Decreases	Decreases
Alkyd	Increases	Increases

Where radical crosslinking reactions dominate then a stiffening and increasing T_g is observed often with a consequent embrittlement (although there may be an initial strengthening of the material on crosslinking, excessive crosslinking inevitably leads to embrittlement). This is typical of alkyds. In either case, the extent of damage to the polymer can be such that material at the surface can be lost through solubilisation or attrition due to inability to sustain even the low strains associated with differential thermal expansion. The net result will be gradual erosion and roughening of polymer from the surface with consequent loss of gloss. Thus, the photo-oxidation reactions ultimately lead to physical changes in the film. Pigment or filler particles may also be exposed as the binder is lost and these particles can be easily removed from the surface – the phenomenon known as chalking.

Although the basic photochemistry is very similar, in many cases there are subtle differences and additional processes which impact upon the detailed behaviour of the polymer networks, leading to differences in the physical behaviour of the materials.

The pigment particles may also be affected by solar radiation, especially the organic types. When mixtures of these pigments are used to produce a specific colour and one of these is more susceptible to radiation, a gradual colour change can take place on exposure as one of the pigments undergoes photochemical reactions. There are some pigments which are themselves photochemically active. A well-known example is titanium dioxide which in the presence of sunlight, oxygen and water will generate hydrogen peroxide and radical species. To reduce this activity, specific forms of titania (rutile is more stable; anatase is photochemically active) with various surface treatments are employed and as a result exterior durable white paints based on titania pigmentation are widely used. Readers should not dwell under the illusion that 'exterior durable' grades of titania actually always improve the durability of a coating. For polymer systems with poor photo-oxidative stability, the scattering of damaging radiation from the coating by the titania particles can lead to enhanced lifetime. However, for polymer systems of good photo-oxidative stability, the addition of titania often leads to reduced resistance to photo-oxidation through the photo-catalytic effect of the pigment.

For some systems, exposure to high humidity or water immersion can cause serious deterioration in the properties of coatings either through chemical attack (hydrolysis) or through the dilational strains caused by water uptake. In many cases the combined effect of water and sunlight represents a more severe environment than either separately. For this reason a number of laboratory accelerated test regimes have been designed to produce exposure of coatings to various cycles of temperature irradiation and humidity. Similarly, a number of test sites, notably Florida, are widely used by the coatings industries and specifiers of coatings as severe natural exposure environments. Coating systems, for exterior applications, are commonly quoted as being 1 year or 5 years Florida which is shorthand for the retention of 50% of original gloss after those periods of exposure in the hot, sunny and humid Florida environment. There still remains the problem of the meaning of 5 years Florida data for coatings which will be used in Northern Europe (say). Almost certainly they will retain their properties for longer than 5 years, but for how much longer? Our inability to answer this question emphasises the value of the reliability methodology. With a well-developed damage model, and characterisation of the exposure environment, it ought to be a relatively straightforward computation to predict the amount of damage a coating can accrue under a given exposure environment. The development of the damage model is the key.

There have been many attempts to produce an accelerated laboratory weathering test for predicting outdoor durability. As described above many incorporate cycles of dry heat, humid heating, irradiated, and dark periods of varying length. It is perhaps unfair to try to summarise the vast amount of effort put into accelerated weathering tests. In general, however, the correlation between these accelerated tests and real time exposure is not very good. To be of any use the test needs to give a result in an acceptably short experimental time say 1–2 months. In order to produce an acceleration factor of approximately 50 (in terms of reproducing 5 years Florida data in 5 weeks), then it is necessary to increase temperature, relative humidity and the intensity of incident radiation. For some time it was common to use radiation of shorter wavelength than occurred naturally in sunlight. This certainly produced rapid results, but they were unrelated to what happened on natural exposure. The emphasis now is on the use of filtered UV sources which cut off at 300 nm (as does natural sunlight), but which deliver higher intensity illumination than sunlight at the shorter wavelengths. Similarly, too high a test temperature may produce erroneous results if, for example, T_g of the coatings is exceeded in the test but not in actual service. It has also been found that dark 'rest' periods are necessary to enable oxygen concentrations in the surface layer to be replenished as happens naturally at night-time. Thus, continuous exposure to light and oxygen may in fact run in oxygen-starved conditions and hence give misleading durability predictions. The 'microenvironment' of the test panel can often be important, *e.g.* a panel at the edge of rack will often be exposed to lower temperatures than a sample in the middle of a rack. This can lead to variability in replicate samples.

Accelerated weathering tests are still very widely used in the coatings industries despite the observations made above and the realisation that they only provide an indication of the performance rather than any predictive capability. Increasing effort has been expended on developing methods of increased analytical sensitivity so that the slight changes occurring after relatively short natural exposures can be monitored accurately. This approach will give quantitative data on some aspect of the weathering process to allow kinetic analysis and the development of models to predict long-term performance. This has been successfully carried out by Ford for automotive paints using electron spin resonance (ESR) spectroscopy to follow free radical processes and also using titrimetric determination of hydroperoxide, an important species in the oxidation scheme. Other techniques examined in this connection are FTIR, chemiluminescence, and most recently ozone depletion measurements.

5.3.1.2 Stabilisation. There are relatively few examples where stabilisation by additives has been used successfully in coatings. For pigmented systems, stabilisers against thermal- and photo-oxidation need to be effective in the surface resin rich-layer which determines gloss levels. Once this layer has disappeared or been damaged, then appearance properties are lost and so any stabilisers need be present only in the surface. These additives will have a finite solubility in the resin and, since they may be consumed in time, protection of the paint will be lost at some stage. Ideally, replenishment of the additive *via* migration from the bulk to the surface at the right rate would provide long-term protection. This has proved difficult to achieve in practice and the use of UV stabilisers, antioxidants, *etc.* in coatings is not as widespread as it is in the plastics and rubber industries. An example where an unexpectedly early failure occurred was in initial tests on Concorde. The skin of this supersonic aircraft became so hot during flight that the stabilisers volatilised out of the coating. Testing of the materials was clearly not rigorous enough and did not cover the extremes of conditions to which Concorde was subject. The problem was quickly rectified. Transparent nanosized particles of titanium dioxide and zinc oxide are currently being promoted as non-leachable, persistent UV stabilising additives for coatings.

One example where stabilisation has proved successful is in automotive paints where in the 1980s a new technology based on the application of a thin (*ca.* 10 μm) layer of unpigmented acrylic polymer over a pigmented primer. This was done for aesthetic reasons. In the early stages of development, the clear topcoat was chosen for its very low UV absorption characteristics and long-term durability. Less attention was paid to the chemistry and durability of the basecoat. Traditional automotive finish coats were pigmented in a conventional way and exhibited a slow but steady loss of gloss as the main failure mode. The new basecoat–clearcoat technology gave much greater gloss stability on field exposure testing. However, the clearcoat was essentially transparent to UV which resulted in degradation of the basecoat. Instead of the expected gloss loss, failure in service was loss of appearance through disruption of the topcoat–basecoat interface. The warranty cost for repairing damaged coating as a result of this unanticipated catastrophic delamination failure mode was very substantial indeed. The use of UV absorbers and stabilising additives in the topcoat now successfully prevents UV penetration to the basecoat and long-term exterior durability for the system as a whole has been achieved.

5.3.2 Physical Ageing

Physical ageing is the name given to the slow changes which occur in glasses with time. In contrast to the chemical ageing processes described above, these changes are reversible when the glass is heated to above its T_g. The origin of this ageing phenomenon lies in the non-equilibrium nature of glasses. When amorphous polymers are cooled through their T_g, they solidify and stiffen to form glasses. In this glassy state, rates of structural relaxation become very slow in comparison to the cooling rate and polymer conformations are held in a non-equilibrium state. If the temperature is held at some value below T_g, the process of relaxation towards equilibrium configuration proceeds and as a result many material properties change with time. The effects are most pronounced on low strain and low frequency properties such as creep and stress relaxation rates which decrease on ageing. There are, however, significant effects on other important properties, notably, elastic moduli, yield stress and density: all increase on ageing and as a result impact resistance, ductility and formability all decrease.

The process of physical ageing is highly temperature dependent. Maximum ageing rates are encountered at temperatures closest to T_g. Similarly, the rate of cooling through T_g is important in terms of the amount of excess free volume trapped in the glass with rapid or quench cooling leaving the material furthest from its equilibrium state. In terms of the mechanical performance of a coating and the measurement of material properties, it is clearly important to know and understand the effect of thermal history, especially in terms of designing test protocols for coatings which are stoved above T_g and those which have a T_g within *ca.* 30 °C of ambient.

The physical ageing phenomenon is well illustrated by the following example in which powder coating of metal pipes of different wall thicknesses gave different performance properties. Here a single powder paint was used to coat a variety of steel pipes under similar conditions of film thickness, stoving times and temperatures, and water quenching. Samples were cut from the pipes soon after coating and bent to assess flexibility of the coatings. Those from thin-walled pipes passed the bend test whereas those from thicker walled pipe failed. When tested the next day, samples from both types of pipe failed the test. The reason for the differences in performance arose from the cooling rates of the two types of pipe in the water quench. The high mass of metal in the thicker walled pipe led to a slow rate of cooling and hence more physical ageing took place on cooling to ambient. The thin-walled pipe coating was essentially quenched under the same conditions and so was in its toughest, most formable condition when tested after coating. Storage overnight led to a similar amount of ageing to that experienced by the coating on thick-walled pipe and this produced failure on bending. The lessons here are that proper sample conditioning and control of thermal history are essential if performance is to be reflected accurately in laboratory tests for glassy materials.

5.4 CHEMICAL EXPOSURE

Some of the most technologically demanding applications for coatings occur in the protection of structures against chemical attack. The range of end uses where this type of performance is required is wide. In terms of size, the structures vary from a food or drinks can holding less than half a litre of liquid to a ship's chemical tank with a capacity of a million litres or more. Similarly, the chemical exposure can be in the form of gas, liquid, or solid, organic or inorganic, hot or cold, acidic, neutral, or basic. The exposure may be continuous or intermittent and in some cases such as chemical cargo tanks there may be a succession of different chemicals.

In general, we can separate the chemical resistance performance of a coating into two broad categories. In the first, chemical resistance is required where coatings suffer occasional exposure to some form of chemical stress. They must nevertheless continue to perform their other functions during and after this exposure. Three examples serve to illustrate this type of chemical attack and are given in order of increasing severity.

Coatings for washing machines need to resist hot or boiling water and solutions of soaps and detergents. Car paints also need to cope with soap and water but resistance to hydrocarbons (as fuels, greases and lubricants) and hydraulic fluids will be required. Finally, aviation finishes need, in addition to many other exacting performance requirements, to be resistant to prolonged exposure to aircraft hydraulic fluids and lubricants.

In the second broad category, the primary function of the coating is to provide corrosion protection for the substrate or structure against long-term exposure to a chemically aggressive environment. The examples of food containers and chemical tankers have already been described, but more common is the protection of steel from simple atmospheric or saltwater attack.

For either category the exposure to some form of chemical environment must not cause the coating to be lost through dissolution or disbondment or by becoming mechanically weak from absorption of fluids. Where prolonged chemical exposure is expected, care in the selection of coating chemistry must be exercised to avoid the possibility of direct chemical attack or reaction. Polyesters would not normally be used, for example, as coatings for structures exposed to strong aqueous bases or acids. Here hydrolysis of main chain ester groups can lead to breakdown and eventual dissolution of the coating. Similarly, hydrocarbon coatings will be attacked by strong oxidising species.

Amorphous thermoplastics will generally be soluble in a particular range of organic solvents dependent on solubility parameters. For this reason thermoset or crosslinked coatings are widely used where resistance to a broad spectrum of chemicals is required. In the absence of direct chemical attack or specific reactions, crosslinked coatings will be insoluble in all solvents. They will absorb solvents, however, which leads to swelling, softening and weakening. The amount by which crosslinked polymers swell depends on a number of factors of which the two most important are crosslink density and the level of thermodynamic interaction between polymer and solvent. Highly crosslinked structures take up less solvent than lightly crosslinked ones and this forms the basis of the familiar solvent swelling technique used to characterise networks. In general, good solvents for the polymers before crosslinking will be the most powerful swelling solvents for the final network and hence taken up in the largest amounts.

The consequence of solvent uptake is that compressive stress is exerted on the adhesive interface between a coating and its substrate through the dilational swelling strains. Disruption of the interface may also occur if solvent–substrate interaction is favoured over that of the polymer. Whilst this may be useful for solvent-based paint strippers in the home, it is clearly going to be a severe nuisance for a coating in a chemical environment.

Solvent uptake will also lower a coating's T_g and, in this softened and swollen condition, tensile and tear strengths will be much reduced and so its resistance to mechanical damage will be impaired. When a coating on a land or ship-based storage tank is exposed to a succession of different liquids, a material retained from a previous filling may be released into the current contents of the tank. This presents obvious problems when such contamination is unacceptable, for example, when foodstuffs are carried after chemical cargoes in a ship's tank. In such cases the sequence of cargoes is strictly controlled.

There have been three traditional approaches to providing resistance to chemical uptake and swelling. Where a single or dominant type of chemical exposure is expected, coatings architectures have been chosen which are markedly different from the chemical, *i.e.* polar polymers for non-polar solvents and *vice versa*. This minimises any swelling. The second strategy is to obtain the highest possible crosslink density (commensurate with other required properties) in the coating for applications where a broad spectrum of chemical resistance is required. For reasons outlined in this chapter, high extents of cure may be difficult to obtain in such systems especially when only ambient curing conditions are possible. In consequence, there may be a significant sol fraction remaining in the coating which will be extractable in some solvents or chemicals. Many cargo tank coating specifications allow for a post-cure treatment, either by hot air, hot water or requiring that the first cargo is held at an elevated temperature. The last approach has been to reduce the concentration of swellable polymer in the coating by the use of fillers and pigments. This has implications for the viscosity of the coating at application and is becoming less attractive with the drive to reduce solvent levels in coatings.

The subject of corrosion protection by coatings remains somewhat controversial despite its obvious importance and the large and continuing research effort devoted to its understanding. In its simplest aspect, coatings provide protection for metal against corrosion by acting as a barrier to the reactants or products of corrosion reactions. Typically, these species would be oxygen, water, hydrogen, ions or electrons.

One body of opinion maintains that organic coatings do not provide a sufficiently impermeable barrier to oxygen or water to prevent corrosion at the metal substrate. The rates of diffusion of these species in perfect, flaw-free organic films would be sufficient to support the observed rates of corrosion of unprotected metals. It further maintains that the fact that coatings can provide very effective corrosion protection is due to the very low permeability of ionic species or electrons in the covalent, organic, low dielectric medium of the coating. To account for the corrosion protection performance of real coatings the presence of flaws, imperfections and structural heterogeneities of various sorts is invoked. These flaws can be macro- or microscopic or at the macromolecular level and are held to be responsible for providing pathways for the species above to sustain local corrosion centres. Some maintain that the absorption of water into the coating causes structural reorganisation of the chains and eventually leads to percolation pathways through the coating which then no longer provides protection. In other words, the necessary flaws are developed during the service life rather than being an intrinsic property of the newly applied film. Many corrosion failures occur at specific sites – typically along weld seams or in areas where the coating has been overapplied (these often go hand in hand), or at sharp edges. These areas have one thing in common: they are generally areas of greater stress, where perhaps there is a lower 'activation energy' for structural reorganisation, or indeed delamination or coating fracture. Much effort is being expended on test methodology which can adequately predict the tendency of a coating to crack under the extremes of application conditions and in-service performance. Most of these methodologies are concerned with cyclic variation in environmental stresses.

However, it is clear that the quality of coating application and film formation is of major importance in corrosion performance. Where this quality is poor and flaws, voids and dewetted areas are detectable by eye or microscope, then corrosion performance is impaired. Multiple paint applications will often improve performance when the chance of coincident flaws is much reduced. Similarly, the addition of non-polar, low dielectric additives such as hydrocarbon waxes and tars has been used to improve corrosion resistance. Tar, however, is considerably out of favour due to the presence of the highly carcinogenic polyaromatics as a constituent, although some grades are available free of this material. The presence of so-called hydrocarbon resins commonly leads to coatings with a tendency to crack if care is not taken during the formulation stage. It should also be said that oxygen/water barrier enhancing fillers such as glass flake, mica, and micaceous iron oxide have also been effective in anti-corrosive coatings. However, in the absence of microscopic flaws, *etc.*, the evidence for heterogeneities at the molecular or supramolecular level being responsible for corrosion pathways is much less convincing – although still a popular opening supposition by corrosion scientists.

5.5 SUMMARY

The emphasis of this chapter has been to describe the links between coating performance in service and the tests and procedures which are used to develop materials to meet a required performance. Attempts to extend these relationships through to polymer structure have also been described. It is clearly an ambitious and daunting task to design coatings for specific end uses starting from a polymer architecture. In practice, of course, it is usual for a succession of people or teams to take a project from materials to commercial products. The key to success is the effective generation and transfer of quality information on what the coatings need to do, what they actually do, what makes them perform the way they do, and hence what can be done to change performance.

5.6 BIBLIOGRAPHY

1. A.J. Kinloch and R.J. Young, *Fracture Behaviour of Polymers*, Elsevier, London, 1983.
2. J. Martin, F. Floyd, S. Saunders and J. Wineburg, *Methodologies for Predicting the Service Lives of Coating Systems*, Federation of Societies for Coatings Technology, 1996.
3. L.E. Nielsen, *Mechanical Properties of Polymers and Composites*, Dekker, New York, 1974.
4. C.H. Hare, *Paint Film Degradation: Mechanisms and Control*, SSPC, 2001.

CHAPTER 6

Binders for Conventional Coatings

A.R. MARRION

6.1 BINDERS AS POLYMERS

Of the four components of a typical coating – binders, fillers, additives and thinners – the character of the binder has the greatest impact on the performance of the coating, and defines its properties. Binders are almost invariably polymeric, if not at the point of application, then certainly when the film is 'cured' and ready to see service. A number of additives are also polymeric materials that exert a significant influence on the binder system.

Polymers are large molecules, usually constructed of one or more repeating units or 'mers' derived from monomers. Their chemistry is extremely diverse (as we shall see) but their character is also heavily dependent on their molecular weight and architecture.

Representing a monomer as M, its tractable polymers can be straight chains (**1**), branched (**2**), radiate (**3**) (stars if the number of branches exceeds about four), hyperbranched (where the branches have branches) (**4**) or even cyclic (**5**). When a second monomer m is introduced it is possible to envisage random (**6**), alternating (**7**), blocky (**8**) or grafted (**9**) copolymers.

(1)

(2)

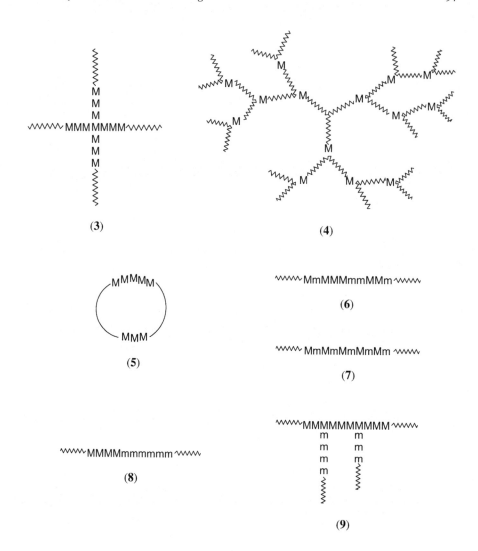

The molecular weights of polymers can vary from a few thousand to several million and it is often more convenient to think in terms of degree of polymerisation (DP), *i.e.* the number of monomer residues.

No known polymer system is composed entirely of molecules of identical molecular weight. The overall molecular weight can be expressed in a number of ways, *e.g.* the number average M_n:

$$M_n = \frac{\sum n_i m_i}{\sum n_i} \qquad (6.1)$$

or the weight average M_w

$$M_w = \frac{\sum w_i m_i}{\sum m_i} = \frac{\sum n_i m_i^2}{\sum n_i m_i} \qquad (6.2)$$

where n_i is the number of molecules of mass m_i and w_i is the weight of such molecules). The ratio M_w/M_n is defined as the polydispersity, d, of a particular polymer and is also an important characteristic.

Another structural class of polymer is the crosslinked network (**10**). Its members defy the above analysis, being of essentially infinite molecular weight, and completely intractable. As discussed in Chapter 4, they are characterised by M_c, the molecular weight between crosslinks, a structural feature that bears strongly on their mechanical properties and chemical resistance.

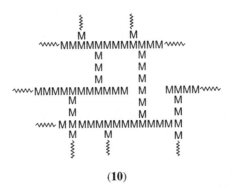

(**10**)

The synthesis of polymers involves either a chain growth mechanism (Scheme 6.1), in which monomers add to a growing chain, or a 'step-growth' mechanism (Scheme 6.2), in which monomers link together randomly in small groups which in turn link to each other producing an exponential increase in molecular weight.

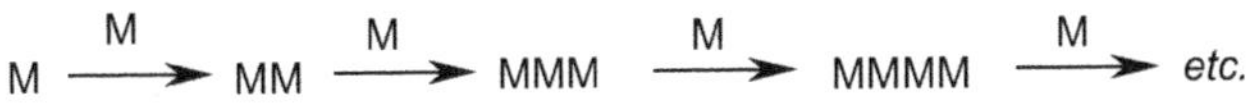

$$M \xrightarrow{M} MM \xrightarrow{M} MMM \xrightarrow{M} MMMM \xrightarrow{M} etc.$$

Scheme 6.1

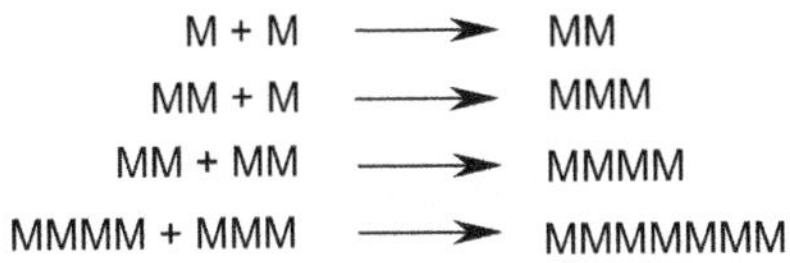

$$M + M \longrightarrow MM$$
$$MM + M \longrightarrow MMM$$
$$MM + MM \longrightarrow MMMM$$
$$MMMM + MMM \longrightarrow MMMMMMM$$

Scheme 6.2

The polyesterification reaction between a diol and a dicarboxylic acid (Scheme 6.3) is representative of the step-growth processes. The product necessarily contains alternating diacid and diol residues and its end-functionality and molecular weight are determined by the starting stoichiometry.

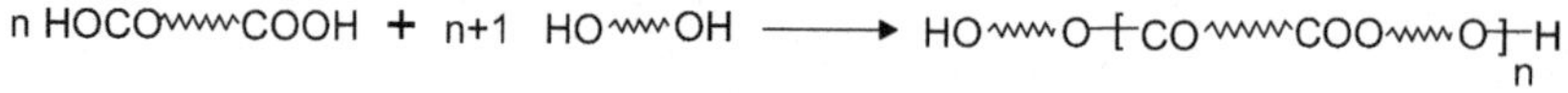

n HOCOᵛᵛᵛCOOH + n+1 HOᵛᵛᵛOH ⟶ HOᵛᵛᵛO⎡CO ᵛᵛᵛCOOᵛᵛᵛO⎤H
                                                        ⌊   ⌋n

Scheme 6.3

At the outset, there is a plentiful supply of reactive groups, and esterification occurs readily. As the reaction proceeds, molecular weight increases and reactive group concentration becomes lower. In practice it is difficult to drive any step-growth process to completion, and high temperatures (250 °C), catalyst, and continuous removal of the condensed species may be needed to achieve a satisfactory conversion.

Combinations of different acid and polyol monomers are frequently used. For example, one of the 'hard' phthalic acids might be combined with the 'soft' adipic acid and a suitable polyol to obtain a satisfactory glass transition temperature (T_g). A degree of branching can be introduced by using low levels of tri- or higher-functional acid or alcohol though the resulting polydispersity is inclined to be high. Formulations are then best derived by computerised iteration, which can also give warning of potentially gellable compositions. Some of the more usual polyester monomers are listed in Table 6.1.

Table 6.1 *Typical polyester monomers*

Group	Structure	Name
'Soft' diacids	$HOCO-(CH_2)_n-COOH$	Alkanedioic Acids ($n = 0$–10)
	$HOCO-(CH_2)_4-COOH$	Adipic acid
	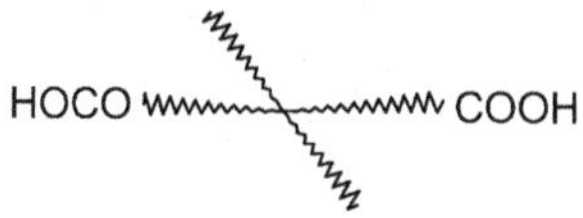 HOCOᵛᵛᵛCOOH	Dimerised fatty acid $n = ca.$ 34)
'Hard' diacids		1,2- $= o$-phthalic acid (anyhdride)
		1,3- $=$ isophthalic acid
		1,4- $=$ terephthalic acid
	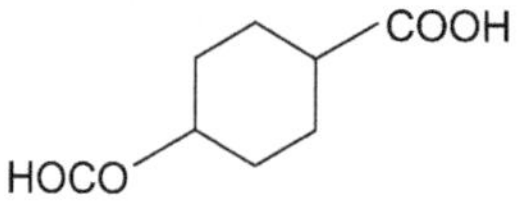	Cyclohexanedicarboxylic acid

(*continued*)

Table 6.1   *Continued*

Group	Structure	Name
Triacids	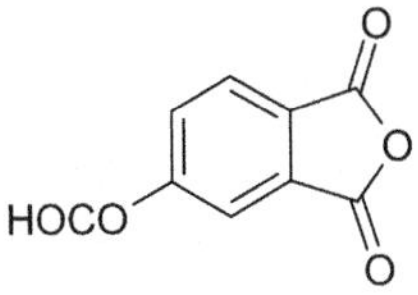	Trimellitic anhydride
Heterofunctional acids	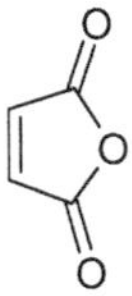	Maleic anhydride
Diols	HO-(CH$_2$)$_n$-OH	Alkanediols ($n = 2$–10)
	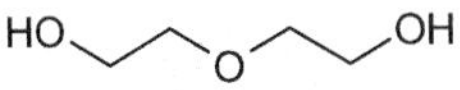	Diethyleneglycol
		Neopentylglycol
	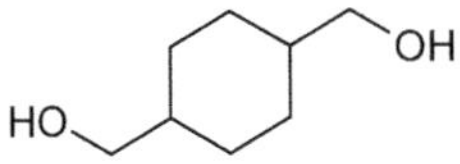	Cyclohexane dimethanol
Triols	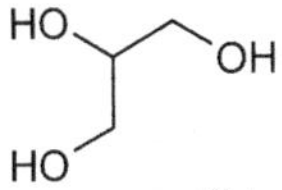	Glycerol
	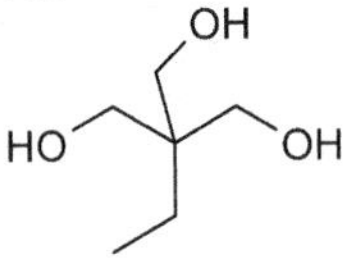	Trimethylolpropane
Tetrols	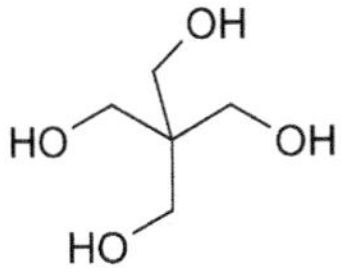	Pentaerythritol

Other step-growth polymerisation processes involving condensation chemistry are the reaction of phenols or amino compounds with formaldehyde, polyamidation of acids with amines and the reaction of epichlorohydrin with diphenols. Important step-growth additions include the reaction of diepoxides with diphenols to produce higher molecular weight epoxy resins, and of isocyanates with diols to prepare polyurethanes.

Polymerisation of 'acrylate' monomers, under free radical conditions, is an example of a chain growth process (Scheme 6.4). Acrylate monomers are olefinic compounds of enhanced reactivity including acrylic acid derivatives and related compounds, and a few 'honorary' acrylics such as styrene or vinyltoluene (*cf.* Table 6.2).

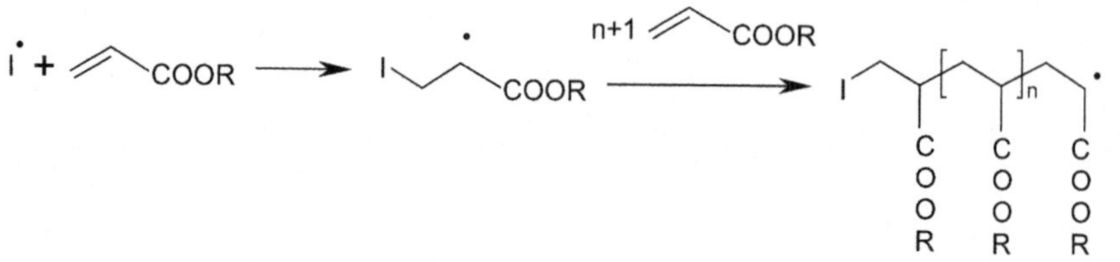

Scheme 6.4

Table 6.2   *Some typical "acrylic" monomers*

Structure	R	R'	Name
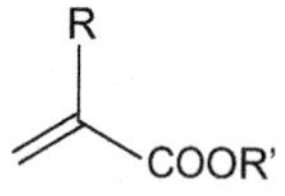 ![structure COOR']	H	H	Acrylic acid
	Methyl	H	Methacrylic acid
	H	Methyl	
		Ethyl	Alkyl acrylate
		Butyl	
		2-ethylhexyl	
	Methyl	Methyl	
		Ethyl	Alkyl methacrylate
		Butyl	
		2-ethylhexyl	
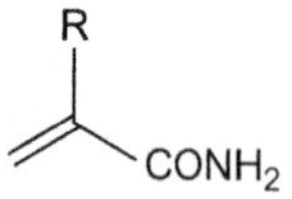 ![structure CONH2]	H	–	Acrylamide
	Methyl	–	Methacrylamide
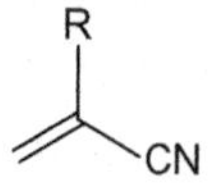 ![structure CN]	H	–	Acrylonitrile
	Methyl	–	Methylacrylonitrile

(*continued*)

Table 6.2 *Continued*

Structure	R	R'	Name
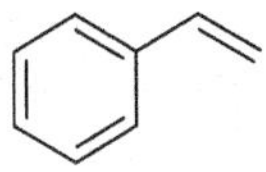	–	–	Styrene
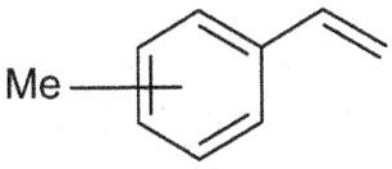	–	–	Vinyltoluene

The initiator, I˙, is a free radical derived by thermal homolysis of a species such as a peroxide or an azo compound, or by a redox reaction (Scheme 6.5). Chains would grow rapidly to a molecular weight determined by the ratio of monomer to initiator, if opportunities for chain termination and transfer did not exist. In reality they usually do. Chain transfer occurs when the growing macroradical abstracts a moiety (usually H˙) from a species in the reaction mixture. A new radical is left behind to initiate a further chain (Scheme 6.6).

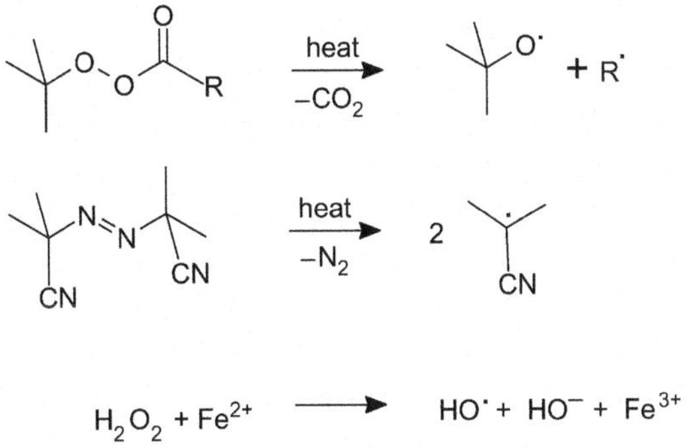

$$H_2O_2 + Fe^{2+} \longrightarrow HO˙ + HO^- + Fe^{3+}$$

Scheme 6.5

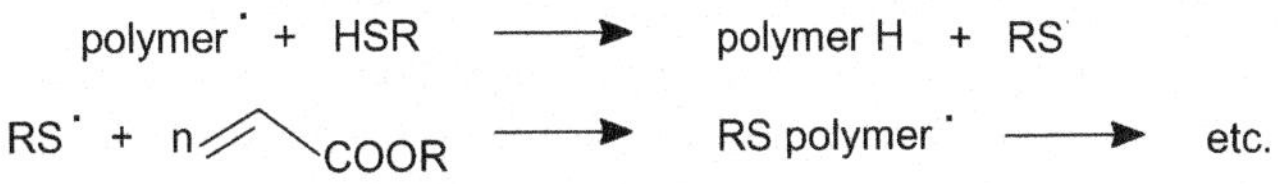

Scheme 6.6

Solvent, monomer, or polymer can act as chain transfer agents but it is common practice to add mercaptans to reduce molecular weight and economise on initiator. Reaction chains can also be terminated by the combination of two macroradicals or by disproportionation, when one macroradical acquires H˙ from another, leaving an unsaturated end group.

Products of free-radical polymerisation of different acrylic monomers are more or less blocky or random in residue distribution, according to the reactivity of each monomer, with growing macroradicals tipped with like or unlike residues according to the 'reactivity ratios' of the new monomer and its predecessor.

Chain growth processes can also be initiated by cationic or anionic species. In some cases chains with 'living' end groups can be produced, and will initiate polymerisation of a different monomer so that well 'tailored' block copolymers can be produced.

The geminal substitution on a polymethacrylate produces a much stiffer backbone than the equivalent polyacrylate, hence methacrylate monomers confer high T_gs on their polymers. A coatings formulation may contain a combination of methacrylates and acrylates to achieve the desired hardness with small amounts of acid or amide to improve adhesion. A wide range of other 'functional' monomers can also be employed to provide crosslinking sites (*cf.* Table 6.4, p. 114).

6.2 POLYMERS AS BINDERS

Even when the liquid coating consists of monomers, a polymerisation process during cure ensures that the binder is ultimately a high molecular weight species. The characteristics of the polymeric binder usually determine the nature of the coating, so that polyurethane paints are noted for their toughness, epoxies for their adhesion, and so on. They also have a major bearing on the method of application and conditions of film formation.

A coating binder in service is required to be indifferent to varying extents, to mechanical abuse, chemical insults and the effects of its environment, yet be applied easily under mild conditions. It is necessary to convert a liquid or fusible solid into an intractable solid on demand, and that transformation is a key feature of any successful coatings system. The physics of film formation was discussed in Chapter 4. Several of the transformation strategies in widespread use are summarised in Table 6.3.

A very wide range of polymers and polymer precursors (prepolymers) has been exploited in coatings, the only essential feature being that they are tractable at the point of application. Thermoplastic acrylic latices, for example, have molecular weights of millions, but are applied in aqueous dispersion. Coalescence of the particles is possible over time. By contrast, solvent-free liquids rely on monomers or oligomers that undergo extensive crosslinking reactions to achieve their film properties. In this chapter, however, we are concerned with 'conventional' coatings that usually

Table 6.3 *Strategies for achieving high performance films from tractable precursors*

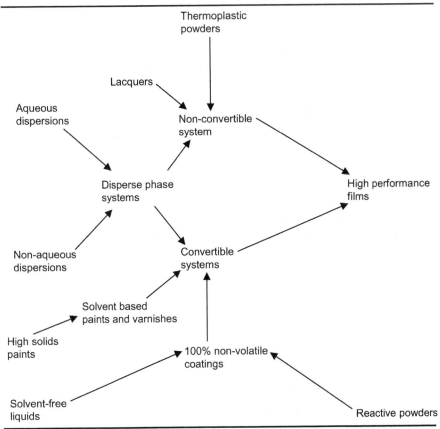

contain volatile organic solvent, and have at least some element of lacquer drying in their film formation.

Many coatings polymers are manufactured 'in house' by the suppliers of coatings, according to their own technology. Others are prepared by specialists, either because they have wide application and the economics of scale can be exploited, or because their manufacture requires particular plant, or control, or is proprietary.

Dissolution of a relatively high molecular weight polymer in a suitable solvent, whose loss by evaporation provides a thermoplastic coating, can give a satisfactory coating known as a lacquer, or non-convertible coating. However, a polymer of sufficient molecular weight tends to be viscous in solution so that high dilution is required. The economic disadvantages of lacquers have long been recognised. Not only is a large proportion of the composition ultimately lost to the atmosphere, but also the film left behind is likely to be much thinner than desired and multiple applications may be necessary. More recently of course, the environmental implications of releasing large amounts of volatile organic material to the atmosphere have been understood (*cf.* Chapter 2). In a similar way, thermoplastic powders are generally too viscous in the melt to coalesce to films of satisfactory integrity and appearance in other than very thick layers, and at high temperatures.

Reactive systems can be based on lower molecular weight precursors which 'crosslink' or 'cure' after application. Such systems are described as 'convertible' or 'thermosetting', and the coatings as enamels. They can be designed with substantially decreased volatile contents and usually show enhanced mechanical and chemical properties as compared with non-convertible systems.

To add a little confusion, there are also 'reactive thermoplastics' in limited use. They exploit reactive systems to provide a degree of molecular weight advancement, stopping short of gelation, so are convertible, yet thermoplastic.

Conventional thermoplastic and thermosetting binder chemistries will be discussed in the present chapter, whilst subsequent chapters deal with high solids, solvent free, powder, waterborne and non-aqueous dispersion media, and inorganic coatings. Each addresses the problem of transforming a liquid film to a tough solid coating in its own particular way.

6.3 THERMOPLASTIC BINDERS

The fact that film formation by lacquer drying is a reversible process, and the polymer can be removed with a suitable solvent is advantageous when temporary coverage is required, as when coating fingernails, and when cleaning or repair of localised damage is called for. It also means that intercoat adhesion can be very great since some re-dissolution on overcoating can occur, whilst the tendency of partially crosslinked systems to swell and disbond is absent.

It is usually necessary to exceed the 'entanglement molecular weight' of the polymer in question to achieve a satisfactory film. The entanglement molecular weight varies from system to system, for some polyesters and epoxies it is in the order of thousands, whilst for acrylic polymers, molecular weights of tens of thousands would be typical. Many practical thermoplastic coatings need molecular weights approaching 10^5 to achieve acceptable performance, and such materials are sometimes diluted to 10–15% non-volatiles with all the problems noted above.

6.3.1   Natural Resins

Some naturally occurring resins and gums are best discussed in the context of thermoplastic coatings, although a number of them possess a certain degree of reactivity.

6.3.1.1   Shellac.   Shellac is the secretion of an insect, *Kerria lacca*, and is gathered from the wild or cultivated commercially on trees in plantations in India, Burma and Thailand. The insects live on the branches in tightly packed colonies and their lac secretions fuse to form a hard protective sleeve 1–3 cm thick over the sedentary insects. The infested branches are cut from the trees and fragmented before the lac is melted and filtered to remove the insect and vegetable debris. Its composition is still somewhat obscure, but it is apparently a low molecular weight polymer or oligomer, richly furnished with hydroxyl groups. Saponification yields 9,10,16-trihydroxypalmitic acid, and a selection of tricyclic acids related to shellolic acid (**11**).

(**11**)

The major usage of Shellac up till about 1940 was for making gramophone records, where its wear-resisting properties were amply demonstrated, but it still has some applications in coatings, including wear resistant floor coatings. Compounded with beeswax, it is the basis of 'French polish', used on high quality wooden furniture. Because of its highly polar character, its ethanol solutions are used by painters to treat the knots in softwood joinery ('knotting'). The knots are rich in rosin, a hydrocarbon-soluble material (*v.i.*) that would otherwise 'bleed' through subsequent coats of paint. Shellac still finds significant application in coatings for pills and sweets, and in hair lacquer.

6.3.1.2   Rosin.   Rosin or 'colophony' is the secretion of certain trees of the genus *Pinus*. It is obtained by tapping mature trees, extraction from stumps left after felling, or as a by-product of paper manufacture. Worldwide production by tapping alone approaches one million tons per annum. It consists predominantly of abietic acid (**12**), pimaric acid (**13**) and related structures.

(12)   (13)

Its principal use is as a component of "oleoresinous" media, where it is processed with drying oils (*v.i.*) or added to alkyds. Its main function is to increase hydrophobicity and lacquer drying. Although non-polymeric, it is able to impart significant film-forming properties when present simply as part of a physical mixture. However, some chemical reaction through the acid group (polyesterification) or the unsaturation (auto-oxidation or Diels–Alder addition) can also occur. Deliberate chemical modification, by esterification with alcohols, formation of salts of sodium, calcium or zinc, or by Diels–Alder reaction with maleic anhydride, has frequently been used. Rosin has been used in antifouling paints to control hydrophobicity whilst permitting slow dissolution in seawater. Another significant application is, in combination with alum, as a sizing agent for paper.

6.3.1.3 Others. Other resins obtained from exotic trees evoke a more romantic era of paint technology, reminiscent of the spice trade, but have very limited use nowadays. Total world production is only a few thousand tons. Examples are manila copal from trees of the genus *Agathis* found in Indonesia and the Philippines and dammar (or damar) from trees of the family Dipterocarpaceae, also widespread in Indonesia and Malaysia. Manila copal and dammar, like other resins are broadly based on terpene chemistry, but copal is soluble in alcohols and ketones whereas dammar is soluble in aliphatic and aromatic hydrocarbons. Another historic example is the fossilised kauri gum from New Zealand, now very scarce.

"Gums" are water-soluble plant materials of polysaccharide constitution, and have never been much used in the conventional coatings industry.

6.3.2 Cellulose Derivatives

Derivatives of cellulose are amongst the most important thermoplastic coatings polymers. Cellulose itself occurs widely in nature. It is a polysaccharide, a polymer of anhydroglucose units (**14**), and exists as crystalline, hydrogen bonded fibres. Esterification of some of the three hydroxyl groups per residue leads to organic solvent soluble polymers.

(14)

"Nitrocellulose" (*i.e.* cellulose nitrate) is prepared by treating cotton waste, wood pulp or other source of cellulose with nitric and sulfuric acids. Levels of nitration varying between 2 and 3 units per residue are attainable and the products are dissolved in strong solvents such as esters. They were originally used in aircraft 'dope' to stiffen and tighten the canvas covering, and in automotive finishes where their relative softness allowed them to be polished to a high gloss. More recently they saw extensive application in car refinishing. In today's rather changed world they are still used in furniture finishes, nail varnish and inks, usually in conjunction with plasticisers and other polymers to provide the required mechanical properties and increased non-volatile content. Though their use has declined considerably they still occupy a significant position.

The mixed cellulose ester, cellulose acetate/butyrate, in several grades according to composition, has widespread importance. It shows unusual compatibility with other polymers and exerts a profound effect on their rheology. In coatings containing leafing metal pigments it promotes alignment of the pigment particles and is frequently added in quite high proportions to automotive "base-clear" base coats.

Other cellulose esters and ethers have limited value as thickeners in water-based compositions, for example.

6.3.3   Acrylics and Vinyls

Thermoplastic acrylics, usually predominantly poly(methyl methacrylate) of molecular weight in the order of 10^5, have been used as automotive topcoats of the "flow and reflow" type. They were sanded to remove imperfections after stoving at a low temperature, then re-stoved at a higher temperature to restore a high gloss.

Alternatively, they can film-form at ambient temperature when sufficiently plasticised either 'internally' with co-monomers such as butyl acrylate or with an external additive such as an ester or certain proprietary chlorinated materials. Such compositions are used as marine topcoats where their retention of the usual white colour and gloss is invaluable. Polystyrene has found limited acceptance in zinc-rich anticorrosive primers. Poly-acrylates have also been the principal components of antifouling coatings for ships. They work by controlled dissolution into the sea.

The best known vinyl polymers are prepared by free radical copoly-merisation of vinyl chloride (15) and vinyl acetate (16) to achieve the required hardness (homopolymers of (15) being too hard ($T_g = 81\,°C$), whilst those of (16) are too soft ($T_g = 32\,°C$). Small amounts of functional monomers such as maleic anhydride are frequently added to promote adhesion, or introduce a low level of crosslinking. Vinyl polymers are typically highly water resistant and have been used alone as can-linings or combined with other thermoplastic or thermosetting polymers such as acrylics, epoxies or hydrocarbon polymers for marine paints. Poly(vinyli-dene chloride) is a crystalline polymer of (17) with remarkable resistance to gas permeation, so is used to line polyester bottles to contain carbonated beverages.

(15)   (16)   (17)

Homopolymers of styrene and vinyltoluene can be used in the same way as acrylics and vinyls, though they are apt to be rather brittle. More effective use is as copolymers with plasticising monomers, or in styrenated or vinyltoluenated alkyds. In the latter examples, a suitable alkyd is processed at 150–170 °C with 30% or more of styrene or vinyltoluene in the presence of a peroxide. Alkyds based on conjugated fatty acids or containing a proportion of maleic acid are preferred. The resulting products consist of hydrocarbon chains, both free and grafted to the alkyd residue. They dry rapidly, mainly by a physical route.

Hydrocarbon or petroleum polymers are low cost materials usually made by cationic polymerisation of unsaturated materials such as indene (19), and various other unsaturated molecules present amongst the by-products of coal tar or petroleum refining. The polymers are usually referred to as coumarone–indene resins but, in point of fact, the levels of coumarone (18), are rarely above a few percent. They have low molecular weights, typically below 1000, and are used as thermoplastic additives to epoxy–amine compositions. Their compatibility with epoxy resins can be enhanced by incorporating 10–15% of phenol during the polymerisation process. It presumably undergoes Friedel Crafts type chain termination reactions.

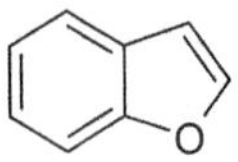

(18) 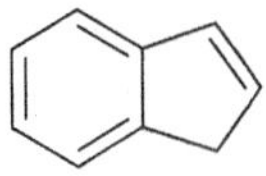   (19)

In much the same way, pitch (coal tar) at levels up to about 50% confers excellent properties on vinyl or epoxy compositions where its black colour can be tolerated, but unfortunately tends to contain mutagenic condensed aromatic hydrocarbons and has lost favour.

Poly(vinyl alcohol) is prepared by hydrolysis of poly(vinyl acetate), since vinyl alcohol exists predominantly as the acetaldehyde tautomer, and cannot be polymerised in the required way itself. The hydrolysis is never complete and the properties of different grades are dependent on levels of residual acetyl groups. It and polyvinyl ethers, such as poly(vinyl*iso*butyl ether), are used as alternative plasticising co-monomers with vinyl chloride.

Polyvinylbutyral is made by treating poly(vinyl alcohol) with butyraldehyde and contains the structural units (**20**) and residual hydroxyl groups, as well as the acetate groups in the original polyvinyl alcohol.

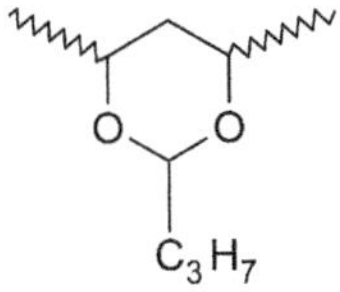

(20)

Vinyl ether polymers (R = Me, Et, isobutyl) are formulated with cellulose nitrate, chlorinated binders and styrene copolymers as coatings for, for example, paper.

Poly(vinyl fluoride) and poly(vinylidene fluoride) are high cost materials with outstanding weatherability and water resistance. They are increasingly used in coil coating for exterior use.

6.3.4 Chlorinated Polymers

Chlorinated polymers are attracting a certain amount of disfavour on the grounds that material involved in fires or incineration of waste gives rise to toxic fumes of hydrogen chloride and possibly also the very dangerous chlorinated benzodioxins under certain conditions. Nevertheless, there is a wide range of chlorinated polymers available and still in use.

Chlorinated rubbers are highly chemically resistant media with broad compatibility, though liable to decompose at only moderately elevated temperatures. They are made by treating rubber with chlorine. Addition to the olefinic bonds takes place first, followed by substitution and more complex reactions with progressive hardening. The most heavily chlorinated systems contain 65% of chlorine. Chlorinated rubber media are used, for example, in high build primers for ships.

Analogous chlorinated systems are chlorinated polyethylene, chlorinated polypropylene, chlorinated ethylene/vinyl acetate copolymers and even chlorinated polyvinyl chloride.

6.4 REACTIVE BINDERS

In conventional thermosetting or 'convertible' coating binders, at least one component is a low molecular weight polymer or prepolymer bearing reactive groups. The groups ultimately form 'crosslinks' in the high molecular weight network that will give the coating its important characteristics. Typically, a second reactive component is a low molecular weight multifunctional molecule or crosslinker. As explained in Chapter 4, both components must have functionality greater than 1, and their average functionality must be greater than 2.

From the perspective of a coatings maker, it may be convenient to manufacture the functional polymer replete with, for example, hydroxyl groups. An isocyanate crosslinker (say) can then be bought from a specialist supplier able to deal with the more demanding chemistry. The crosslinker is likely to form a relatively small fraction of the composition since it will probably have a much lower equivalent weight than the main film-forming polymer. The economics of the arrangement can thus be satisfactory to all parties.

One or both components can be polymers, and in some cases a single component is selfcrosslinking. The important thing is that multiple intermolecular reactions take place to form a network. It is also possible to form polymeric coatings *in situ* from precursors that are essentially monomeric, as described in Chapter 7.

A crosslinked network is defined by its chemistry, its vitrification temperature (T_g) and its crosslink density (M_c), as discussed in Chapter 4. From the point of view of design, we have to consider the materials to be incorporated, the nature of the crosslinks to be formed and the functionality and equivalent weight of the precursors. A further problem is to ensure that the required reaction takes place as fully as possible under given conditions.

Low molecular weight curing agents can be simple molecules, di-, tri- or multifunctional in the required reactive group. There are several examples in the next section. However, if two such molecules were persuaded to react with one another, the resulting network would probably be excessively dense, and consequently brittle. A typical coatings network might have an M_c between a few hundred and a few thousand to achieve the necessary mechanical properties. Accordingly, one component is often polymeric, with well-spaced functional groups, and its backbone is the major contributor to the overall properties of the network.

In an idealised linear polymer represented by (**1**), there are broadly two ways of arranging functional groups. They can be 'terminal' as in (**21**) or 'pendent' as in (**22**). Obviously the range of possibilities and combinations in the structural types (**1**)–(**9**) is enormous, and its impact on the final properties of a crosslinked network is incalculable.

F-MMMMMM-F

(**21**)

$$\begin{array}{c}\text{MMMMMMMM}\\ |\quad|\quad\quad|\\ \text{F}\quad\text{F}\quad\quad\text{F}\end{array}$$

(**22**)

6.4.1 Polyesters as End-Functional Binders

It is easy to make linear polyesters, functional in either hydroxyl or carboxyl, simply by combining diacids and diols in the required ratio in a step-growth polycondensation. The functionality is terminal.

As an example, 4 mol of isophthalic acid could be pictured reacting with 5 mol of neopentyl glycol, releasing 8 mol of water to give a difunctional molecule of molecular weight 1040, therefore equivalent weight 520 (Scheme 6.7).

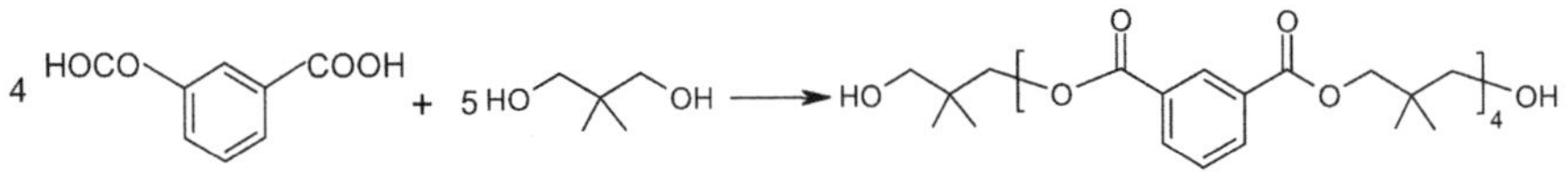

Scheme 6.7

In reality, the reaction would be to some degree incomplete and the molecule would be terminated with a statistical distribution of both acid and hydroxyl groups. The molecular weight would be an average value, and some side reactions would have taken place. Nevertheless, the product would be rich in terminal hydroxyl groups. It would cure with a trifunctional isocyanate to give excellent mechanical properties, since most of the polyester backbone would be elastically effective. Being a polyester, it would still be vulnerable to conditions under which the ester groups could be hydrolysed.

If alternative end groups were needed, it would be necessary to carry out a separate tipping procedure, or introduce a chain-terminating monomer containing a third type of functionality that would not be involved in the polyesterification process. Examples are tipping with a diisocyanate or with a phenolic acid (Scheme 6.8).

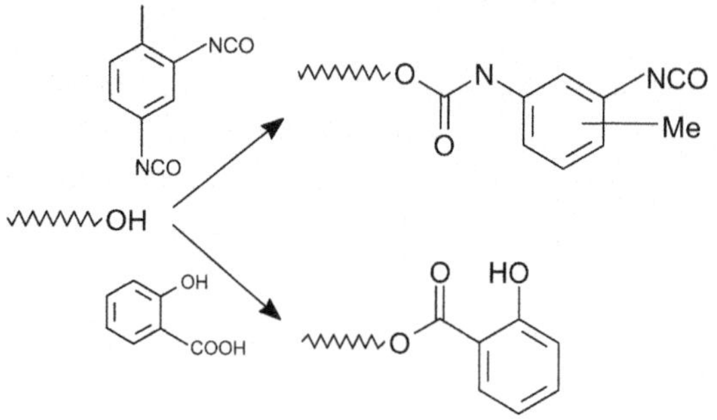

Scheme 6.8

If a polyester with functionality above two is needed, it is possible to introduce a low level of multifunctional monomer, as described above. There would be however a strong tendency towards premature gelation, and functionalities above about four are difficult to achieve.

6.4.2 Acrylics as Binders with Pendent Functionality

Acrylic polymers, our exemplars of chain growth, are quite different. Functionality is introduced by replacing some of the monomers with others bearing additional reactive groups. A selection of the commonly available "functional monomers" is given in Table 6.4.

The resulting product approximates to a linear backbone with pendent functional groups. When crosslinked through functional groups along the backbone, lengths of polymer are inevitably left 'dangling', and full use is not made of the elastic strength of the molecule. Consequently, acrylics have a tendency to show rather poorer mechanical strength than polyesters under similar conditions.

The average equivalent weight is easily estimated, for example a polymer containing 15% by weight of hydroxyethyl acrylate ($MW = 130$) would have a theoretical equivalent weight of $100/15 \times 130 = 867$. (It would of course be prudent to measure the hydroxyl value of the finished polymer.) Its functionality, on the other hand would be determined by molecular weight, whose measurement is a significant technical task. Moreover, as we have seen, the polymer would contain a distribution of molecular weights, with a further spread of functionalities superimposed on the distribution

Table 6.4 *Ambifunctional "acrylic" monomers (R=H or methyl)*

Structure	Name	Reactive group
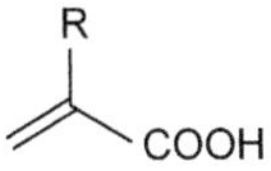	(Meth)acrylic acid	—COOH
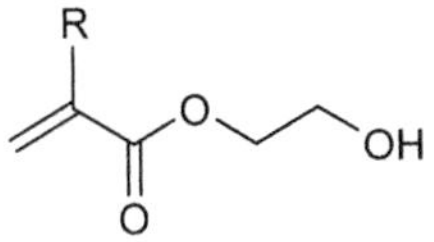	Hydroxethyl (meth)acrylate	—OH
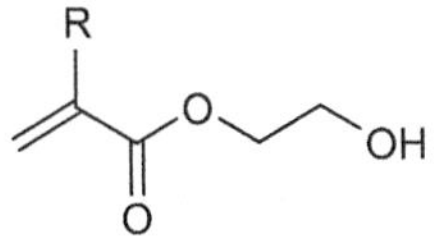	Hydroxypropyl (meth)acrylate	—OH
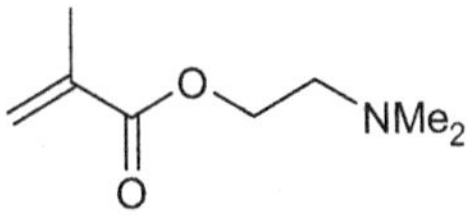	Dimethylaminoethyl methacrylate	—NR$_2$
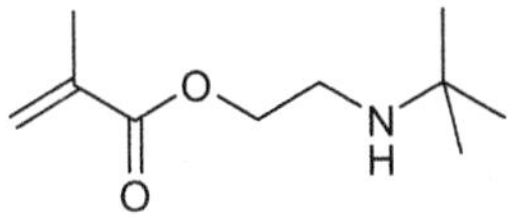	*tertiary*-butylaminoethyl methacrylate	>NH
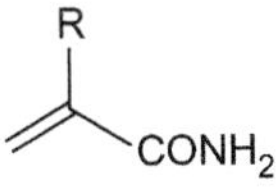	(Meth)acrylamide	—CONH$_2$
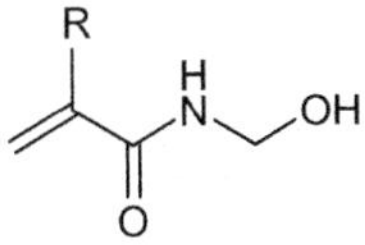	Methylol (meth)acrylamide	Activated—OH
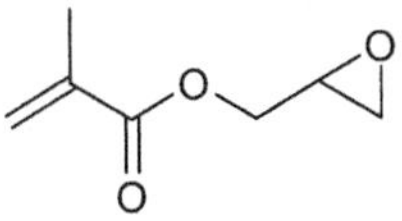	Glycidyl (meth)acrylate	Epoxy

(*continued*)

Table 6.4 *Continued*

Structure	Name	Reactive group
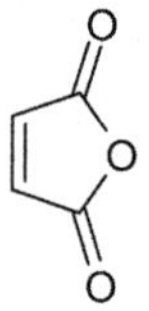	Maleic anhydride	Cyclic anhydride
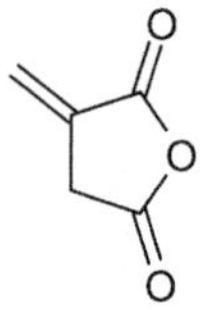	Itaconic anhydride	Cyclic anhydride
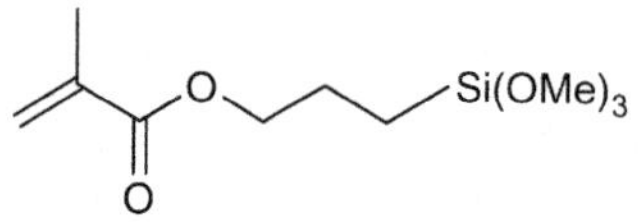	Trimethoxysilylpropyl methacrylate	Hydrolysable silyl ether
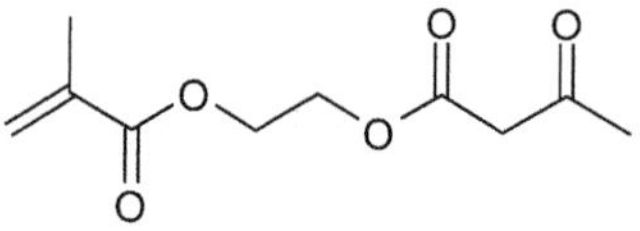	Acetoacetoxyethyl methacrylate	Acidic carbon
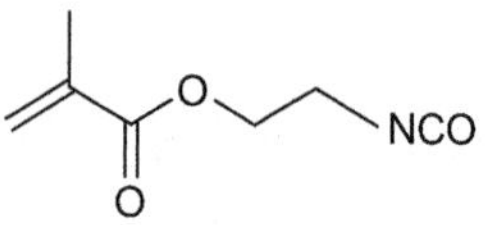	Isocyanatoethyl methacrylate	Isocyanate
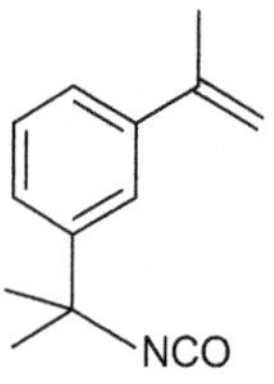	Dimethylisocyanatomethyl-3-isopropenylbenzene	Isocyanate

consequent on the molecular weight. By way of illustration, the hydroxy acrylic described above would contain some chains of molecular weight 8670 amongst all the other possibilities. However, they would not all contain 10 hydroxyl groups: there would probably be representatives of every variation from completely non-functional chains to hydroxyethyl methacrylate homopolymer ($f = 66$). All the other fractions would be similarly complex.

Further complexity arises from the variable copolymerisability of different pairs of monomers, giving rise to more or less statistical distributions of functionality. The consequences for network structures and the mechanical properties of a coating, of introducing blocky precursors are all too apparent.

In the next section, we will see that alkyds are polyesters with reactive groups distributed along their chains, that the most useful epoxies are end-functional step-growth oligomers, and that polyamidoamines used to cure them contrive to have terminal and intermediate reactive groups.

Amino resins and phenolics, to the extent that they are polymers or oligomers, probably also have chain end and intermediate reactive sites.

6.5   CROSSLINKING CHEMISTRY – BINDING THE BINDERS

The crosslinking process is essentially a polymerisation, although one whose monomers may themselves be polymeric, and whose product is an intractable network. It follows that crosslinking reactions can be characterised by their chemistry, and as chain growth- or step-growth processes.

It is also important to view them in the context of their application. The most important consideration is whether the coating has to develop its properties at ambient temperature, or whether it can be stoved. Rather different chemistries have usually been selected for crosslinking the different binder types described in this and subsequent chapters, as Table 6.5 attempts to show.

In very general terms, processes driven by condensation and the loss of a volatile component are well suited to most stoving applications, where removal of a small organic molecule or water is accommodated by the low initial viscosity of the hot film. However, they can be troublesome in some, such as powder coatings, where they may contribute to a 'popping' or 'pinholing' phenomenon in thick films. Also of course, an organic condensate will compromise the low volatile organic status of such a coating.

Ambient cured systems require more energetic chemistry, and many systems rely on the attack of a nucleophilic agent on a suitable electrophile such as a strained ring or double bond. If the electrophile happens to be susceptible to attack by water, its use in aqueous systems is somewhat circumscribed.

Table 6.5 *Classification of important crosslinking processes*

Coating type	Heterolytic			Homolytic	
	Step growth		Chain growth	Step growth	Chain growth
	Ring opening	Condensation			
Conventional ambient (Chapter 6)	Epoxy-amine, Isocyanate-hydroxyl, Acrylate or maleate-amine			Autooxidation	
Conventional stoved (Chapter 6)		Amino-formaldehyde, Phenol-formaldehyde-hydroxyl			
High solids (Chapter 7)	Isocyanate-thiol, Amine, maleate-thiol, Acrylate or maleate-acetoacetate		Cationic epoxy cure	Thiol-ene addition, Autooxidation	Radiation cured acrylates, Peroxide-cored unsaturated polyesters
Powder (Chapter 7)	Epoxy-acid, Epoxy-amine, Hydroxy-azlactone, Oxazoline-acid	Acid-ethanolamide, Blocked isocyanate-hydroxyl	Epoxy-anhydride		
Water-borne (Chapter 8)	Epoxy-amine, Aziridine-acid, Carbodiimide-acid	Amino-formaldehyde or Phenol-formaldehyde-hydroxyl, Carbonyl-amine, Blocked isocyanate-hydroxyl		Autooxidation	
Inorganic (Chapter 9)		Silanol condensation		Oxidative coupling of thiols	

Finally, it is interesting to note that some of the most successful solvent-free liquid coatings are based on chain-growth free-radical chemistry. Reasons for their suitability are explored further in Chapter 7.

We will next review a number of important crosslinking chemistries, deferring consideration of those that are particularly associated with high solids, solvent free, water based and inorganic systems to the appropriate chapters. It is important to recognise though, that the distinctions are not rigid, and a chemistry favoured for water-borne systems may also find use in powder coatings, and so on.

6.5.1 Auto-oxidative Curing–Ancient and Modern Chemistry

Auto-oxidative crosslinking of allylically unsaturated products is probably the oldest curing chemistry, as well as the most widely used on metal and wood, in primers and topcoats. Its mechanism is complex, but Scheme 6.9, where RH represents the molecule activated to H$^{\bullet}$ abstraction, gives an approximate picture of the most important processes.

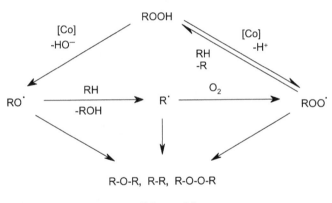

Scheme 6.9

Molecules, from which H can be abstracted easily, are susceptible to auto-oxidation. Allylic molecules are typical examples since the resulting radicals are stabilised by resonance, and they find the widest use in the coatings industry. 'Doubly allylic' groups are more reactive by one or two orders of magnitude since they can participate in more extensive resonance (Scheme 6.10).

Oxygen, which exists as a diradical, can react with adventitious radicals to form peroxyl radicals. They in turn abstract hydrogen from the allylic site to generate new carbon-centred radicals. Radicals due to doubly allylic systems exist predominantly in the conjugated form, which persists in the crosslinked network. Various transition metals, most notably cobalt,

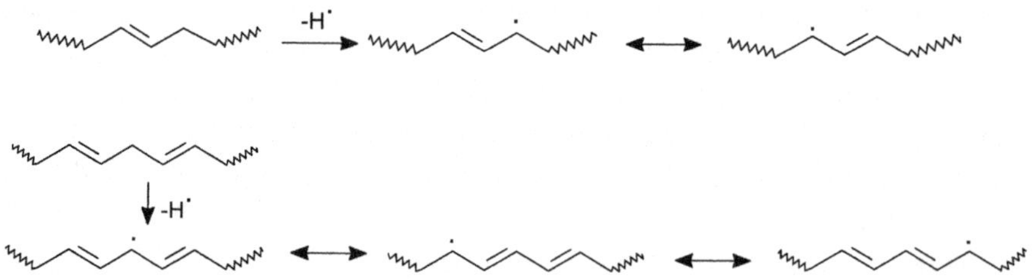

Scheme 6.10

catalyse the decomposition of the resulting hydroperoxide in a redox process (Scheme 6.11).

$$ROOH + Co^{2+} \longrightarrow RO^\cdot + HO^- + Co^{3+}$$
$$ROOH + Co^{3+} \longrightarrow ROO^\cdot + H^+ + Co^{2+}$$

Scheme 6.11

Overall, in the presence of air, a cycle generates a continuous supply of oxygen- and carbon-centred radicals that can form crosslinks by combination. When conjugated double bonds are present in the starting materials, a new stabilised radical is produced by addition to one of the radical species, and can participate in further chemistry (Scheme 6.12). With suitable prepolymer architecture, the crosslinks formed can lead rapidly to a thermoset network. The radical addition process is independent of the oxygen supply, so cure of conjugated systems can proceed throughout thick coatings. Non-conjugated systems have a marked tendency to 'skin'.

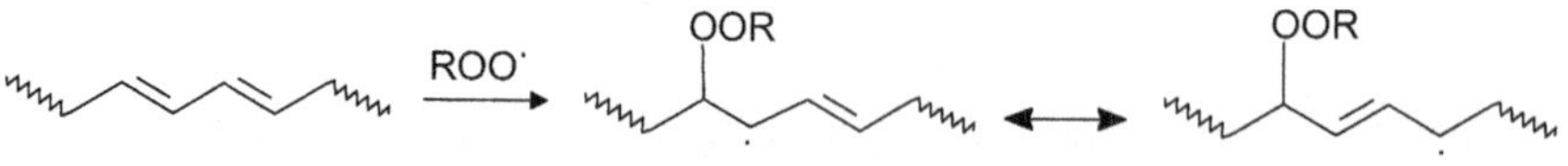

Scheme 6.12

6.5.1.1 Natural Oils. The exploitation of linseed oil in coatings may have begun as early as Roman times in the West, and Tung oil has been used for at least as long in China. Since then many other 'drying oils' have been recognised. All are triglycerides corresponding to (**23**) where OCOR represents the residue of a mixture of long chain fatty acids, some at least containing the allylic unsaturation required for curing. The great majority of the important fatty acids are based on chains of 18 carbon atoms.

(23)

As noted above, isolated unsaturation, conjugated pairs of double bonds and 'methylene interrupted conjugation' (*i.e.* CH_2 attached to two olefinic groups) can all participate in auto-oxidation reactions, and all are available in various natural oils. Most of them contain predominantly the more reactive *cis*-olefinic bonds. The triene-acids produce the most rapid cure, because they introduce two reactive sites per residue but they are often associated with a tendency to yellow. Oils with a significant proportion of triene-acid are described as 'drying oils' as they have sufficient functionality to crosslink in their own right. Those containing mainly diene-acids are known as 'semi-drying oils' as it is necessary to convert them to a more highly functional form to achieve efficient drying.

Some commercially and historically important fatty acids and their sources are given in Table 6.6. Linseed, soya bean, tall oil and castor oil are the mainstays of the modern coatings industry. Soya bean oil contains low levels of linolenic acid, so can be used in relatively colour stable coatings. Poppy seed oil is exceptionally low in linolenic residues, so was traditionally favoured for artist's paints, but the cultivation of opium poppies for their oil is no longer encouraged! Tall oil is a by-product of the paper industry. It originates as the usual kind of triglyceride in the bark of pine trees, but is hydrolysed as the wood pulp is processed, and isolated as a mixture of fatty acids and variable amounts of rosin. The ricinoleic residues in castor oil can be dehydrated at elevated temperatures to introduce a second olefinic group (now predominantly *trans*-). Roughly equal amounts of the conjugated ('ricinoic') and 'methylene interrupted' dienes are produced, and resulting binders have unusual auto-oxidative activity. Tung oil is sometimes known as China-wood oil, confusingly shortened to 'wood oil', although it is actually extracted from the nuts. Its triple conjugation makes it remarkably susceptible to auto-oxidation, and it is used for example in wrinkle finishes. The wrinkling effect is an expression of rapid surface drying. Perilla oil and oiticica oil are, respectively, cultivated in the Far East and gathered from the wild in Brazil. Both offer high reactivity, but are not significant in international trade.

Suitable natural oils are mobile liquids that can crosslink to satisfactory coatings. However, heat treatments known as 'bodying' were developed in

Table 6.6 *The origins of fatty acids of commercial and historical importance*

Acid	Systematic name	Oil	%	Plant source
Stearic	Octadecanoic acid			Mainly animal sources
Oleic	Octadec-9c-enoic	Palm oil	50	*Elaeis guineensis*
		Olive oil	83	*Olea europaea*
		Safflower oil	45–78	*Carthamus tinctorius*
Linoleic	Octadeca-9c,12c-dienoic	Soyabean oil	54	*Glycine max*
		Tall oil	35	*Pinus* species (*via* paper manufacture)
		Poppyseed oil	65	*Papaver somniferum*
Linolenic	Octadeca-9c,12c,15c-trienoic	Linseed oil	50	*Linum usitatissimum*
		Perilla oil	70	*Perilla frutescens*
Eleostearic	Octadeca-9,11,13-trienoic	Tung oil	80	*Vernicia fordii*
Ricinoleic	12-hydroxyoctadec-9c-enoic	Castor oil	90	*Rinicus communis*
Ricinoic	Octadeca-9,11-dienoic,	Dehydrated Castor oil	40	–
	Octadeca-9,12-dienoic*		50	
Licanic	4-keto-octadeca-9,11,13-trienoic acid	Oiticica oil	73	*Licania arborea*

*Dehydration of the ricinoleic acid residues produces comparable amounts of the conjugated and methylene interrupted species.

the early days of the coatings industry to improve their film forming character and film properties. The oil was held at a high temperature in contact with air, and possibly catalysts, until the required viscosity was achieved. Optionally, various natural resins such as rosin were incorporated to promote hardness (*v.s.*). Such 'oleoresinous' media find only limited modern application.

6.5.1.2 Alkyd Resins. Auto-oxidative cure chemistry advanced considerably in the 1920s with the invention of "alkyd" resins. An alkyd is a polyester with fatty acid residues attached along its backbone. Though saturated alkyds are known, and are cured through terminal acid or hydroxyl functionality, the term is usually understood to refer to polyesters furnished with auto-oxidisable side chains and providing much higher functionality than any oil, bodied or otherwise.

In modern practice, a variety of diacids, polyols and fatty acids are reacted, and viscosity monitored, until the required product is obtained. The traditional method involved preparation of a monoglyceride from the chosen oil and additional glycerol, completion being assessed by the miscibility of the product with methanol. The monoglyceride was then processed with phthalic anhydride and optionally more oil or glycerol. The product typically contained the residue (**24**). The theory of poly-condensation predicts that most alkyd resin formulations will gel before completion of the reaction. For that reason great care has to be taken towards the end of the process to ensure that the viscosity does not exceed the desired level.

(**24**)

Alkyds are described as long, medium or short oil according to the triglyceride content. There is little agreement amongst authorities on the precise definition of the various oil lengths, but > 60, 60–40 and $< 40\%$ oil, respectively, can be taken as a rough guide. Oil length is a rather meaningless concept in alkyds based on polyols other than glycerol, and it is more usual to describe them in terms of fatty acid content. Increasing oil length leads to increasing solubility in aliphatic hydrocarbons and easier manufacture. Short oil alkyds have increased potential to gel during manufacture. Typically, they dry most rapidly, but with a higher element of lacquer drying.

The auto-oxidation reaction is catalysed by transition metals and it is usual to add oil-soluble metal compounds such as cobalt naphthenate or octoate as 'driers' or 'siccatives'. Their action is understood in terms of catalysed decomposition of hydroperoxides as described above. Regret-tably, some concerns about the health effects of cobalt have surfaced recently, and efforts are in hand to replace it with rather less effective metals such as manganese. Several other metals also influence drying. Their action is less easy to explain, but lead, zirconium, aluminium, bismuth and neodymium are sometimes described as 'polymerising' driers and promote 'through drying'. Calcium, potassium, lithium and zinc are also used as 'auxiliary' driers to modify the properties of the final film. Driers for auto-oxidising systems are discussed in more detail in Chapter 10.

Unfortunately, the auto-oxidation process that leads to curing is essentially the same as that which leads to degradation, and does not stop when the required level of cure is obtained (*cf.* Chapter 5). All alkyds tend to embrittle through their service lifetime, more so when high levels of drier have been used. They are, nevertheless, extensively used in architectural and heavy duty markets.

Alkyd resins have probably been subject to more blending, modification and formulation than any other binder type. Much of that chemistry is beyond the scope of this chapter, but it is worth recalling a few major types.

Silicone alkyds were designed for enhanced weathering resistance. A fraction of the diacid is typically replaced with a disilanol, so that the backbone contains both ester and Si–O–C linkages. They are discussed further in Chapter 9. Styrenated, vinyltoluenated or acrylated alkyds are prepared by processing a pre-formed alkyd with the specified monomer under radical conditions, and were described in Section 6.3. They dry by solvent loss initially but retain some potential for oxidation. Modification of alkyds with phenolic resins (*q.v.*) also deserves mention.

6.5.1.3 Alternative Oxidising Systems. There have been many attempts to incorporate oxidising character into alternative film formers. They include epoxy esters, made from bisphenol A or epoxy novolac polymers esterified with fatty acids, and urethane alkyds and oils made by polymerisation of fatty acid monoglycerides with diisocyanates. Acrylic alkyds have been made *inter alia* by esterifying glycidyl functional acrylics with fatty acids, but have not achieved the popularity of epoxy esters or urethane alkyds. The products show property enhancements characteristic of their components – adhesion, toughness and weatherability, respectively.

6.5.1.4 Oriental Lacquer. The earliest, and most complex, example of an auto-oxidising coating was probably Chinese lacquer, the refined sap of a small tree, *Rhus verniciflua*, also known as *Toxicodendron vernicifluum*, in use in the Orient some 5000 years ago. Chinese or Japanese lacquer is a mixture containing urushiol (**25**) and related materials.

(**25**)

It cures by a combination of auto-oxidation of the unsaturated side chain, and oxidative coupling of the catechol residue, promoted by the phenoloxidase enzyme, laccase, present in the mixture (Scheme 6.13). It is not clear whether the phenyl radicals have any role in the reactions of the side chain. Curing of the coatings traditionally took place overnight in damp cellars.

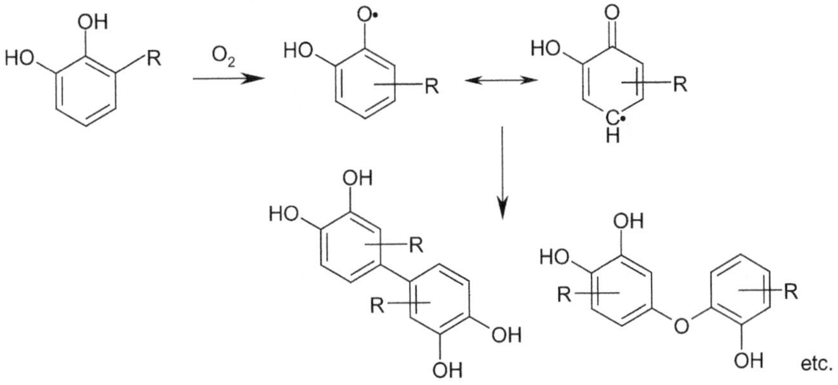

Scheme 6.13

Oriental lacquer has a claim to be treated as a solvent-free coating as most volatiles are removed by prolonged gentle heating, but also has a foot in the water-based camp since the raw latex is sometimes used as a primer.

6.5.1.5 Further Variations on the Auto-oxidation Theme. Attempts to synthesise alternative auto-oxidisable species have yet to provide film formers with the combination of performance, convenience and economy which derivatives of natural fatty acids can provide. None has become widely used except as a reactive diluent (*q.v.*). Numerous attempts have also been made to upgrade the reactivity and performance of standard alkyd resins. Apart from the extensive opportunities to vary the fatty acid, polyol and diacid components of the basic polymer, several additives can produce dramatic effects. Additions of radically polymerisable materials such as methacrylates, *e.g.* (**26**) are particularly useful. The radicals produced by auto-oxidation can initiate chain growth polymerisation which continues in the absence of air, in marked contrast with auto-oxidation itself, whose need of oxygen is often manifest as skinning. Thick coating films containing methacrylates can 'through dry' efficiently, thus overcoming one of the principal shortcomings of alkyds.

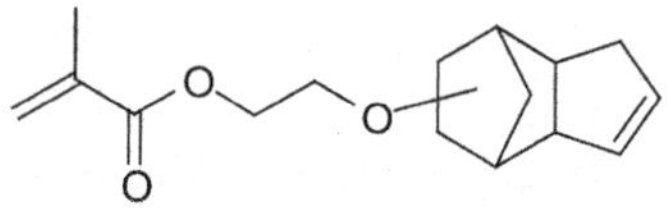

(26)

A further variant involves addition of thiol functional molecules. As shown in Scheme 6.14, thiyl radicals are produced in the auto-oxidation process and can give rise to additional crosslinks by adding to the double bonds in adjacent molecules.

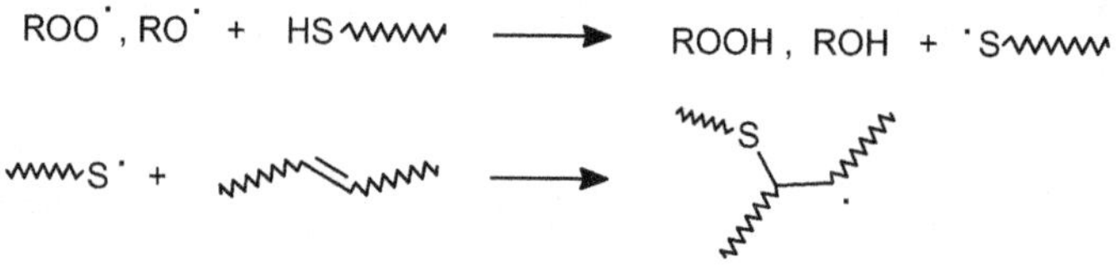

Scheme 6.14

6.5.2 Curing Primers at Ambient Temperature

6.5.2.1 Epoxide Chemistry. Epoxy resins are amongst the most satisfactory media for high performance primers for metal objects such as ships and aircraft. They are of particular value as anticorrosive primers for steel, perhaps because dense networks with excellent adhesion and no hydrolysable sites can be produced.

Any molecule containing two or more oxirane groups can be described as an 'epoxy', but the term epoxy resin is usually understood to denote the oligomers obtained by condensation of 'bisphenol A' or similar polyphenol with epichlorohydrin (Scheme 6.15). Bisphenol A itself is prepared by condensing phenol with acetone.

The $n = 0$ species (**27**) is a solid (m.p. *ca.* 45 °C), but when $n = ca.$ 0.14 the small amount of secondary –OH containing oligomer inhibits the crystallisation and a metastable syrup results. It is used in its own right for high and 100% non-volatile coatings. It is also used as a feedstock to prepare solid grades ($n = 1$–4) by 'advancement' with additional bisphenol A. Alternatively, the solid grades can be prepared directly from bisphenol A and epichlorohydrin in appropriate proportions (the Taffy process). The latter procedure produces twice as many oligomer fractions in a similar molecular weight envelope. A parallel series of resins can be based on bisphenol F (from phenol and formaldehyde). They have lower viscosity but are more inclined to crystallise.

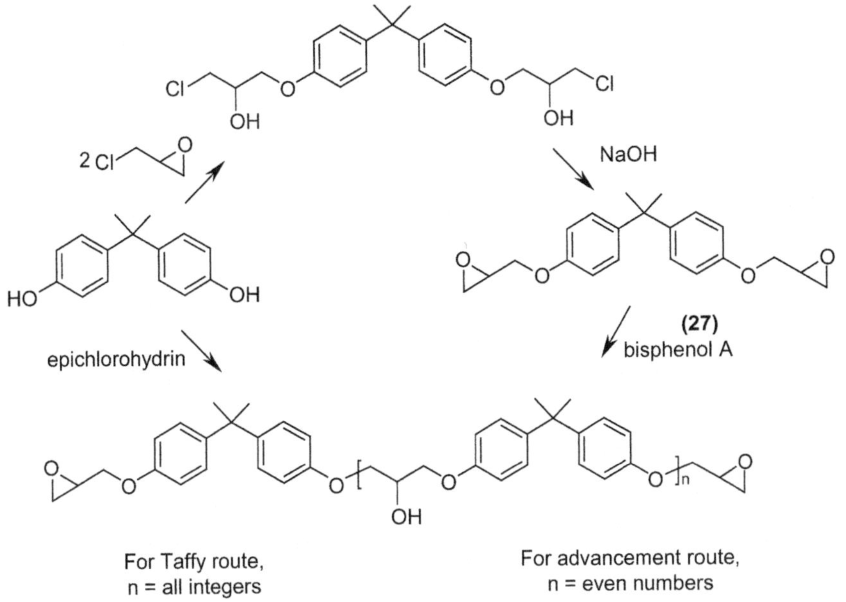
Cl
OH
O
O
Cl
OH
2 Cl
O
NaOH
HO
OH
O
O
O
O
(27)
bisphenol A
epichlorohydrin
O
O
O
OH
O
O
n
O
For Taffy route,
n = all integers
For advancement route,
n = even numbers

Scheme 6.15

Epichlorohydrin–bisphenol A polymers, cured for example with amines, yield coatings characterised by hardness, chemical resistance and excellent adhesion. Unfortunately their weathering resistance is very poor and exterior exposure leads rapidly to a condition known as 'chalking' – the polymer is eroded and powdery pigment is left on the surface. Amongst other available aromatic epoxies are epoxidised 'novolacs' (**28**), the polymeric counterparts of bisphenol F, and epoxidised amino phenols (**29**). They offer increased functionality and, consequently, greater hardness in the cured film. Epoxidised aromatic amines are widely used in aerospace composites.

Improved weathering performance can be obtained by using aliphatic epoxies, though usually at the expense of hardness. They include hydrogenated versions of liquid bisphenol A epoxide, nominally (**30**), and copolymers of glycidyl methacrylate and related monomers (*cf.* Table 6.4). Epoxidised oils or polybutadienes and materials such as (**31**), (**32**) and (**33**) also have interesting properties. However, 'internal' epoxy groups have greatly reduced reactivity towards nucleophiles and are more susceptible to self-condensation induced by Lewis or Brönsted acids.

For ambient temperatures, amines are the curing agents of choice. Simple aliphatic polyamines such as ethylenediamine, diethylenetriamine and

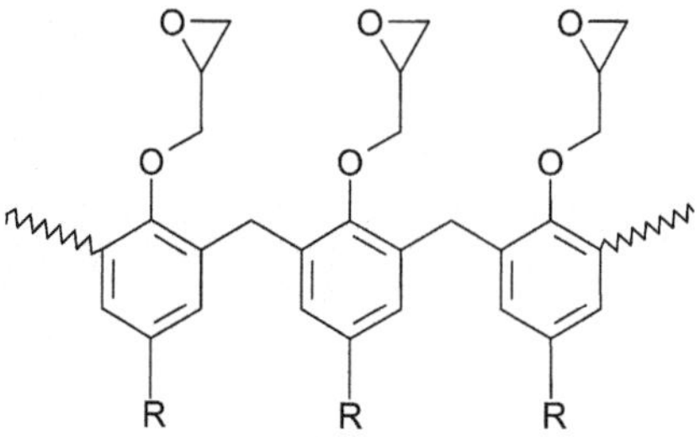

(28)

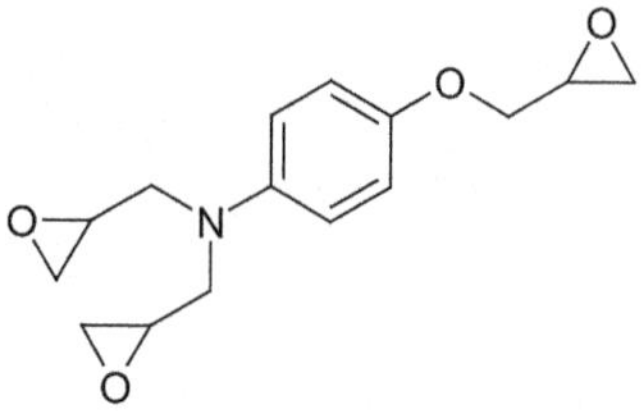

(29)

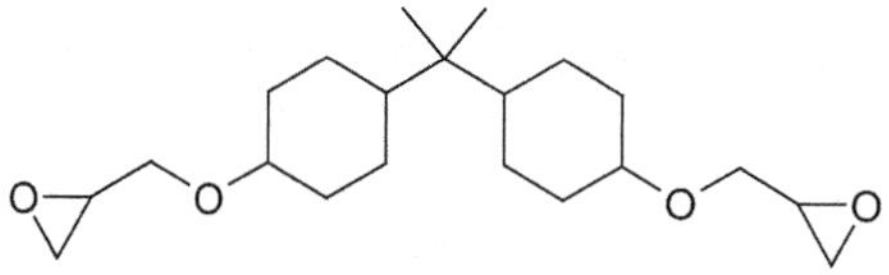

(30)

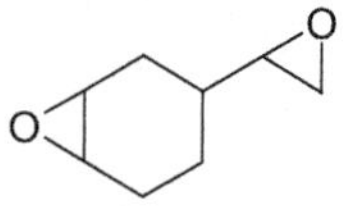

(31)

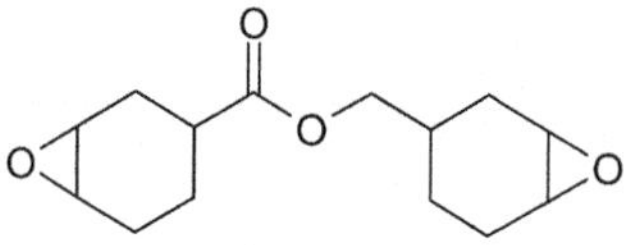

(32)

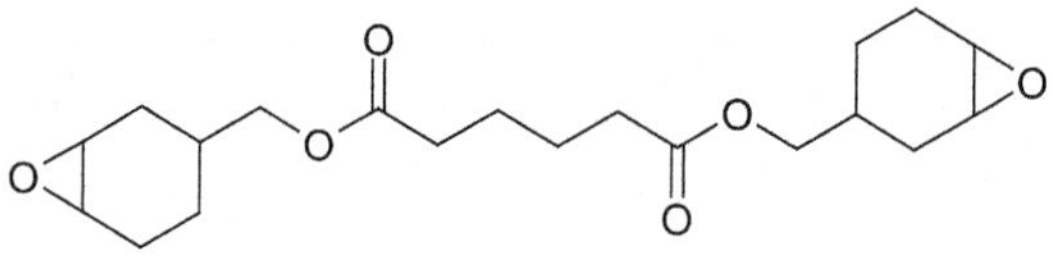

(33)

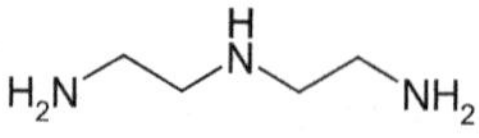

(34)

triethylenetetramine are used. All active hydrogens participate so that diethylenetriamine (34), for example, is pentafunctional.

The generated tertiary amines catalyse the reaction of epoxies with hydroxyl groups, and anionic epoxy homopolymerisation which often contributes significantly to the curing process (Scheme 6.16).

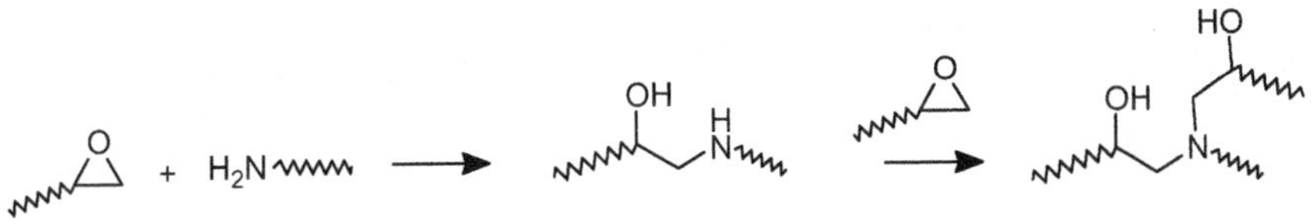

Scheme 6.16

The epoxy–amine reaction is catalysed by hydroxylic species such as phenols and possibly alcoholic solvents. Since hydroxyl groups are generated by the ring opening of the epoxide ring, epoxy–amine reactions show a degree of autocatalytic character.

Simple amines are rather volatile and toxic and, being strongly basic, tend to react with carbon dioxide at the coating/air interface with consequent cure inhibition and surface stickiness, the so-called 'amine bloom' (Scheme 6.17).

$$RNH_2 + CO_2 + H_2O \longrightarrow RN^+H_3 \ HCO_3^- \xrightarrow{RNH_3} RN^+H_3 \ ^-OCONHR$$

Scheme 6.17

Cycloaliphatic amines such as isophoronediamine (**35**), bicyclic amines such as (**36**), amine tipped polyethers and prepolymers prepared from simple polyamines and epoxide resins are all used to address the above problems.

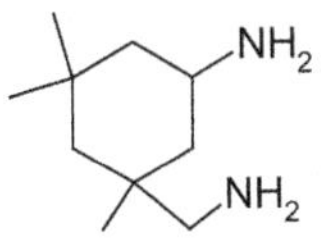

(**35**) 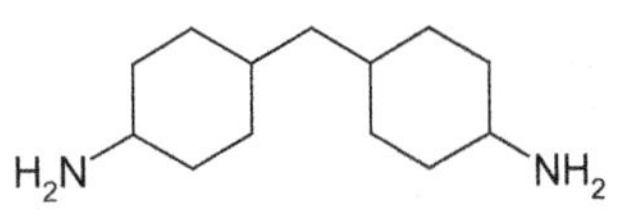   (**36**)

Amine adducts can either be 'isolated' or 'partial'. In the latter case an excess of low molecular weight amine is used, and remains in the composition to vitiate its toxicological and volatility advantages. However, partial adducts are much easier to make.

Amino functional polyamides formed from simple amines and long chain diacids including dimerised fatty acids (*cf.* Table 6.1) (*e.g.* (**37**)) are effective epoxy curing agents.

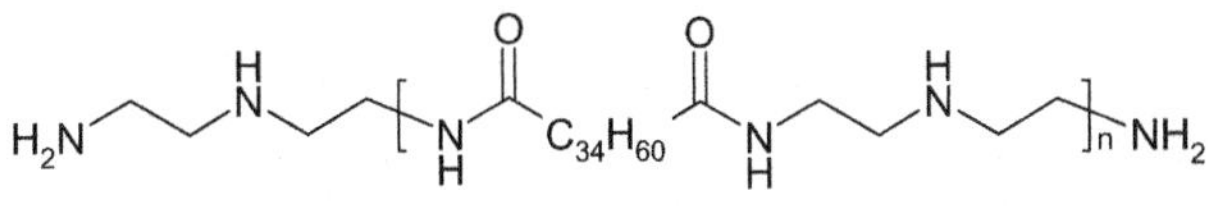

(**37**)

The so-called polyaminoamides can be made to contain various levels of imidazoline by dehydration of chain segments to varying extents (Scheme 6.18).

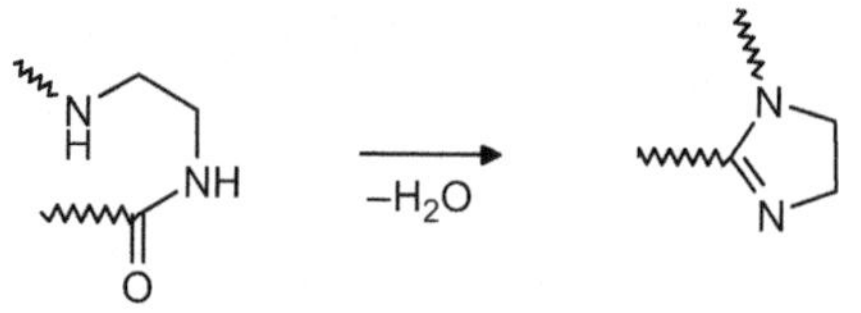

Scheme 6.18

High levels of imidazoline give lower reactivity since some amine groups are not immediately available for reaction, being released by slow hydrolysis on contact with atmospheric moisture. Polyamides provide highly flexible and corrosion resistant thermosets. Alternatively, polyamines can be partially reacted with mono-functional fatty acids to give amidoamines. Mannich bases such as those derived from amine, phenol and formaldehyde (Scheme 6.19 is a simplified representation) show enhanced reactivity, perhaps due to the catalytic effect of the phenolic hydroxyl, and are often used in 'winter workable' compositions.

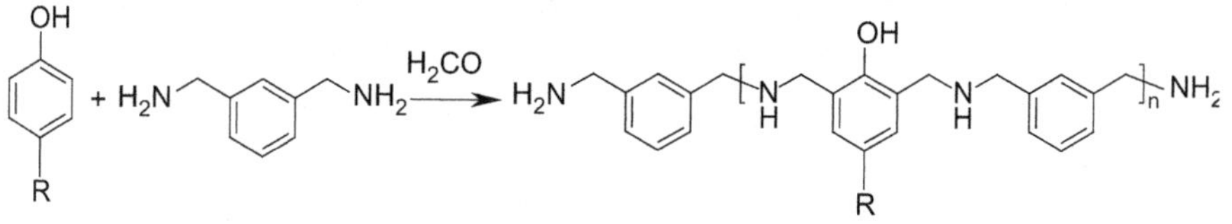

Scheme 6.19

It is also possible to present amines in protected form, such as ketimine, so as to provide a latent cure dependent on atmospheric moisture.

Secondary amines are generally observed to be less reactive than primaries, despite their higher nucleophilicity. However, secondary amines with the minimum of steric hindrance, such as those where one substituent is methyl, or especially cyclic secondary amines containing piperidine or pyrrolidine residues can show exceptional reactivity.

Various aromatic polyamines are effective curing agents at stoving temperatures but they are generally more toxic than aliphatic amines so find little application in coatings. Their chemical resistance, however, is exceptional.

Epoxy–thiol chemistry (Scheme 6.20) has not been much exploited in coatings, perhaps because of the malodorous character of most thiols, and their limited commercial availability, but *cf.* Chapter 7. However, their reaction with epoxies can be much faster than that of amines, given a basic catalyst to generate the thiolate ions. Technically excellent thermosets free of carbonation problems can be prepared using suitable polythiols and tertiary amine catalysts.

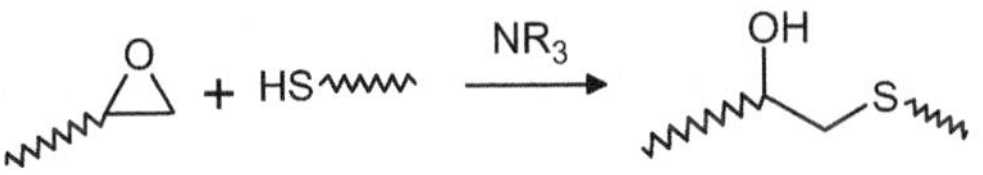

Scheme 6.20

Reactions between epoxies and carboxylic acids (Scheme 6.21) are relatively sluggish at ambient temperature, and polyacid curing agents find their most important application at elevated temperature in powder coatings (*q.v.*).

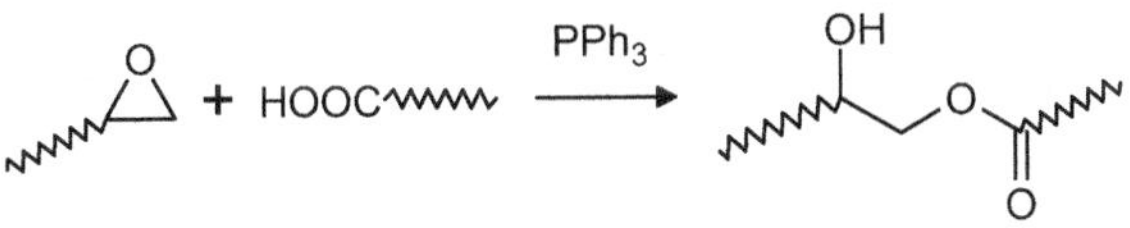

Scheme 6.21

That said, a number of successful proprietary chemistries achieve ambient cure using aliphatic epoxies with suitable acid-functional systems.

Homopolymerisation of epoxy resins can be initiated by anionic or cationic species. Anionic polymerisation (Scheme 6.22) as noted above is an almost inevitable concomitant of crosslinking by amines, and is a generally helpful phenomenon since it leads to higher connectivity than the simple step-growth process. It can be initiated by tertiary amines, including those formed during epoxy cure, but species such as imidazoles and pyridines are particularly effective. Some formulations are designed with more than the usual stoichiometric excess of epoxy over amine to allow for a significant level of homopolymerisation.

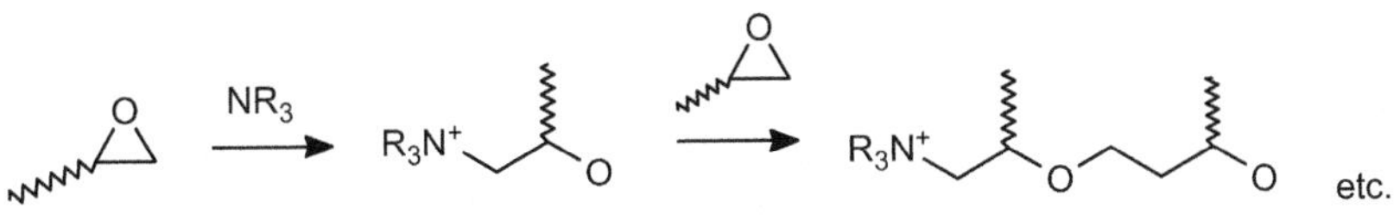

Scheme 6.22

Cationic homopolymerisation is initiated by Lewis and Brönsted acids such as BF_3, $SnCl_6$, H_2PF_6 and CF_3SO_3H. 'Internal' epoxies, especially the cycloaliphatic varieties, are particularly susceptible because of the carbonium ion stabilisation available to them. They are widely used in electrical potting compositions, often in combination with polyol additives that can act as chain terminators and increase the connectivity of the system. However, they have found limited application in coatings because of the sensitivity of the initiators to moisture and other adventitious nucleophiles. There is though, significant interest in photoinitiated cationic cure, providing a route to radiation cured epoxies (*cf.* chapter 7). Besides epoxies, other cyclic ethers, vinyl ethers, lactones and cyclic sulfides can be polymerised cationically.

Anhydride and epoxy residues have a strong tendency to undergo alternating chain growth which greatly exceeds the affinity of carboxylic acids for epoxides or hydroxyl groups for anhydrides, though all four functional groups are frequently formulated together. The process can be initiated by a tertiary amine and the mechanism of Scheme 6.23 has been proposed. The technology has been most frequently used in powder coatings.

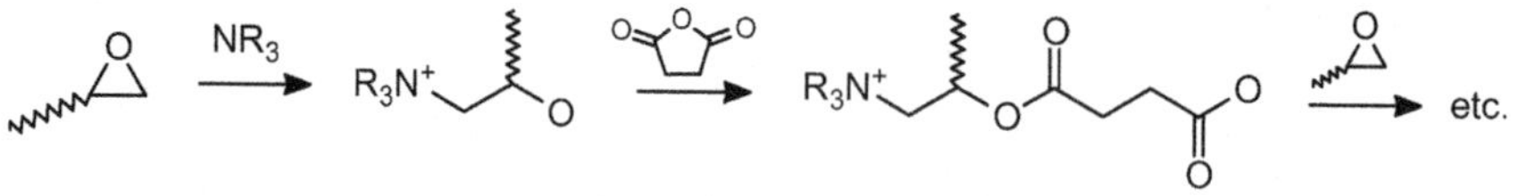

Scheme 6.23

6.5.3 Curing Topcoats at Ambient Temperature

6.5.3.1 Isocyanate Chemistry. Isocyanate chemistry was investigated by Bayer in the 1930s with a view to producing polymers with the attractive hydrogen-bonding characteristics of polyamides. Incorporation of urethane groups into polymeric systems produced marked enhancements of their properties, attributed to secondary crosslinking through their hydrogen bonds (**38**) to the extent that a pure polyurethane polymer would be quite intractable, and it was sufficient to incorporate a few urethane linkages per molecule, as when crosslinking.

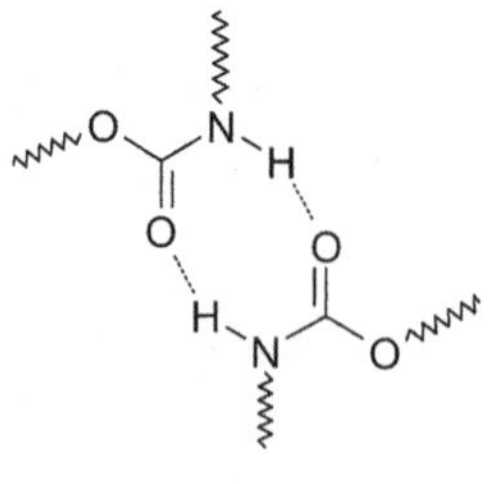

(**38**)

The various linkages produced from isocyanates and active hydrogen species are also highly resistant to hydrolysis – ureas even more so than urethanes.

Some of the reactions of isocyanates with active hydrogen compounds are given in Schemes 6.24–6.28. At elevated temperatures, the products of the reactions can react with further quantities of isocyanate (Schemes 6.29–6.31).

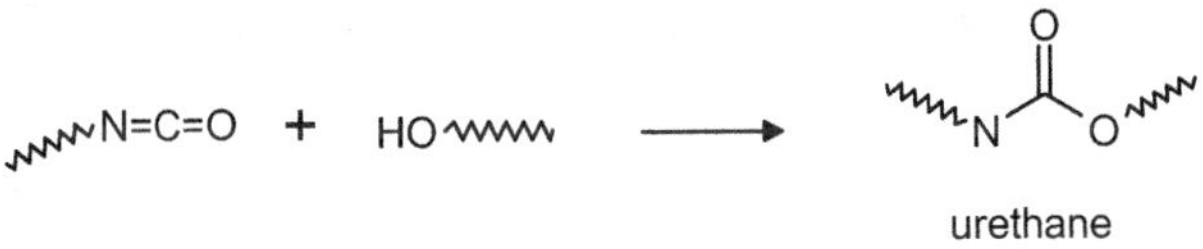

Scheme 6.24

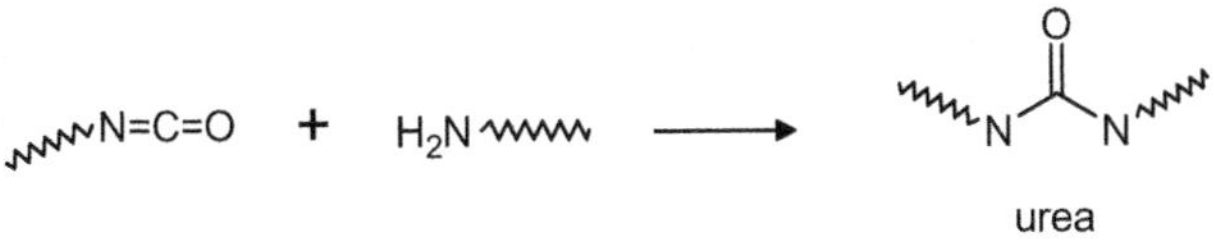

Scheme 6.25

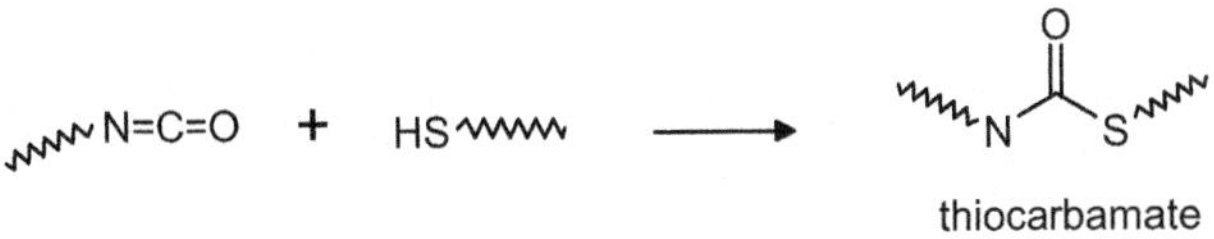

Scheme 6.26

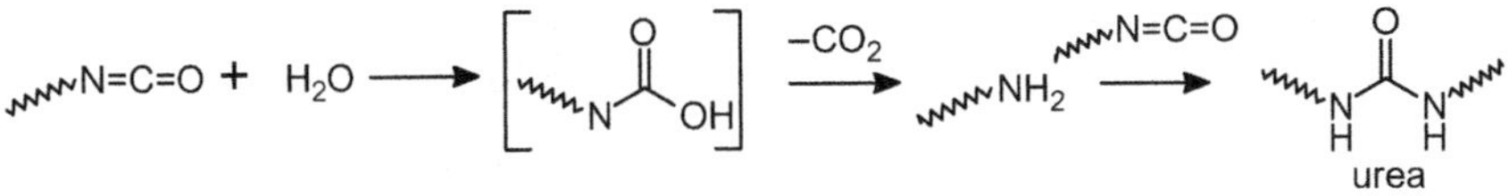

Scheme 6.27

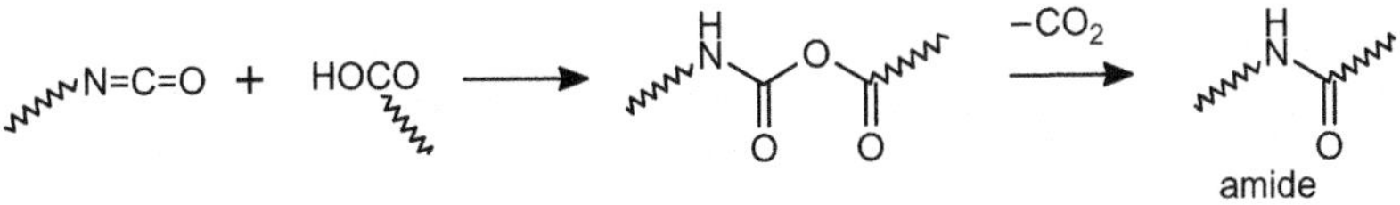

Scheme 6.28

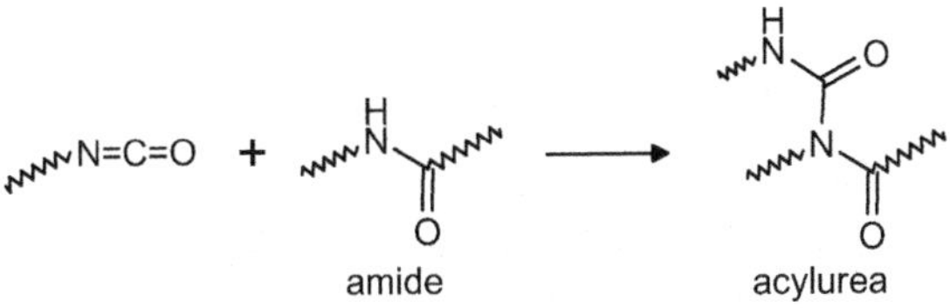

Scheme 6.29

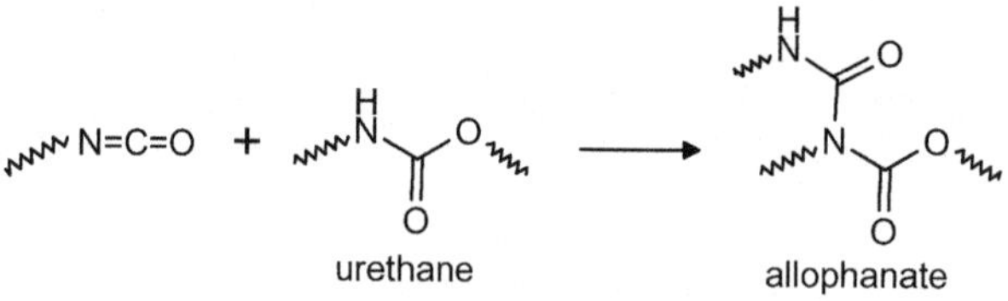

urethane   allophanate

Scheme 6.30

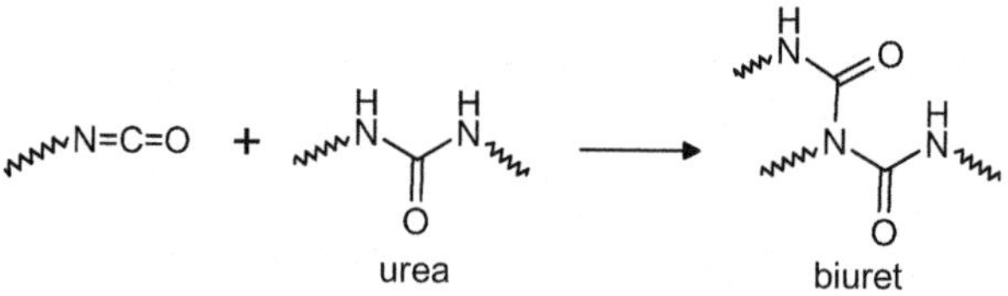

urea   biuret

Scheme 6.31

Since aliphatic isocyanates can yield highly weatherable coatings with excellent mechanical and chemical resistance properties, they have become the ambient cured topcoats of choice in many high performance applications. Some difunctional isocyanate monomers are listed in Table 6.7.

Table 6.7   *Some common difunctional isocyanates*

Structure	Name
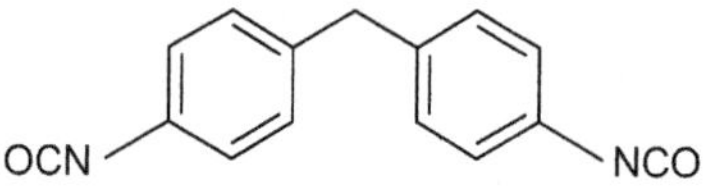 (39)	Methylenediisocyanate (MDI)
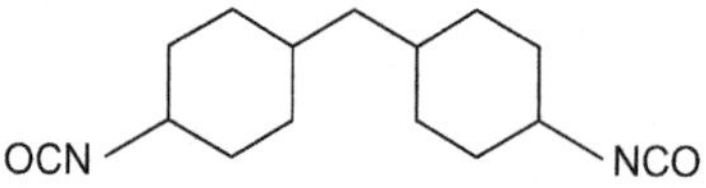 (40)	Hydrogenated MDI (HMDI)
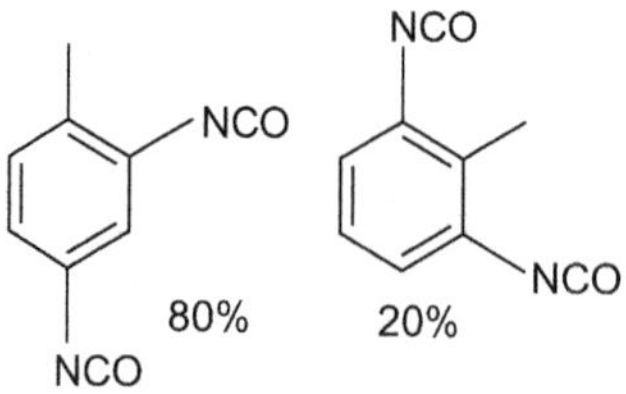 (41)	Tolylenediisocyanate (TDI)

(continued)

Table 6.7   *Continued*

Structure	Name
OCN–(CH₂)₆–NCO (**42**)	Hexamethylenediisocyanate (HDI)
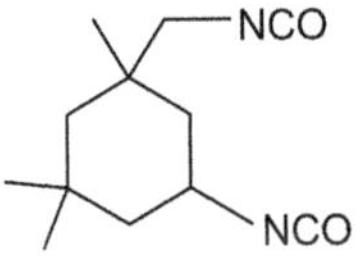	Isophoronediisocyanate (IPDI)
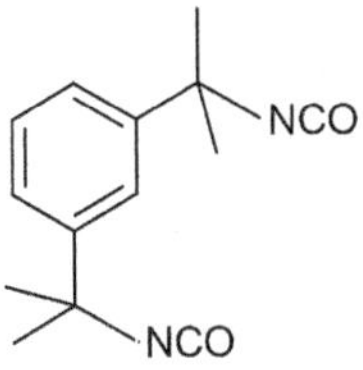	*m*-Tetramethylxylylenediisocyanate (TMXDI)

They are usually prepared by the action of phosgene (carbonyl chloride) on the corresponding diamines (Scheme 6.32).

Scheme 6.32

An alternative synthetic approach uses the addition of isocyanic acid (HNCO) to isopropenylbenzenes (Scheme 6.33).

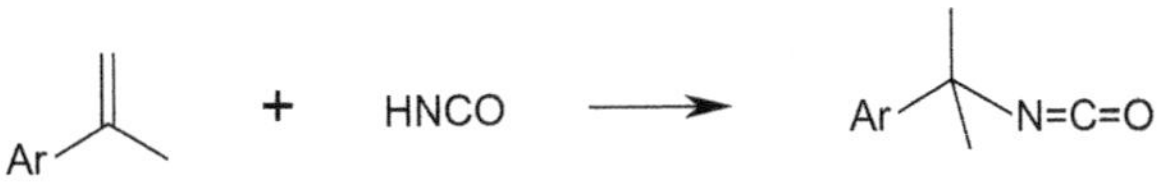

Scheme 6.33

Most isocyanates produce a very marked physiological reaction in certain individuals. It usually manifests itself as a form of respiratory distress known as industrial asthma, whose unfortunate victims become sensitised to very low levels of airborne isocyanate. Accordingly, great care must be taken when using isocyanates in coatings, where they have the maximum opportunity of entering the atmosphere, either as vapour or aerosol (due to spraying). The hindered isocyanates prepared from isopropenylbenzenes are of reduced reactivity, and have been claimed to be free of pulmonary sensitisation.

Diphenylmethanediisocyanate (**39**) and its hydrogenated equivalent (**40**) find limited use in coatings compositions, in view of their volatility and consequent health risk. Tolylenediisocyanate (**41**) and hexanediisocyanate (**42**), however, are the most widely used in coatings curing agents. Both are powerful sensitisers, and sufficiently volatile to constitute a hazard. It is necessary to use them in 'prepolymer' form, which has the further advantage of increasing their functionality to three. (**41**) is usually adducted 3:1 with a triol such as trimethylolpropane. (**42**) is either reacted 3:1 with water to produce a biuret (**43**) or catalytically trimerised to the isocyanurate (**44**), which is generally preferred for its hydrolytic stability and additional 'hardness'.

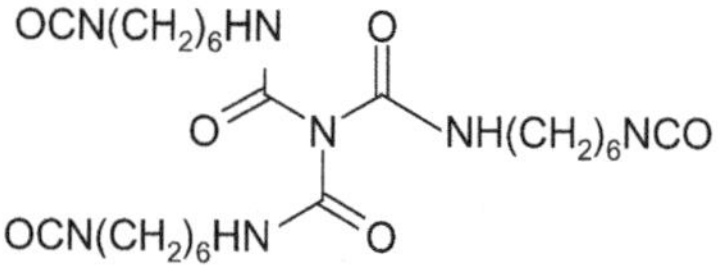

(**43**) 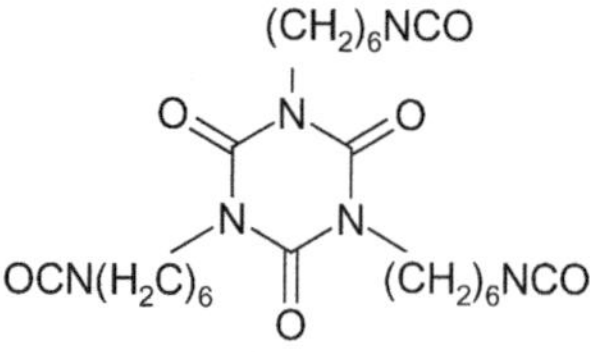   (**44**)

Addition polymers from the unsaturated isocyanates, noted in Table 6.4, have some specialised uses.

HDI prepolymers are used to crosslink hydroxyacrylics or hydroxy-polyesters at ambient temperature to films of excellent weathering resistance and application properties. They are used as topcoats, for example in car refinish and aircraft coating. The reaction of isocyanates with hydroxyl compounds is sluggish in the absence of catalyst and various organotin compounds and tertiary amines are used. The addition of pentane-2,4-dione (acetylacetone) moderates reactivity and can be used to extend pot life without impairing the properties of the final film.

Another interesting approach to 'cure latency' involves the use of tertiary amine catalyst in the form of a vapour. It has been introduced with the spray propellant, or the newly coated article can be passed through a vapour-filled chamber so that very rapid curing takes place.

As indicated in Scheme 6.27, isocyanates can be cured with water. The carbamic acid formed by reaction of water and 1 mol of isocyanate decarboxylates spontaneously to form a primary amine. Reaction with a further mole of isocyanate leads to a urea linkage. Moisture curing of isocyanates is somewhat limited in its application, since the evolution of carbon dioxide is likely to disrupt a thick film. However, it has the advantage of being 'latent' – the liquid paint is stable indefinitely but cures rapidly as a film on contact with atmospheric moisture. Moisture cured systems find most frequent use as wood varnishes. It is also frequent practice to formulate high performance urethane aircraft finishes with an excess of isocyanate over polyol so that a proportion of urea linkages forms on exposure to atmospheric humidity. The resulting coatings are more resistant to the hydraulic fluids that frequently find their way onto the outsides of aeroplanes than pure urethanes.

Curing with primary and secondary amines is finding increasing favour because the urea linkage generated is even more hydrolytically stable than urethane, and because the reaction is extremely rapid without needing any catalysis. Ultra-rapid curing coatings have been developed for application by twin-fed spraying equipment, and can be walked on virtually as soon as the paint has made contact with the substrate. If more control is needed the process can be restrained by derivatising the amine as ketimine, enamine or oxazolidine which release amine only after hydrolysis by atmospheric moisture (Schemes 6.34–6.36).

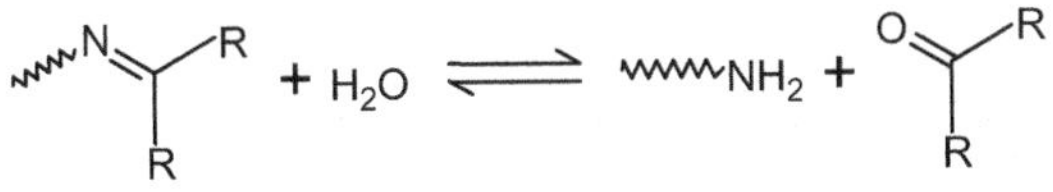

Scheme 6.34

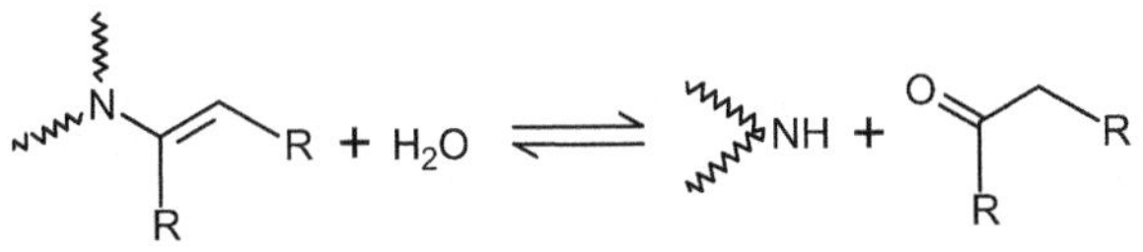

Scheme 6.35

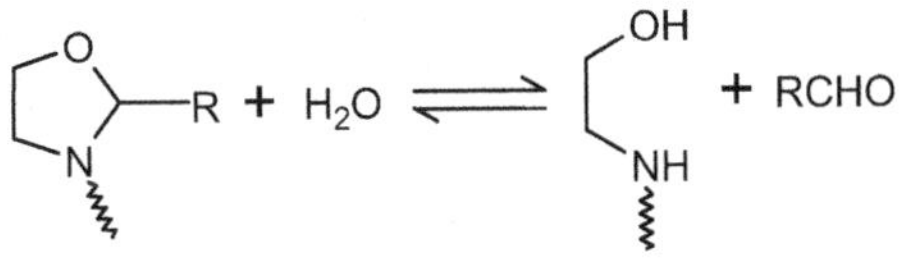

Scheme 6.36

Alternatively, highly hindered systems such as aspartic esters find some use, though mainly as reactive diluents (*cf.* Chapter 7).

Aromatic amines are less reactive than most aliphatic equivalents and have been used in elastomeric compositions. However, they present the same severe health hazard, as when they are used with epoxy resins.

A further aspect of isocyanate chemistry is the possibility of catalytic dimerisation and trimerisation (Schemes 6.37 and 6.38). The dimer (uretidinedione) exists as reactive four membered rings, used as blocked isocyanates, whereas the trimer (isocyanurate) is very stable. There have occasionally been proposals to use the trimerisation process as a curing reaction.

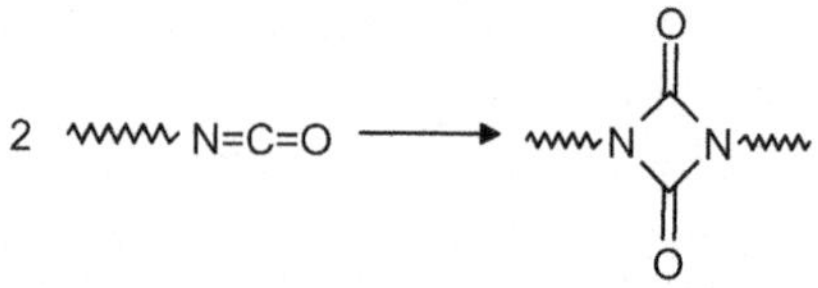

$$2 \quad \text{\small www} N{=}C{=}O \longrightarrow$$

Scheme 6.37

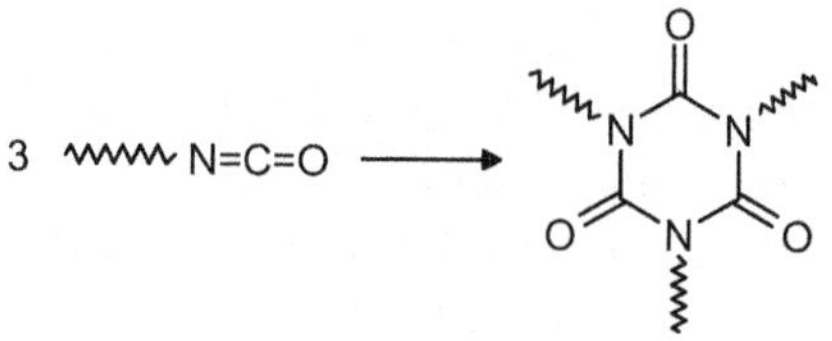

$$3 \quad \text{\small wwww} N{=}C{=}O \longrightarrow$$

Scheme 6.38

The cyclic trimer of hexamethylenediisocyanate is an ideal trifunctional curing agent in view of its structure and stability. The earlier biuret system tended to contain sufficient monomer to create an unacceptable atmospheric concentration.

6.5.3.2 Isocyanate-free Binders. Because isocyanate chemistries combine such excellent properties with a significant health hazard, there have been many attempts to produce equivalent ambient-cured crosslinkers with reduced toxicity. The strong and growing market for isocyanates gives a pointer to the limited success of such attempts, but a number of important 'niche' chemistries have emerged. Some of them will be discussed in the following chapters as being more directly relevant for high solids, water based, or inorganic hybrid coatings, but the following examples are more appropriate to conventional coatings.

The reaction of multifunctional cyclic carbonates with primary amines has attracted considerable attention, and the resulting β-hydroxyurethanes have been claimed to be significantly more resistant to hydrolysis than conventional urethanes because of internal H-bonding (Scheme 6.39).

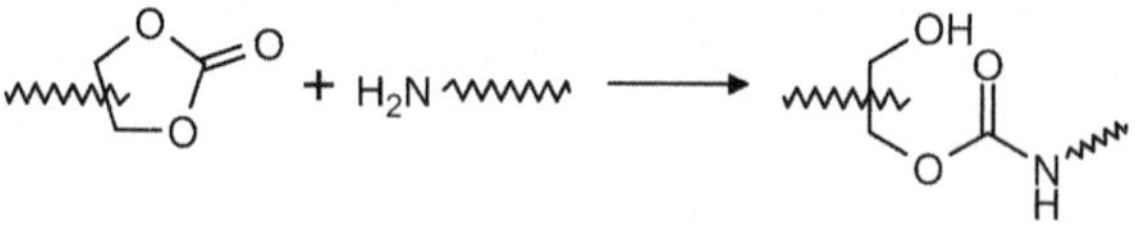

Scheme 6.39

Epoxy–amine chemistry can be adapted to topcoat use as, for example, when a copolymer of glycidyl methacrylate is cured with an acrylic polymer containing primary amine groups. A primary amine functional monomer, if it could be prepared, would hompolymerise at once by Michael type addition, so it is necessary to use an indirect route to prepare the amine reagent. A methacrylic acid polymer can be reacted with aziridine to produce aminoethyl methacrylate residues (Scheme 6.40).

Scheme 6.40

A number of proprietary technologies exploit the cure of epoxy binders, such as copolymers of glycidyl methacrylate, with *tertiary* amines. The amines can be copolymers of dimethylaminoethyl methacrylate, often containing comparable levels of polymeric carboxylic acid. It is likely that both the ammonium and the carboxylate ions react with epoxy groups, but that the most important curing mechanism is homopolymerisation of the epoxy groups, catalysed by the quaternary ammonium salts generated *in situ*.

Another early approach used a polymer containing cyclic anhydride groups, typically a copolymer of maleic or itaconic anhydride. It was cured with a polyol in the presence of catalytic amine (Scheme 6.41). The formation of water sensitive half-ester residues did not necessarily degrade its properties. Anhydride cured polyols have been used in car re-finishing coatings.

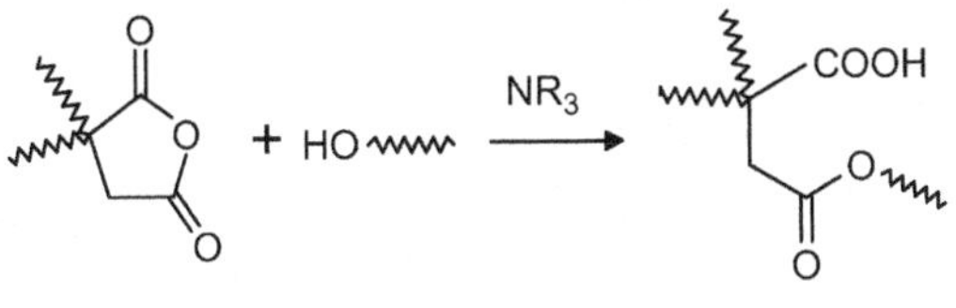

Scheme 6.41

In similar vein, a polymer containing the rather less reactive residues from vinyl azlactone reacted at a convenient rate with primary amines (Scheme 6.42). Unfortunately, vinyl azlactone is a rather costly material.

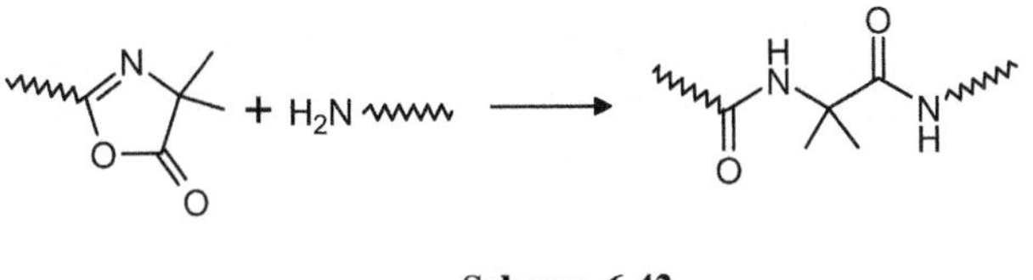

Scheme 6.42

The Michael and Michael-type reactions of acrylates and related activated olefins such as maleates and fumarates have been the subjects of intensive study. In the former case, the nucleophiles are the anions derived from carbon acids such as acetoacetate or malonate (poly)esters (Scheme 6.43). A strong base such as sodium methoxide or an amidine is required to deprotonate the esters. The Michael type reaction exploits the addition of primary or secondary amine or thiol to the acrylate residue. In the case of the thiol, a tertiary amine catalyst is needed to bring about the reaction (Scheme 6.44).

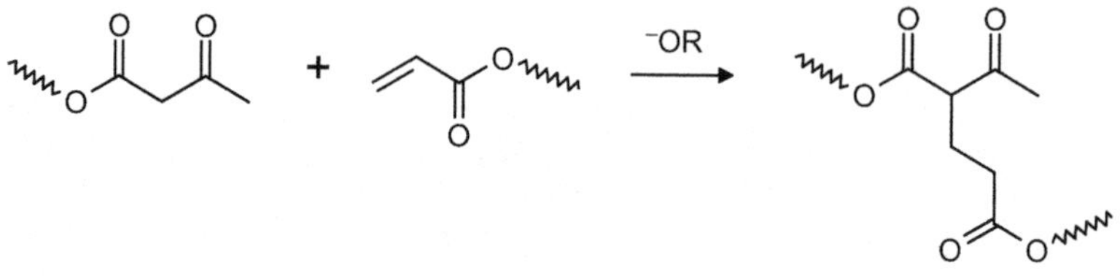

Scheme 6.43

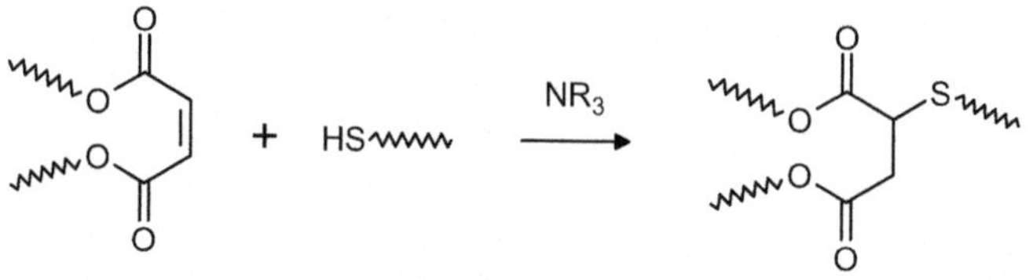

Scheme 6.44

All the Michael related chemistries are capable of fast reaction at low temperatures, but still more reactive Michael acceptors are the alkylidene-malonates made from malonic esters and aldehydes. They are most effectively crosslinked with thiols (Scheme 6.45).

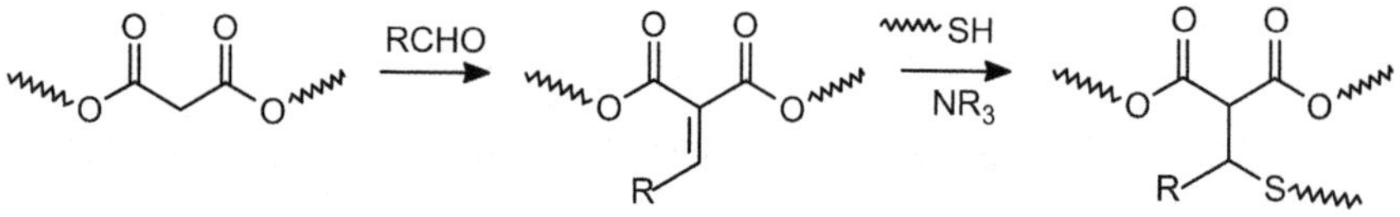

Scheme 6.45

Although condensation processes are less well suited to ambient cure than addition, a novel chemistry promoted in the 1980s involved copolymers of a proprietary monomer 'MAGME' (**45**), the methyl ether of acrylamidogly-colic acid. The activated ester groups transamidated readily with primary amines leading to cured networks.

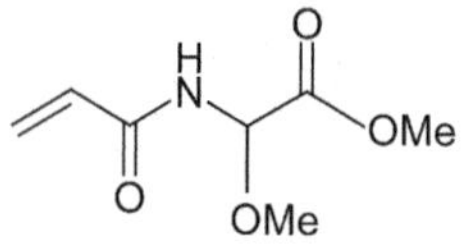

(**45**)

6.5.4 Stoving Topcoats

6.5.4.1 Amino Formaldehyde Crosslinkers and Polymers. The most important chemistry used in stoved coatings is based on optionally alkylated, amino-formaldehyde condensates. The principal 'amino' precursors are urea (**46**), thiourea (**47**), melamine (**48**), benzoguanamine (**49**), glycoluril (**50**) and copolymers of acrylamide and substituted acrylamide (**51**). Other amides and urethanes have been exploited occasionally.

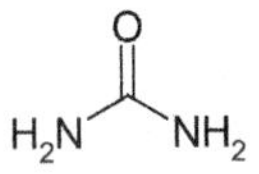

(46)	(47)	(48)

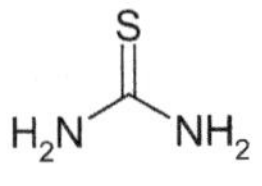

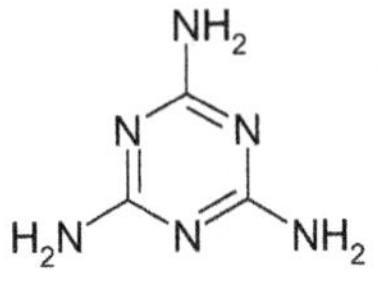

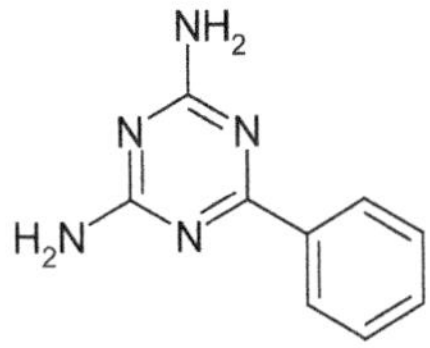

(49)	(50)	(51)

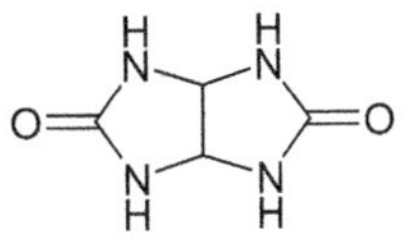

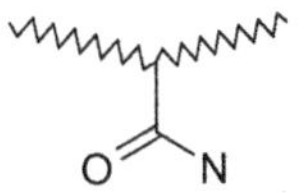

Melamine–formaldehyde (MF) products, introduced in the 1940s, are the most widely used amino resins, and their manufacture and crosslinking chemistry can be used to illustrate the more general reaction. As shown in Scheme 6.46, melamine reacts with formaldehyde in alcoholic solvent using acid or basic catalysts to introduce alkoxymethyl substituents and methylene or ether bridging groups. Methylene bridging groups, N–CH$_2$–N, appear to be rather labile unless grouped together in hexahydro-triazine rings, and ether bridges, N–CH$_2$–O–CH$_2$–N, may play a more important part in the structures than has often been supposed.

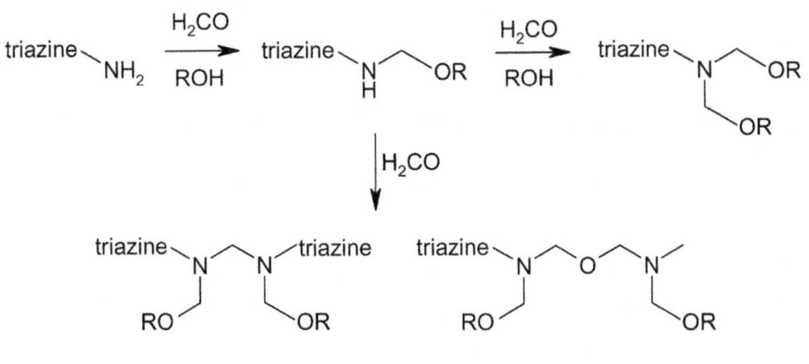

Scheme 6.46

The resulting entities can be monomeric or oligomeric, and one or both hydrogens on the amino groups can be substituted. Non-alkylated systems are also known, but the hydroxymethylamino groups tend to be excessively reactive. The alcohols used are typically methanol and butanol, sometimes in combination. Butanol gives a more tractable and lower viscosity product. Alcohol is usually present in the solvent, and aids stability by displacing the equilibrium away from decomposition. When the solvent is lost from the film, reaction can go ahead swiftly.

A major distinction exists between fully and partially alkylolated melamines. Hexamethoxymethylmelamine (**52**), for example, is commercially available as an essentially discrete substance.

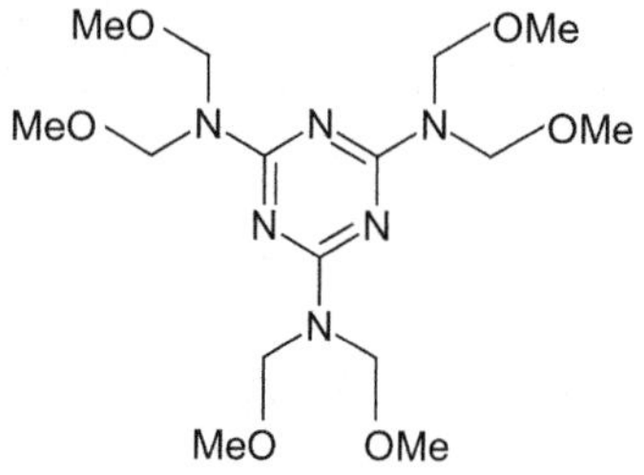

(**52**)

It and similar materials require strong acid catalysis to undergo a relatively straightforward condensation with hydroxyl, carboxyl, amide or carbamate functional film formers (Scheme 6.47). They can also react with activated CH, as in acetoacetates or aromatic CH groups as in phenols. A typical catalyst is *p*-toluenesulfonic acid, and transetherification usually takes place between 120 and 150 °C.

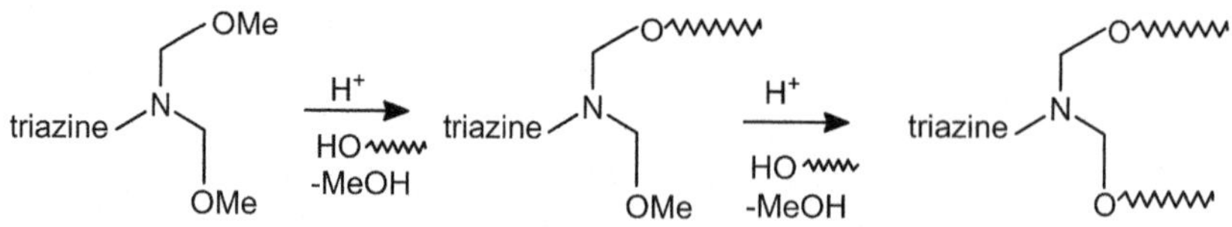

Scheme 6.47

Incompletely alkylated versions react with similar functional groups, but in two stages. Alcohol is first lost to produce a formaldimine intermediate. It is followed by the addition of the active hydrogen group to create a linkage (Scheme 6.48). The overall effect is still transetherification. Partially alkylolated systems require weaker acid catalysis, the carboxylic acid present in many film formers often being sufficient. A much wider range of reactions, including self-condensation, becomes possible and it follows that the species, as prepared, will tend to be oligomeric. Continued condensation in the absence of other materials leads to densely crosslinked thermosets used as surface coatings for laminates.

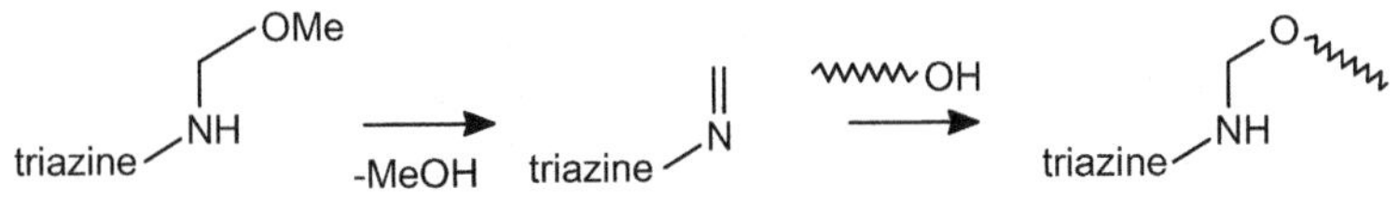

Scheme 6.48

It has generally been held that partially methylolated aminos are more reactive than fully methylolated versions. However, Jones (1994) showed that rigorous removal of the partially methylolated material from a melamine resin could yield a system that cured at temperatures as low as 50 °C or even ambient, with the usual sulfonic acid catalyst. He attributed the improvement to the removal of basic materials that would otherwise counteract the protonation brought about by the acidic catalyst.

Both types of system yield some self-condensed material containing methylene and ether bridges and it is usual to formulate coatings to contain considerably more MF than suggested by stoichiometry. MF crosslinkers are used with polyester and acrylic polyols as high quality stoving topcoats for motor cars.

Urea formaldehyde mouldings were in use in the 1920s and urea remains the cheapest 'amino' precursor. By comparison with melamine, it gives highly reactive polymers with rather poor water resistance, being susceptible to hydrolysis. They can be cured at room temperature using a strong acid catalyst, such as hydrochloric, and are combined with hydroxyl functional polyesters in furniture finishes.

Benzoguanamine gives increased flexibility but reduced solvent and weathering resistance because of its lower functionality, and the presence of the conjugated phenyl group to absorb energy from solar radiation. Benzoguanamine formaldehyde crosslinkers are used, like MF, in automotive topcoats. There have been a few attempts to prepare guanamine derivatives with alternative side groups, but none appears to have achieved widespread acceptance.

Glycoluril resins emit relatively little formaldehyde and are said to be useful in high solids coatings. Because the secondary amino groups can only participate in full alkylolation, monomeric curing agents can be made under a variety of conditions. Perhaps for the same reason, they demonstrate relatively high reactivity.

Alkoxyacrylamide polymers naturally reflect the properties of the co-monomers, but can be cured at low temperatures and are used, for example, in domestic appliance finishes where good colour and alkali resistance are important. A variety of N-substituted acrylamide monomers is also available.

6.5.4.2 Phenol Formaldehyde Polymers and Crosslinkers.

Amongst the oldest synthetic polymer materials, 'phenolics' are prepared by condensing phenols with formaldehyde. Phenol is reactive at the 2, 4 and 6 positions and will condense with formaldehyde ultimately forming a densely crosslinked network that was the basis of the early plastic 'Bakelite', developed by Leo Baekland in 1907. In the coatings industry it is more usual to employ 4-substituted phenols so that linear materials are produced. Other aldehydes can form polymers with phenols, but are less reactive than formaldehyde and are rarely used.

When formaldehyde is in excess, short chains terminated with methylol groups known as 'resoles' (**53**) are formed. If the reaction is carried out in an alcohol the resole hydroxyl groups are alkylated. An excess of phenol provides 'novolacs' (**54**) with reactive aromatic sites at the chain ends. Branches can be introduced by judicious incorporation of unsubstituted phenol. In both cases a certain amount of ether bridging also takes place.

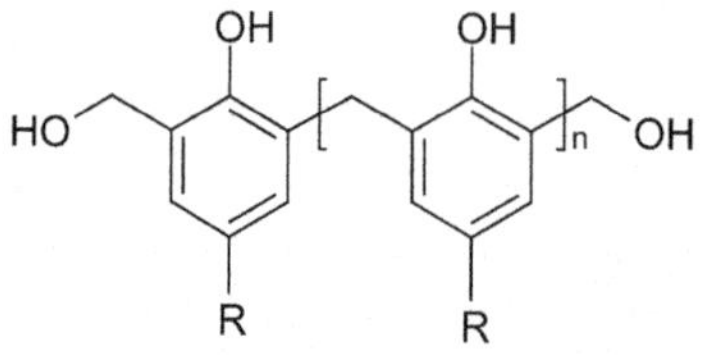

(**53**) 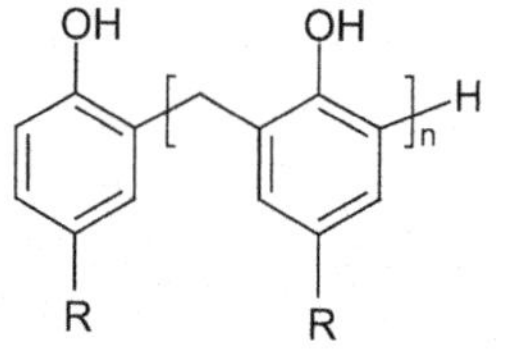　　　　　　　　　(**54**)

Novolacs will condense with further formaldehyde or resoles, but otherwise are inert except with respect to their phenolic groups which can be used to cure epoxy polymers or converted to epoxy groups. The methylol groups on resoles, whether alkylated or not, are reactive with polyols on stoving under basic, acidic or neutral conditions, though acidic catalysis is most effective. The reaction proceeds by addition to a quinomethide intermediate (Scheme 6.49). Addition of a further resole group to the quinomethide can occur either directly to form an ether bridge, or by a Diels–Alder reaction between two quinomethides to form a spiro-linkage. The overall effect is self-condensation.

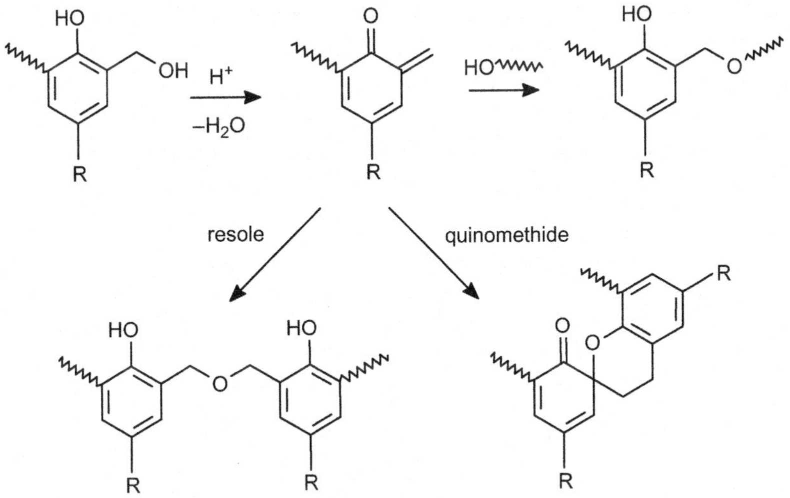

Scheme 6.49

Resoles are sometimes cured and processed with alkyd resins, when addition of the olefinic sites to the *o*-quinomethide intermediate can yield chroman ring structures (**55**).

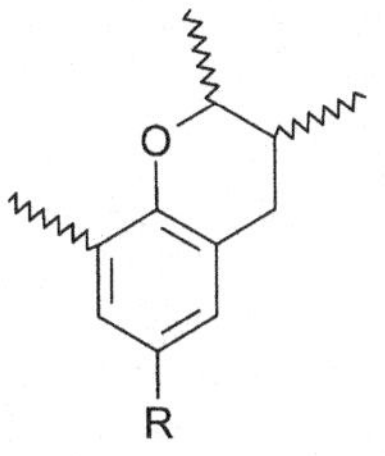

(**55**)

Oxidation of phenolic residues produces highly conjugated groupings of intense red or yellow colour (Scheme 6.50). The dark colour inevitably associated with phenolic polymers prevents their widespread use in coatings. However, they are valued in can linings for their great chemical inertness, and are often used in conjunction with epoxies. They are also used as adhesion promotors in, for example, epoxy–amine aerospace primers.

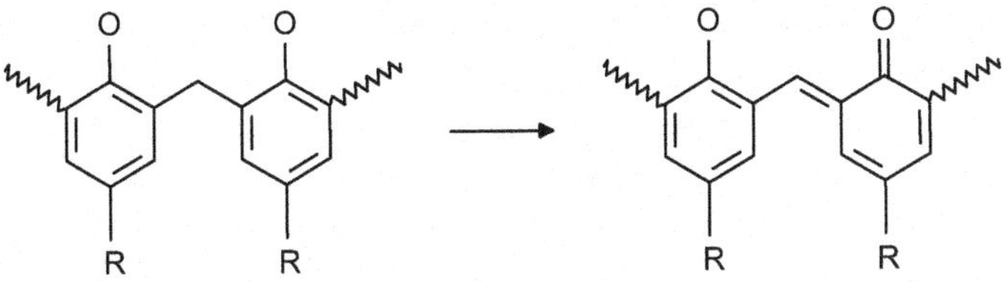

Scheme 6.50

6.5.4.3 *Blocked Isocyanates.* The bond formed from certain mono-functional active hydrogen compounds and isocyanates is labile, and the so-called 'blocking agent' can be displaced by a polyol or polyamine to form a crosslinked network. The process frequently appears to be a mixture of elimination–addition, and addition–elimination, as shown in Scheme 6.51. Effective blocking agents can be alcohols, phenols, amines or amides.

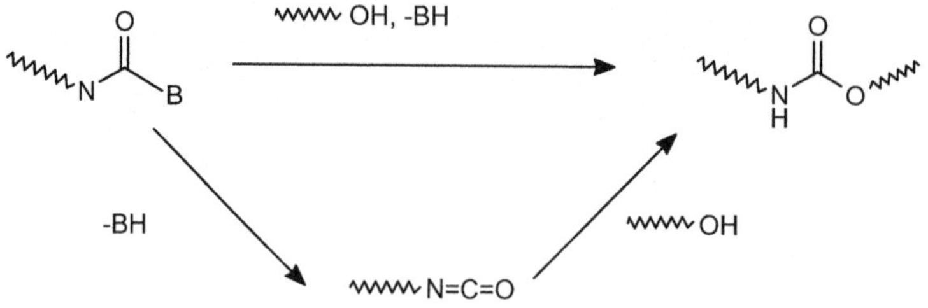

Scheme 6.51

The rate of reaction with a functional film former at a given temperature is influenced by the nature of the attacking species and the blocking group, as well as the solvent and any catalyst used. At one extreme, a phenol-blocked isocyanate can be cured at ambient temperature with a polyamine in the presence of amidine catalyst. However, blocked isocyanates are usually formulated as more or less stable, single package compositions that cure on stoving.

Some blocked isocyanates, and the temperatures at which their reaction with polyols occurs at a convenient rate, are listed in Table 6.8, but the 'reaction temperatures' should be treated with considerable reserve.

Blocked isocyanates are used most frequently in powder, but also in water-borne compositions. Alcohol-blocked isocyanate curing agents, for example, have been used extensively in cathodic electrodeposition formulations.

A number of proprietary blocking agents are said to provide crosslinkers effective at low temperatures, dibutylglycolamide systems, for example, have been reported to crosslink polyols at 120 °C, but the stability of such a composition at ambient temperature must be in some doubt. Malonic and acetoacetic esters were regarded as useful low temperature isocyanate blocking agents, but investigations by Wicks have established that the acylated keto-ester was instead activated to transesterification by an E1CB mechanism involving a ketene intermediate (Scheme 6.52).

Isocyanates blocked with Meldrum's acid also react through the blocking agent itself, eliminating carbon dioxide and acetone to form a mixed malonic acid amide–ester. The blocked isocyanate is so acidic that it can be

Table 6.8 *Some typical curing temperatures for blocked aliphatic isocyanates*

Blocking agent	Structure	Typical cure temperature (°C)
Aliphatic alcohol	–	>180
Phenol	–	175–180
Triazole	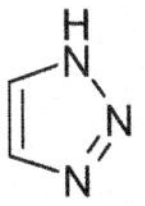	180
Glycol ether	–	160
Caprolactam	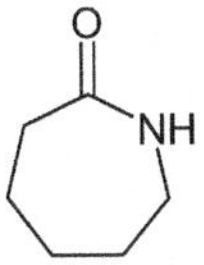	150–160
Imidazoline	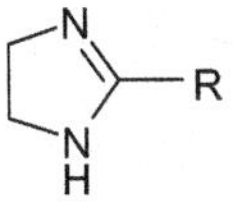	160
Methylethylketoxime	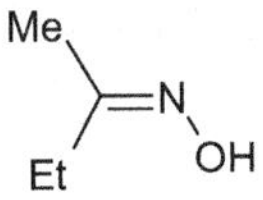	150
Formate ester		110

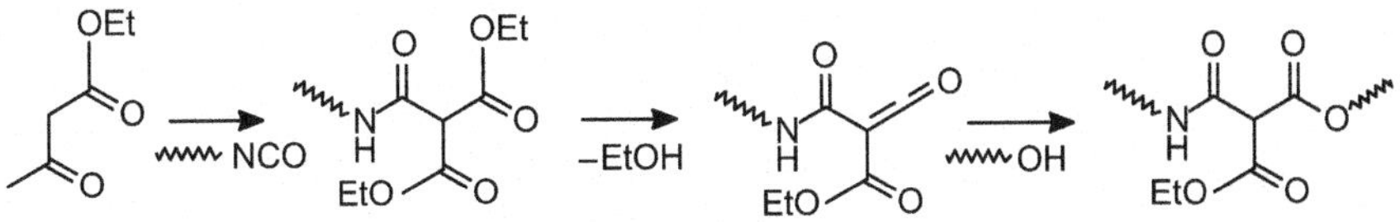

Scheme 6.52

neutralised with an amine and dispersed in water. Isocyanates blocked with sodium hydrogensulfate are intrinsically water soluble and have been proposed for use in water-borne coatings.

The possibility of obtaining 'blocked isocyanates' without ever handling free isocyanates has undoubted appeal, and a number of technologies have been proposed. The reaction of cyclic carbonates with amines to produce a β-hydroxyurethane was described above. Alternatively, a multifunctional amine can be reacted with ethylene or propylene carbonate to provide a blocked isocyanate deblocking by loss of glycol, and with reactivity comparable with that of phenol-blocked isocyanates (Scheme 6.53).

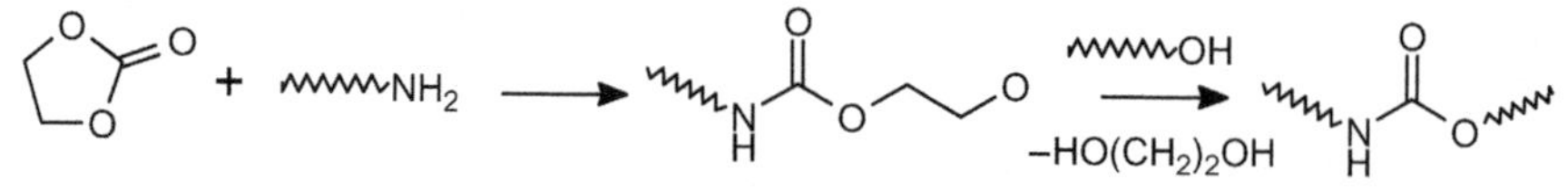

Scheme 6.53

A hydroxamic ester can undergo a Lössen rearrangement to isocyanate at the required rate, at temperatures as low as 100 °C (Scheme 6.54). Hydroxamic esters are normally used in two-pack compositions.

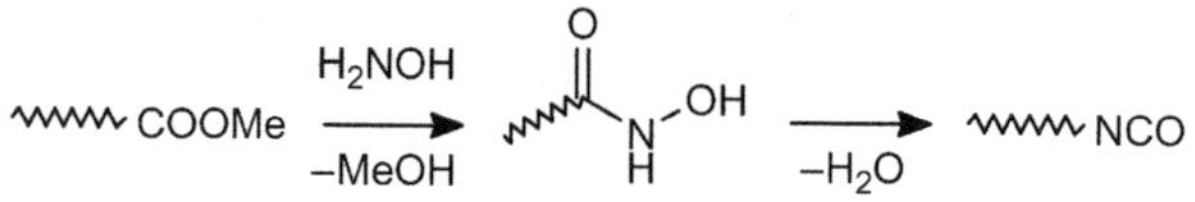

Scheme 6.54

Aminimides prepared from esters and *N,N,N*-trisubstituted hydrazinium salts, or hydrazine in the presence of an epoxide, eliminate tertiary amine at about 150 °C with formation of isocyanates (Scheme 6.55).

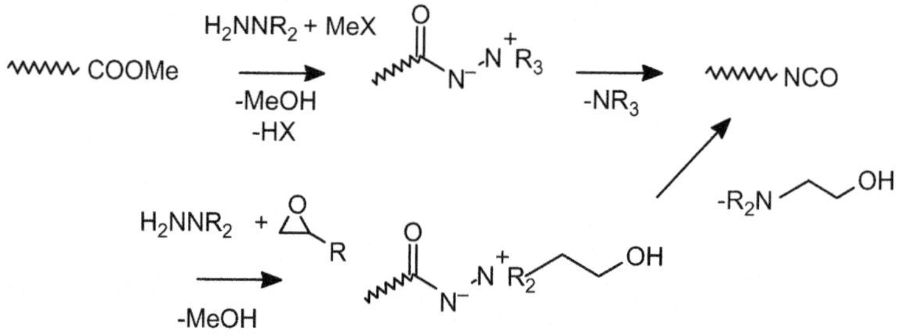

Scheme 6.55

Amongst other *in situ* routes to isocyanates might be mentioned. The cleavage of bicyclic furoxan species (Scheme 6.56), the ring opening of bis-cyclic ureas (Scheme 6.57) and the degradation of low molecular weight polymers containing uretidinedione groups (Scheme 6.58) to yield diisocyanates (*cf.* Chapter 7).

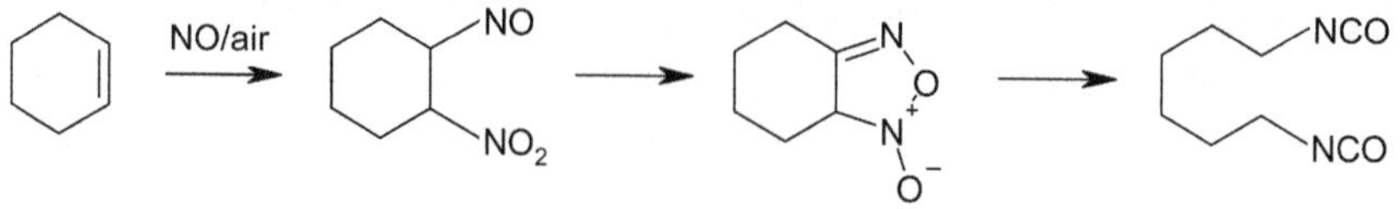

Scheme 6.56

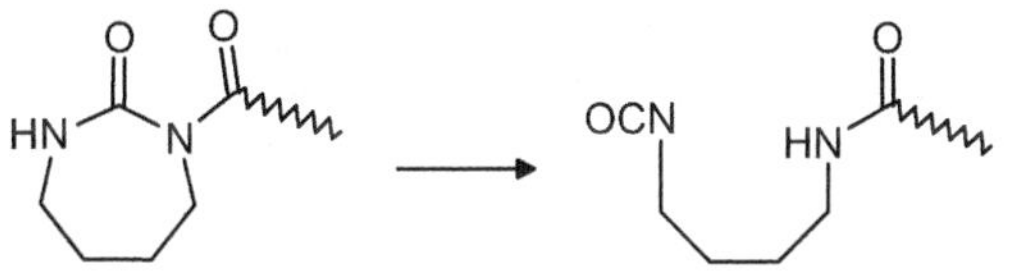

Scheme 6.57

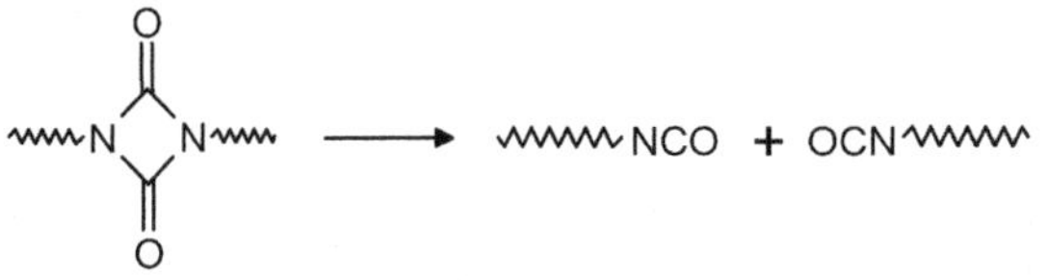

Scheme 6.58

Unfortunately, although they do not contain free isocyanates, blocked isocyanates may nevertheless react like isocyanates *in vivo*. Accordingly it is usually considered appropriate to treat them with the same severe precautions as isocyanates themselves, so their application in some areas is inhibited.

6.6 POLYMER BLENDS

The role of additives in formulations is properly the province of the paint formulator, and is dealt with in Chapter 10. However, in a number of cases, such large amounts of polymeric material are used to modify coatings properties, or for some similar purpose, that they have to be regarded as co-binders, and the overall system as a polymer blend.

Since different polymers are not usually miscible, miscibility is achieved by using an additive of low molecular weight, or furnished with groups designed to interact with those on the main film former. It is necessary both to ensure the required miscibility in the fluid state (when the presence of solvents may be helpful), and in the cured system, when a thermoplastic is all too likely to be ejected from the network by syneresis. The glass transition temperatures of the resulting mixtures are determined by the Fox equation [(13) in Chapter 7]. The usual effect of adding a low T_g thermoplastic material to a composition is to plasticise it and reduce its viscosity.

A case in point is the addition of coal tar to epoxy–amine systems to achieve increased water resistance. The combination has been in use for over 50 years and gives outstanding properties. The coal tar is nominally a mixture of aromatic hydrocarbons, heterocyclic and phenolic materials and colloidal carbon. The excellence of the combination is often attributed to a slow reaction between the epoxy and the phenolic groups in the coal tar. In view of the carcinogenic potential of components of coal tar, it has largely been superseded by hydrocarbon resins made from indene and related olefins (*cf.* Section 6.3). Even so, greatly improved properties have been observed when the polymer contains a few phenolic groups to ensure compatibility.

It has already been noted, in Section 6.3.2, that substantial amounts of cellulose acetate–butyrate are used in leafing aluminium car finishes, amongst other applications. Its effect is to modify the rheological properties of the binder so as to improve 'metal control'. Microgel, introduced in Chapter 8, is used in a somewhat similar way. It also has the effect of reinforcing most networks since it introduces regions of dense crosslinking.

Another technology that introduces substantial amounts of an alien polymer to a range of different generic binder systems is that of universal tinter systems. Pigment pastes are prepared in standard dispersion resins that are required to be compatible with all the different compositions to be tinted. Depending on colour, the amounts of such materials present in the final composition can be very substantial. Examples of 'universally compatible' resins are chlorinated polymers, acrylic polymers containing high levels of cyclic monomers such as isobornyl acrylate (**56**) and aldehyde–urea

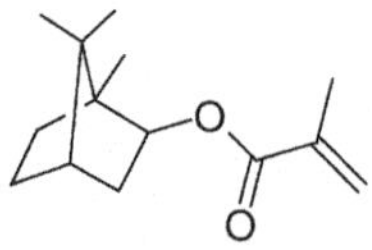

(**56**)

resins typically prepared by the Mannich reaction of urea with formaldehyde and isobutyraldehyde. The nominal intermediate first formed is represented by (**57**), though in fact the structure is much more complex and contains significant amounts of cyclised material.

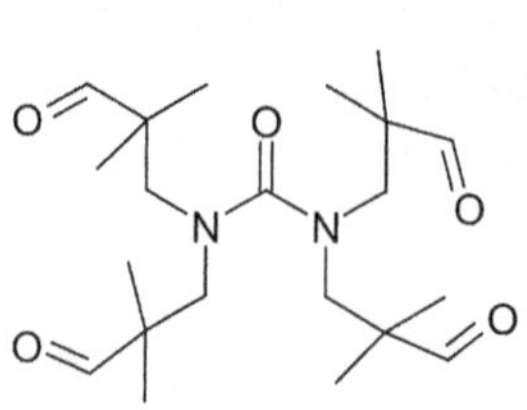

(**57**)

On treatment with base, (**57**) yields a resinous material with very broad compatibility.

A different approach embraces the general immiscibility of polymers. There has been great interest in self-stratifying coatings in which a polymer of relatively low surface energy, such as an acrylic, collects at the surface of, for example, an epoxy coating. In that way a two-coat system achieved in a single application has been envisaged.

In general terms, blends of miscible polymers enable their components to bring their different qualities to the combination, but blends of immiscible polymers may offer altered morphologies, and consequently access to totally unexpected properties.

6.7 CONCLUDING REMARKS

Chapter 6 was intended to give an impression of the role of polymeric binders in coatings. It should be apparent that they are crucial to the delivery of a remarkable range of properties. Where water, light, heat, mechanical abuse or the degradation of its substrate threaten the integrity of a coating, the correctly chosen binder, properly cured, and supported in a well-designed formulation will not be found wanting.

6.8 BIBLIOGRAPHY

1. J. Bentley, 'Surface coatings and inks' in *Oleochemical Manufacture and Applications*, F.D. Gunstone and R.J. Hamilton (eds), Sheffield Academic Press, 2002.
2. B. Ellis (ed), *Chemistry and Technology of Epoxy Resins*, Blackie Academic and Professional, Glasgow, 1993.
3. F.N. Jones, G. Chu and U. Samaraweera, Recent studies of self-condensation and co-condensation of melamine–formaldehyde resins in cure at low temperatures, *Prog. Org. Coat.*, 1994, **24**, 189–208.
4. J.W. Nicholson, *The Chemistry of Polymers*. Royal Society of Chemistry Paperbacks, The Royal Society of Chemistry, Cambridge, 1997.
5. G. Oertel (ed), *Polyurethane Handbook*, Carl Hanser Verlag, 1993.
6. P. Oldring and G. Hayward (eds), *Resins for Surface Coatings*, SITA Technology, London, 1987.
7. T.C. Patton, *Alkyd Resin Technology*, Wiley, New York, 1961.
8. S. Paul in *Comprehensive Polymer Science*, Vol. 6, G. Allen and J.C. Bevington (eds), Pergamon Press, Oxford, 1988.
9. T.A. Potter, J.W. Rosthauser and H.G. Schmelzer, Blocked isocyanates in coatings, *Proc. Waterborne and Higher Solids Coatings Symposium*, February 5–7, 1986, New Orleans.

CHAPTER 7

Binders for High Solids and Solvent-free Coatings

A.R. MARRION and W.A.E. DUNK

7.1 INTRODUCTION

We have seen that the presence of organic solvents in coatings is something of a liability, as they impart flammability, toxicity and cost yet are (almost) entirely lost from the final coating. It is little wonder that a generation of coatings chemists has busied itself trying to reduce their level or remove them altogether.

There are broadly two strategies for solvent removal, as distinct from replacement with a more benign liquid such as water (*cf.* Chapter 8). Either the binder can be made more fluid, so that the application viscosity can be reached with less solvent or none at all, or it can be made more solid, so that it can be applied as a powder. In both cases, severe demands are placed on its design. Binders for high solids liquids have to be highly fluid, but still cure to an intractable solid, whilttst those for powder coatings must be solid enough to retain their particulate identity at high ambient temperatures, yet be fluid enough to coalesce into a smooth film on stoving.

It is also possible to identify an intermediate position, where solid particles are dispersed in a liquid that may or may not be organic. The viscosity of the composition is in that way decoupled from the viscosity of the binder, as explained in Chapter 8, and high molecular weight materials can be used. However, the geometry of particle packing ensures that solids of above about 70% (including any pigments) cannot be exceeded without severe loss of fluidity.

7.2 HIGH SOLIDS AND SOLVENT-FREE LIQUIDS

7.2.1 Binder Design

A typical conventional coatings binder has T_g and molecular weight high enough to form films with quite limited chemical crosslinking. It is therefore a solid or highly viscous liquid, and requires substantial dilution to achieve application viscosity.

7.2.1.1 Molecular Weight.   It is intuitively obvious that the viscosity of a polymer melt or solution will increase with molecular weight, but the relationship can be conveniently expressed by the empirical expression (7.1), of which the familiar Mark Houwink Sakurada equation is a specialised version applicable to dilute solutions

$$\eta = k M_w^{\alpha} \qquad (7.1)$$

where η is the low shear viscosity, and k and α are constants. α is usually between 1 and 2 below the entanglement molecular weight, and about 3.4 above it, so that a strong dependence of viscosity on weight average molecular weight becomes a dramatic one above a certain degree of polymerisation.

7.2.1.2 Polydispersity.   Relationship (7.1) points up another important consideration. Because viscosity has an exponential relationship with molecular weight, the highest fractions in the distribution have a dominating effect. The consequence of high polydispersity was demonstrated very clearly in a classic experiment by Paul Flory: He mixed two polyesters of high and low molecular weight, but otherwise identical composition, and heated the mixture to allow ester exchange reactions to take place. The viscosity at the melt temperature fell as polydispersity decreased towards the equilibrium value, although the average degree of polymerisation was unchanged.

It is therefore important in designing a high solids binder to reduce molecular weight as far as possible, but also to reduce polydispersity.

7.2.1.3 Vitrification Temperature.   Another parameter of fundamental significance for design purposes is glass transition temperature, T_g. The relationship of the melt viscosity of a linear polymer to its T_g is described by the Williams, Landel, Ferry (WLF) equation that has its basis in the Doolittle theory of polymer free volume (7.2):

$$\ln \eta = \ln A + B \frac{V_o}{V_f} \qquad (7.2)$$

where A and B are system specific constants, and V_o and V_f are occupied and free volume, respectively.

An approximate form of the WLF equation (7.3), that neglects the impact of density changes, was given by Wicks and Hill in their seminal 1982 paper on the design of high solids coatings:

$$\ln \eta = \ln \eta_{T_g} - \frac{40.16(T - T_g)}{51.6 + (T - T_g)} \tag{7.3}$$

T is the temperature at which viscosity is measured, and η_{T_g} is the viscosity of a polymer as it enters the glassy state and is usually in the range of 10^{10} Pa s, though dependent on the molecular weight.

The values obtained by applying the WLF equation quantitatively are determined by the arbitrary selection of T_g, and must be treated with considerable reserve. However, they do show that a low T_g is required to achieve low viscosity. The T_g of a polymer is controlled both by its composition and by its molecular weight. A rule of thumb (7.4) relates T_g and number average molecular weight.

$$T_g = T_{g\infty} - \frac{K}{M_n} \tag{7.4}$$

$T_{g\infty}$ is the vitrification temperature of the polymer at a molecular weight approaching infinity, and K is a system-specific constant, typically 10^5 for acrylic polymers, but a much lower figure for polyesters.

7.2.1.4 Functionality.

The ideal binders for high solids systems might seem to be low molecular weight, highly flexible oligomers with low polydispersity. But, if they are to form crosslinked films with the same T_gs as could have been delivered by conventional coatings, there is a considerable shortfall in properties that must be made up entirely by the crosslinking process. A further requirement, that bears strongly on the design of the binder, is to maximise crosslinking, but minimise interactions between the prepolymers resulting from the presence of polar groupings.

Network formation is always imperfect (*cf.* Chapter 4), but it is at least possible to minimise imperfections by using well 'tailored' precursors. It is helpful, for example, when using end-functional binders, if they are as nearly 'telechelic' as possible. In a truly telechelic molecule, all the chain ends bear the required reactive groups.

When a terminally functional polymer such as a polyester is reduced in molecular weight, its functionality is unchanged, but its equivalent weight is reduced as shown in Scheme 7.1. A higher density of reactive groups is obtained automatically.

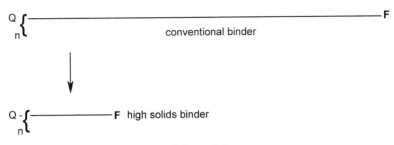

Scheme 7.1

By contrast, when a polymer with pendent reactive groups, such as an acrylic, is reduced in molecular weight, its functionality is lowered but its equivalent weight remains unchanged. It is necessary to introduce additional functional groups to restore its functionality as shown in Scheme 7.2. Equivalent weight is reduced and even so there may be a significant number of chains with only one functional group, or none at all.

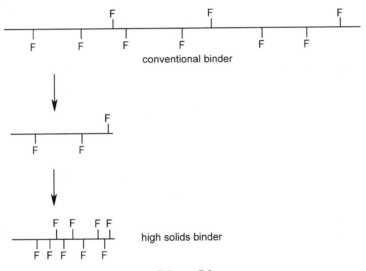

Scheme 7.2

7.2.1.5 Radiate and Star Structures. As noted in the previous section, it is often inappropriate to reduce molecular weight or T_g beyond a certain point and alternative approaches may have to be adopted.

Radiate or star polymers offer several attractive features. A star shaped molecule, even fully relaxed in a melt or solution, extends over a much smaller volume than the corresponding linear species, and entanglements and friction are consequently reduced. It also has, by definition, a larger number of chain 'ends'. Since the ends of a polymer chain have more freedom of movement, and consequently a lower T_g than its centre, the overall T_g and viscosity due to the star structure is further reduced.

Star structures also offer the possibility of introducing more than two terminal reactive groups, so allowing for low equivalent weight combined with high functionality. Such molecules are described as polytelechelic.

In broad terms, the most effective low viscosity star polymers should have a large number of arms, and the arms should all be of the same molecular weight. In other words, the molecules should be as compact as possible, and have low external and internal polydispersity, subject to the problems noted below.

7.2.1.6 Hyperbranched Molecules. Star structures are taken to their logical conclusion in dendrimers, where the arms branch repeatedly in a perfectly regular way to provide spherical molecules with their termini packed closely together in the surface. Remarkable fluidity has been observed in dendrimers of quite high molecular weight, and the discontinuity in α in formula (7.4) that corresponds to the onset of entanglement at a certain MW has not been observed. They have been used at quite low levels as processing aids in structural polymers and the apt description 'macromolecular ball bearings' has been applied. Synthesis of dendrimers is usually quite laborious as they are built up a 'generation' at a time, so they have not proved cost effective in coatings. However hyperbranched polymers, simply made by polymerising $A–B_2$ type monomers (where A reacts with B), are beginning to be used quite routinely. They fall short of the perfection found in dendrimers but, provided the 'degree of branching', (7.5), is above about 0.5, can give usefully low viscosities.

$$\text{Degree of branching} = \frac{\text{dendritic moieties} + \text{terminal moieties}}{\text{total units}} \quad (7.5)$$

7.2.1.7 Other Special Architectures. Other structures that have potential to yield viscosities lower than would be expected from an equivalent linear polymer are cyclic polymers, polymeric rotaxanes and specific anisotropic phases due to the introduction of liquid crystalline moieties (mesogens). Cyclic polymers have no chain ends, so tend to show higher T_g than their linear counterparts, but in the molecular weight range of interest their compact structure provides more than adequate compensation. Polymeric rotaxanes are physically bonded copolymers containing essentially friction-less linkages. Meanwhile, the ordered structures of certain assemblies of mesogens allow layers of molecules to glide past one another.

Since none of the above molecular arrangements has achieved major technical importance in the coatings industry, they merit no further discussion.

7.2.2 The Corollaries of High Solids Binders

The very significant flow-related defects often associated with high solids coatings were discussed in Chapter 3.

7.2.2.1 Mechanical Properties. The mechanical properties of high solids coatings can often be disappointing, because the network is constructed from relatively short, less extensible molecular components, and because it may be difficult to prepare sufficiently well-tailored precursors. The dangling chains resulting from monofunctional prepolymers may have a highly adverse effect on the cured network – increasing softness without contributing to its elastic strength.

There is also some evidence that highly polydisperse prepolymers produce the best mechanical properties, so that the monodisperse systems designed for fluidity may prove unexpectedly disappointing.

7.2.2.2 Reaction Rates. The overall effect of reducing prepolymer molecular weight, and adding further reactive groups if necessary, is to increase the concentration of functional groups in a curing composition. The concentration is further increased by the removal of the solvent that would be present in a conventional coating. Bearing in mind the typical dependence of reaction rate on the product of the concentration of both reactants, (Equation (7.6)) it is clear that the necessary changes to binder structure can lead to an unacceptably fast reaction in high solids compositions. Even so, the absence of any physical drying may mean that it takes much longer for a satisfactorily cured film to develop.

$$\text{Rate} = k[\text{A}][\text{B}] \qquad (7.6)$$

7.2.2.3 Toxicology. High solids binders often show significantly increased toxicity as compared with conventional materials, because they contain reactive components whose molecular weights fall below the value of 300–500 where easy absorption through the skin takes place. They also tend to contain high concentrations of powerfully electrophilic groups such as epoxy, isocyanate and acrylate that are particularly reactive with nucleophiles in the human metabolism. The lipophilic backbones with polar reactive groups also tend to create the amphiphilic structures that are best able to penetrate the lipid bilayers present in living cells.

Furthermore, some of the high solids materials may possess significant volatility and be available for inhalation.

To some extent, the toxic problems associated with high solids binders can be mitigated by careful selection of cure chemistry and consequently the functional groups required. It is, for example, prudent to ensure that the most dangerous (electrophilic) groups are attached to the highest molecular weight components.

7.2.3 Some Specific High Solids Binders

Polyester and acrylic binders again provide a useful contrast as in Chapter 6.

7.2.3.1 Polyesters. Polyesters and similar condensation polymers are very amenable to changes that lead to low viscosity. The molecular weight of the usual type of polyester can be reduced to the point where it consists of a single monomer unit, and its chemical composition can be adjusted until it is simply a polymethylene chain, interrupted occasionally by ester groups. Polymethylene chains offer low T_g as well as the minimum of frictional resistance to flow. Even the presence of pendent methyl groups will increase the viscosity significantly. Polyester molecules can, furthermore, be made almost perfectly telechelic by using an excess of the monomer that is to provide the end groups, a volatile diol, for example, and stripping off the unreacted material.

Unfortunately, the polydispersity will still be at least two for conventional polycondensation, functionality will be limited to two and the end groups will, by definition, be limited to either hydroxyl or carboxyl, and have a strong hydrogen bonding effect. All aliphatic polyesters of the sort described are also rather too susceptible to hydrolysis to provide weatherable coatings. Functionality is low and attempts to increase it by adding a multifunctional monomer are unlikely to be rewarding. Although a degree of branching is expected to be helpful, any benefit would be greatly outweighed by the production of much higher molecular weight components with a dominating effect on viscosity.

Air drying alkyds represent a special case of polyester since they contain pendent fatty acid chains, which provide the crosslinking sites. The terminal groups are therefore of no interest and a monofunctional chain stopper such as benzoic acid or *p-tert*-butylbenzoic acid can be introduced in a controlled amount to define the molecular weight and reduce polydispersity. There is no advantage in a branched structure, but because of the unusual combination of monofunctional, difunctional and trifunctional monomers, a statistical degree of branching is inevitable. The C_{18} pendants also contribute to a bulky structure.

Established alkyd technology provides scope for increasing solids by moving to 'longer' oil alkyds as described in Chapter 6. The consequences are as described there: viscosity is reduced as the fatty acid content increases. More fatty acid residues are present for curing, but the potential for physical drying is diminished. The importance of selecting a fatty acid mixture with a high proportion of the most reactive kinds of allylic unsaturation is clear, if high solids are to be achieved.

Because alkyds were traditionally based on phthalic anhydride, fatty acids and glycerol (or phthalic anhydride and a triglyceride with extra glycerol), the literature abounds with high solids advances based on monomer changes. Adipic acid is a favourite for reducing T_g, whilst cycloaliphatics such as cyclohexane-1,4-dicarboxylic acid can be used in the same way without an undue penalty in terms of hydrolytic stability. Alternative polyols avoid the severe processing needed to esterify the secondary hydroxyl in glycerol.

The best known star polyesters are derived from caprolactone by a chain-growth process as indicated in Scheme 7.3. The strained caprolactone ring can be opened by a hydroxyl group catalysed by a tin compound, the resulting product is itself hydroxyl functional, so can attack a further caprolactone molecule.

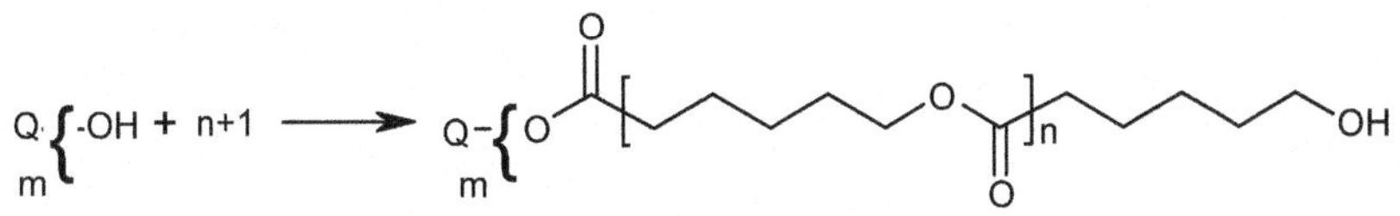

Scheme 7.3

It is only necessary to process a polyol with a suitable amount of caprolactone to produce a polytelechelic star polymer with the same number of arms and functionality as the starting polyol. The inter-arm dispersity settles to an equilibrium value after prolonged processing. It is also possible to react polyacids with caprolactone to obtain polytelechelic acids.

Polycaprolactones are of low viscosity and apparently greater hydrolytic stability than comparable polyesters derived from (say) adipic acid and hexane-1,6-diol. Unfortunately chains of more than about three caprolactone units tend to be crystalline, making it difficult to present homogeneous workable mixtures to the painter. Nevertheless they are used, for example, in high solids aerospace polyurethane topcoats.

Another approach uses the alternating ring opening of epoxies and cyclic anhydrides to achieve a polyester with acid or hydroxyl tips (Scheme 7.4). If the process is conducted stepwise under mild conditions polydispersities as low as 1.1 can be achieved. Unfortunately, the diol and diacid residues are particularly susceptible to hydrolysis because of the β-orientation of hydroxyl and carbonyl groups allowing anchimeric assistance.

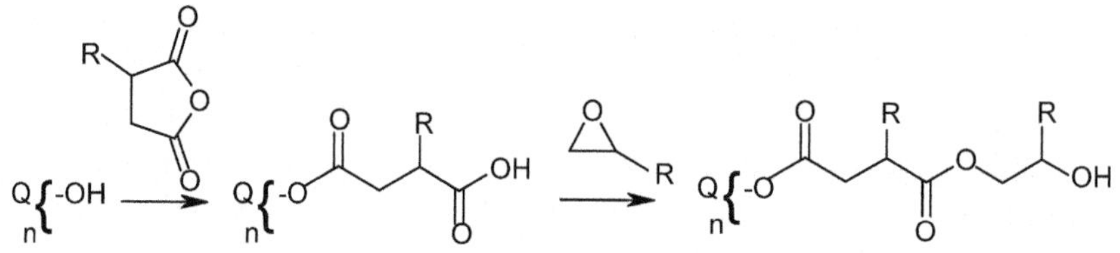

Scheme 7.4

Interest in hyperbranched polyesters is growing rapidly. The favourite monomer is dimethylolpropionic acid. Despite the high hindrance of the acid group, it can be polyesterified under conventional conditions, when it naturally leads to molecules with a preponderance of hydroxyl groups (Scheme 7.5)

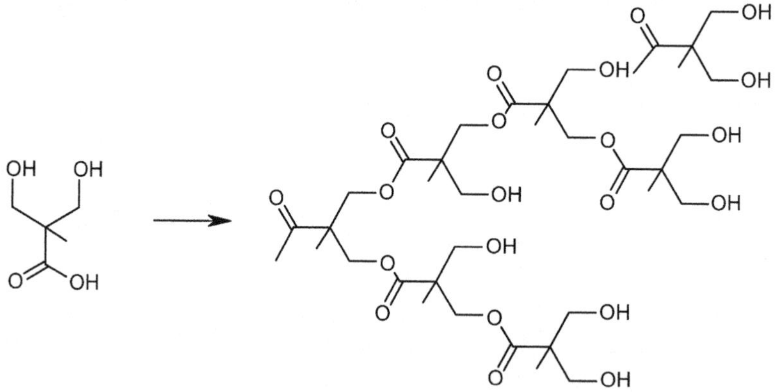

Scheme 7.5

The products have low viscosity and high functionality. The hydroxyl groups are relatively reactive, because they tend to be on the outside of a molecule tending towards spherical, and the hindered ester groups ensure good hydrolytic stability. The hydroxyl groups can be modified, for example with drying oil-fatty acids to provide high solids alkyds. The products are again of low viscosity and high reactivity.

Other AB$_2$ type monomers claimed in the patent literature are reactive products of ethanolamine with methyl acrylate (**1**) or dimethyl maleate (**2**). The methyl ester tips on the products of polymerisation can readily be transformed into other functional groups

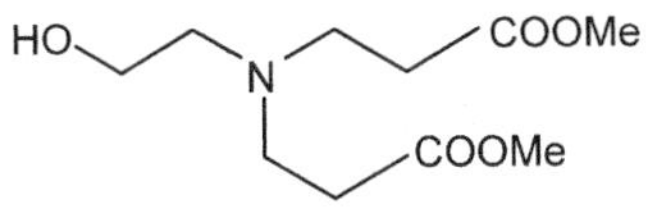

$$(1) \qquad\qquad\qquad (2)$$

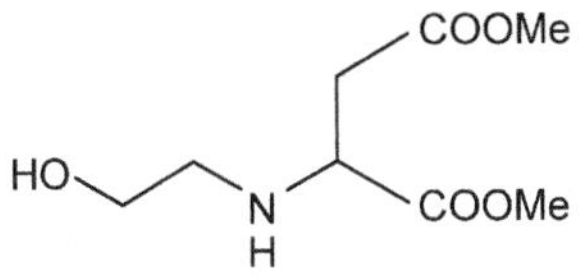

7.2.3.2 Acrylics. Acrylic polymers stand in striking contrast to polyesters in that they are, necessarily, made by a chain-growth process, and it is inherently difficult to produce them with very low molecular weight. Moreover, when low molecular weight is achieved, the functionality may drop so low as to compromise the structure of any network formed, as indicated in Scheme 7.2. The level of non- and mono-functional species in an acrylic polymer can be calculated using statistical techniques, for a given degree of polymerisation and level of functional monomer. For ordinarily low molecular weights and functional monomer concentrations it can be alarmingly high as indicated in Table 7.1.

Although the large number of different 'acrylic' monomers can be assembled in an almost infinite number of combinations, the scope for structural variations remains rather limited. The chains of methylene groups are interrupted at every third carbon atom with pendent alkoxycarbonyl groups or, in the case of methacrylates, with a methyl group as well as the ester. There are of course analogous molecules such as acrylamides and acrylonitrile, and some activated olefins substituted on the 2-position, but

Table 7.1 *The occurrence of non- and mono-functional molecules in acrylic polymers, calculated assuming conditions for Flory's most probable distribution*

DP_n	Reactive monomer (mol%)	Mol% (f = 0)	Mol% (f = 1)
20	5	49	26
	10	31	24
	20	17	17
	30	10	13
50	10	15	14
	20	7	9
	30	4	6
100	10	8	8
	20	4	5
	30	2	3

Table 7.2 *Glass transition temperatures of homopolymers of some (meth) acrylates*

Side-chain alkyl group	Acrylate (°C)	Methacrylate (°C)
Methyl	10	105
Ethyl	−24	65
Butyl	−54	20
Hexyl	−57	−5
Dodecyl	−3	−65
Tetradecyl	24	−9
Hexadecyl	35	15

they generally produce stiffer chains than the corresponding acrylates. T_g can be tailored to the desired value by selecting acrylates or methacrylates and the appropriate alkyl group. Increasing the length of the side chains initially reduces T_g, but excessively long side chains start to show crystallinity, and actually raise the T_g of the backbone. Table 7.2 illustrates the situation.

Interestingly, bulky monomers can be used to produce high T_g polymers that nevertheless show reduced viscosity. The monomers are intrinsically 'hard' but because of their bulk, force the polymer chains apart. Examples are cyclohexyl methacrylate, isobornyl methacrylate and styrene.

Commercial vinyl esters based on highly branched C_9, C_{10} and C_{11} *neo*-acids are remarkable for the T_gs of their homopolymers. They range from −40 °C (*neo*-undecanoate) through −3 °C (*neo*-decanoate) to +70 °C (*neo*-nonanoate) and provide an alternative approach to T_g control. Though much less reactive than acrylic monomers, vinyl esters can undergo free radical polymerisation, particularly in water-based systems (*cf.* Chapter 8).

The use of acrylates rather than methacrylates provides reduced T_g, but at a price. The polyacrylate backbone contains a tertiary carbon for every monomer residue. The hydrogen atoms on those sites are relatively easily abstracted, with the result that polyacrylates are more susceptible than methacrylates to photooxidative degradation, leading to reduced weathering performance in the finished film. For the same reason, adventitious grafting is more likely during manufacture, leading to a degree of branching in the polymer, and consequently increased viscosity.

Also, whereas methacrylate radical chains tend to be terminated by disproportionation, the radicals on a polyacrylate chain are more reactive and more likely to terminate by combination, leading to a doubled molecular weight (Scheme 7.6).

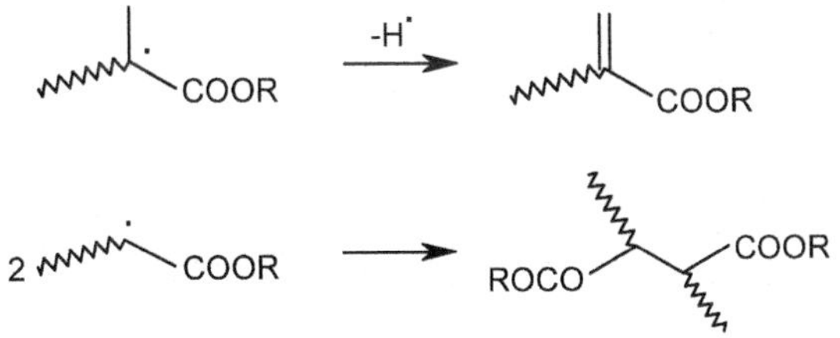

Scheme 7.6

The problem is usually mitigated by chain transfer to the solvent, to monomer or to deliberately added chain transfer agent. Thiols are the most frequently used chain transfer agents, and overall the most effective. They terminate growing chains and initiate new ones as indicated in Scheme 7.7. The addition, preferably in a controlled way, of a few percent of a thiol such as dodecylmercaptan to a radical polymerisation is sufficient to ensure lowered molecular weight and polydispersity, and would be the traditional recourse in the design of a high solids acrylic.

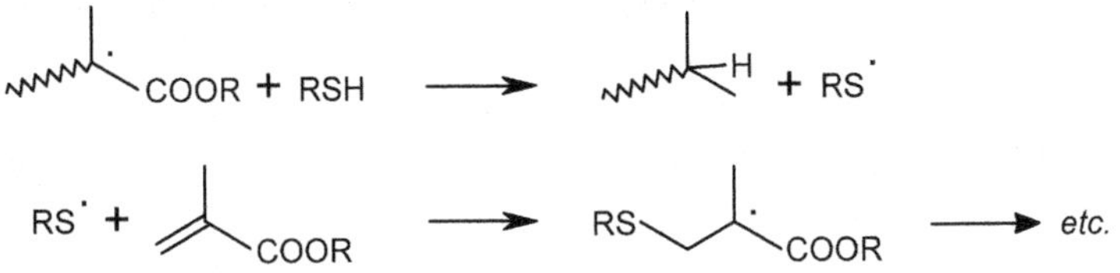

Scheme 7.7

The efficiency of a chain transfer agent, Q, is defined by the chain transfer coefficient:

$$C_Q = \frac{k_{transfer,Q}}{k_{propagation}} \qquad (7.7)$$

where the k symbols represent the respective rate constants. It can be deduced that

$$\frac{1}{DP} = \frac{k_{transfer,Q}R_p}{k_{propagation}^2[M]^2} + \frac{k_{transfer,Q}[Q]}{k_{propagation}[M]} \qquad (7.8)$$

where DP is the degree of polymerisation, R_p the propagation rate and [Q] and [M] the concentrations of chain transfer agent and monomer, respectively. Equation (7.8) shows the impact of chain transfer on molecular weight: C_Q greater than 1 leads to a marked reduction.

Thiol chain transfer agents offer other possibilities: a hetero-bifunctional thiol such as mercaptoethanol or mercaptopropionic acid can add one extra group to nearly every chain, converting monofunctional oligomers to a bifunctional form. Polythiols can also be used to produce approximations to star polymers. For example, additions of pentaerythrityl tetramercaptopropionate can yield a four-armed radiate acrylic with lower viscosity than its linear equivalent.

There has been enormous interest in the control of free radical polymerisation over the last two decades and a great many technologies have offered very low molecular weight, polydispersity approaching 1, special architectures at will, and the placement of functional groups exactly where they are required. All, however, have made stringent demands on experimental conditions, requiring the absence of moisture, high purity monomers and strict temperature control. They are also often limited to specific monomers.

Anionic polymerisation routinely provides molecular weight control in polystyrene-*b*-polybutadiene rubbers, for example, but has not been much applied to acrylics because nucleophilic groups, including those most in demand for curing chemistry, and traces of water, destroy the active anionic centres. Nevertheless, anionic polymerisation has provided the model to which controlled radical polymerisation concepts aspire.

Free radical polymerisation has a tremendous industrial importance, being used for over 50% of the world's production of polymers. The procedure is reasonably tolerant to moisture and impurities and only oxygen needs to be excluded from the vessel. Monomers do not require extensive purification. Furthermore, reaction occurs under mild conditions and does not need elaborate reactors or strong cooling. However, precise control over molecular weight and polydispersities is not available since all the polymeric chains are not initiated at the same time, and the active-ended polymer radicals undergo nearly diffusion-controlled termination by bimolecular processes. The diffusion control of termination also limits the preparation of block polymers.

The concept of bringing living character to free radical polymerisation (here referred to as 'controlled' polymerisation) seems to have been suggested by Bianchi *et al.* in 1957. Various other strategies were subsequently proposed but none showed clear evidence of living character. In 1982 Otsu and his co-workers proposed a model for living radical polymerisation where a propagating polymer chain end might dissociate into a polymer A and a small radical B, which would be stable enough not to initiate a new chain (Scheme 7.3). On this basis Otsu conceived the idea of initiator, transfer, terminator agents that he called 'iniferters' (Scheme 7.8).

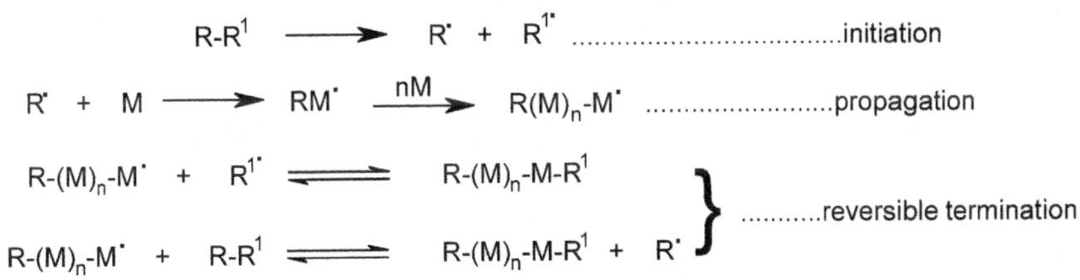

$$R-R^1 \longrightarrow R^{\bullet} + R^{1\bullet} \quad \text{.................................initiation}$$

$$R^{\bullet} + M \longrightarrow RM^{\bullet} \xrightarrow{nM} R(M)_n\text{-}M^{\bullet} \quad \text{.........................propagation}$$

$$R\text{-}(M)_n\text{-}M^{\bullet} + R^{1\bullet} \rightleftharpoons R\text{-}(M)_n\text{-}M\text{-}R^1$$
$$R\text{-}(M)_n\text{-}M^{\bullet} + R\text{-}R^1 \rightleftharpoons R\text{-}(M)_n\text{-}M\text{-}R^1 + R^{\bullet}$$

$$\left.\begin{array}{c} \\ \\ \end{array}\right\} \quad \text{...........reversible termination}$$

Scheme 7.8

Iniferters may be cleaved either thermally or photolytically to provide radicals. Examples of thermally cleaved systems are triphenylmethylazobenzene ($Ph_3CN=NPh$) in which the phenyl radical initiates polymerisation whilst the triphenylmethyl radical acts as the radical trap, and 1,2-disubstituted 1,1,2,2-tetraphenylethanes that can act as reversible capping agents.

Diaryl and dithiuram disulfides are typical photochemical iniferters. Particularly useful examples are tetraethylthiuram disulfide (**3**), benzyl *N,N*-diethyl dithiocarbamate (**4**) and *p*-xylylene-*bis*(*N,N*-diethyldithio-carbamate) (**5**). Termination of polymer chains may occur either by recombination with primary sulfur radicals or chain transfer to initiator.

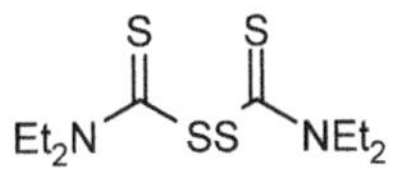

(3)

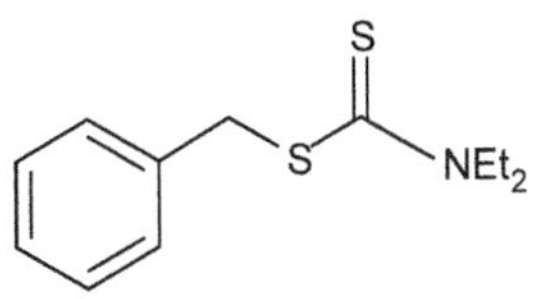

(4)

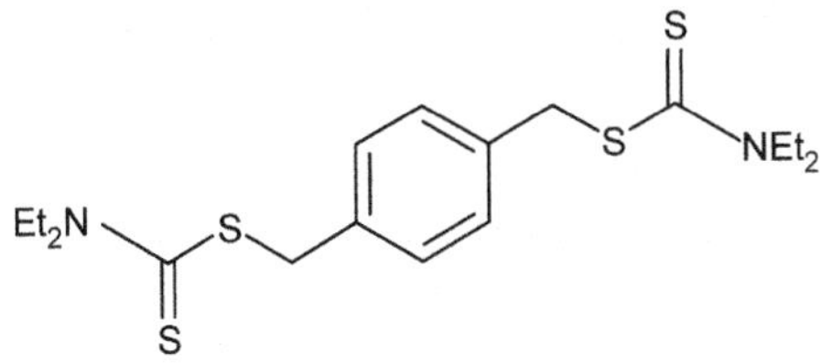

(5)

There have been many attempts to gain control over free radical polymerisations but only in the past 20 years has sufficient progress been made to bring close to living character to the reactions. The most promising approaches seem to be nitroxyl mediated polymerisation, atom transfer radical polymerisation (ATRP) and reversible addition-fragmentation chain transfer (RAFT). Although the literature is quite liberally scattered with suggestions for controlling free radical polymerisations, most fall short of anything like the levels of control that are available in truly living reactions, or for that matter the best achieved by the methods cited above. The only other technique that has found favour in industrial use is cobalt mediated polymerisation described as catalytic chain transfer. It proves to be an effective way of making acrylic oligomers, and DuPont and ICI both have inventions and commercial practice in the area.

Cobalt mediated polymerisation utilises low spin Co^{2+} complexes as chain transfer agents. The mechanism is believed to involve the repeated disturbance of each propagation step by abstraction of hydrogen atoms from the propagating polymer to yield unsaturated chain ends and hydrogen transfer agent adducts, Co^{3+}–H. Subsequent transfers of hydrogen radicals to the growing chain result in reinitiation of the processes. Cobalt is usually chelated as the phthalocyanine, porphyrin or cobaltoxime.

Nitroxyl mediated polymerisation of styrene was first reported in 1985, but it was not until 1993 when Georges and co-workers reported their findings on the polymerisation of styrene in the presence of the 2,2,6,6-tetramethylpiperidinyl-1-oxy radical (TEMPO) (**6**), that the method was extensively investigated.

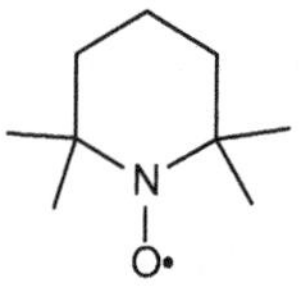

(**6**)

In the earliest experiments, bulk polymerisation of styrene was carried out in the presence of benzoyl peroxide and TEMPO at 127 °C but it took 70 h to reach >85% conversion. Later it was found that the addition of camphorsulfonic acid and acids of similar strength substantially reduced the time (88% conversion in 5.5 h).

Since 1993 there have been many reports of work using nitroxyl mediation. Styrene and substituted styrenes generally work well, but (meth)acrylates give only moderate results. At present, efforts are being made to find new types of stable radicals that do not suffer from the tendency of nitroxyl radicals to abstract β-hydrogen from initiating or propagating radicals. The tri-*p-tert*-butylphenylmethyl radical has been proposed.

Metal catalysed atom transfer radical addition (ATRA) is an efficient method for C–C bond formation in organic chemistry, but in 1995 Wang and Matyjaszewski proposed its application to radical polymerisation processes to control the growth of chains. They utilised a copper/dipyridyl complex as a halogen transfer agent that functions between dormant and active chains. Early work was carried out at temperatures >100 °C with an alkyl halide as initiator and CuCl complexed with 2,2-dipyridyl ligands that acted as an efficient promoter of halide atom transfer. Using 1-phenylethyl chloride, well-defined polymers were produced with controlled molecular weights and polydispersities in the range 1.3–1.45. Subsequently, modifications led to the preparation of polymers having pre-determined molecular weights, M_n *ca.* 10^5 and polydispersities as low as 1.05.

In parallel with Matyjaszewski's disclosure, a similar approach was published by Sawamoto and co-workers who used ruthenium, rhenium and other transition metal complexes to induce one electron redox cycles through interaction with halogens at polymer terminals. This approach has continued to be investigated but the simplicity of the Matyjaszewski technique and the use of less expensive metals are more attractive.

One of the most recent techniques in the field of controlled radical polymerisation is the RAFT process developed by Moad and his colleagues at the CSIRO establishment in Australia. Like ATRP the origins of this method lay in organic synthesis, and the formation of carbon–carbon bonds using radical addition-fragmentation, *e.g.* in the allylation of aldehydes. For chain transfer to operate effectively by this mechanism, the process must compete with propagation. To achieve this, the transfer agent must have a double bond with reactivity toward propagation radicals comparable to that of the polymerising monomer, fast fragmentation of the adduct radical formed and efficient re-initiation of the polymerisation by the expelled radical. The generic structure of RAFT agents is (**7**).

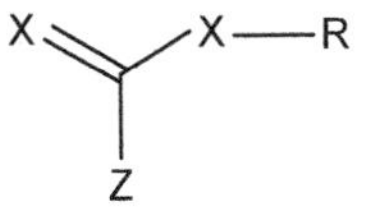

X=C	-	reactive double bond
X–R	-	weak single bond
R	-	radical leaving group
Z	-	modifies addition and fragmentation rates

(**7**)

Effective RAFT agents include dithioesters (**8**), trithiocarbonates (**9**), dithiocarbamates (**10**) and dithiocarbonates (xanthates) (**11**). The odour problem when working with these thio-compounds may be a limitation for the moment, but studies with methacylate macro-monomers and organic iodides have been conducted to overcome this problem.

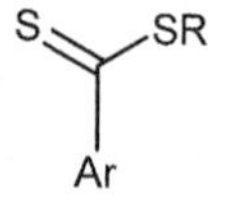

$$
\begin{array}{ccc}
\underset{Ar}{\overset{S}{\diagdown}}\text{--SR} & \underset{SR}{\overset{S}{\diagdown}}\text{--SR} & \underset{R\diagdown NH}{\overset{S}{\diagdown}}\text{--SR} \\
(8) & (9) & (10)
\end{array}
$$

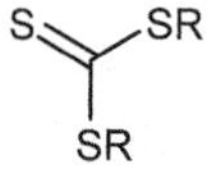

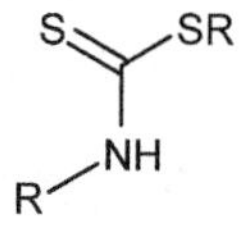

$$
\underset{OR}{\overset{S}{\diagdown}}\text{--SR'}
$$

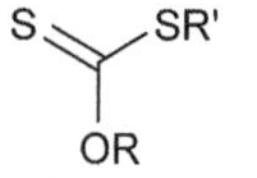

eg R = C$_2$H$_5$, R' = CH(CH$_3$)$_2$

(11)

The RAFT process shares with ATRP an ability to accommodate a wide range of monomers and functionalities as well as a similarly wide range of preparative techniques.

Some control of radical polymerisation can be achieved through reactor engineering. Acrylic polymers prepared in a strong centrifugal field have shown reduced polydispersity. Evidently some of the usual chain termination processes were prevented by the high shear and divergent alignment of the growing chains.

Controlled free radical polymerisation is a very active area of research that commands serious industrial interest and is growing rapidly. It is not yet at a stage of extensive industrial exploitation particularly in the coatings industry, but its potential value in high solids systems is evident. Undoubtedly, however, the most effective solvent-free approach to acrylic coatings is to form the polymers from monomers *in situ* as we shall discuss in Section 7.2.5.

7.2.3.3 Other Binder Types. The principles illustrated by polyesters and acrylics can be applied more widely: polyethers such as polypropylene glycol and polytetramethylenediol offer low T_g, telechelic species, often used in elastomeric urethane systems in place of the hydrolytically sensitive aliphatic polyesters.

Aliphatic polycarbonates made by polymerising ethylene or propylene carbonate contain a mixture of ether and carbonate linkages, the latter being more hydrolytically stable than other ester groups and have been used for much the same purposes. There are also several end-functional polybutadienes made by anionic polymerisation that have found use in elastomeric compositions.

Epoxy resins based on epichlorohydrin are now mainly used at molecular weights approaching those of their monomer. Liquid epoxies contain only about 1.1 repeat units, since the pure diglycidyl ether of bisphenol A is a solid crystalline material and therefore unsuitable. Additional backbone flexibility, hence lower viscosity, is obtained in the corresponding derivative of bisphenol F.

Still greater fluidity is obtainable by using intrinsically more flexible segments. The hydrogenated counterpart of a liquid epoxy would be an obvious candidate. However, an exact analogue is not available in view of the different reactivities of the starting materials. Alternative commercial offerings decorate a standard epoxy with highly flexible in-chain and lateral chains, *e.g.* (**12**).

(**12**)

If the quest was purely for a polymer with the lowest viscosity for a given molecular weight, a polydimethylsiloxane would be the obvious choice with its long Si–O bonds, free rotation and consequent low T_g. Other things being equal, however, siloxanes alone are unpromising candidates for film formation, as will be discussed in Chapter 9.

7.2.4 Crosslinking Chemistries for High Solids

Many high solids systems are crosslinked using the same chemistry as was described in Chapter 6 for conventional coatings. However, the minimum solvent requirement is best achieved by considering the design of all parts of a composition, including crosslinkers and crosslinking chemistry. Certain chemistries are particularly suited to the purpose.

Functional groups on the film former and crosslinker should have the minimum potential for interactions with each other before reaction, as any such physical effects tend to increase viscosity. However, it is essential that the crosslinking chemistry be very effective, so that the unpromising starting materials can be transformed into a hard coating. Also, since the linkages formed during cure will constitute a high proportion of the material present, it is important that they contribute the correct physical and chemical properties to the network. The ideal chemistry involves functional groups that are not very polar at the outset, and cannot participate in hydrogen bond donation or acceptance. After reaction, they should constitute a linkage that *does* hydrogen bond, and contains hard structures such as rings.

Crosslinks formed by a chain-growth process are of especial interest, because each functional group can generate a connectivity of (approaching) two and crosslinking is both dense and efficient. It also allows the use of species with a single functional group to act as reactive diluents by constructing polymers *in situ*.

As noted in Chapter 6 condensation processes are generally undesirable if they contribute a significant amount of volatile organic product to the overall emissions, but there are numerous cases where they can be tolerated. They may have the additional advantage of introducing a degree of 'latency' to control the pot lives of over-reactive systems, since the condensate remains in the bulk mixture and inhibits the forward reaction. After application, volatiles are lost from the film and the curing process can proceed rapidly.

7.2.5 Photochemically Initiated Chain-growth Crosslinking

For the purposes of the present coatings industry, chain-growth crosslinking is mainly concerned with the ionic polymerisation of epoxies and related materials, and radical polymerisation of acrylates, and similar activated olefins. In both cases, the most important technologies use photochemical processes to initiate the reaction.

The use of UV radiation for the curing of coatings is seen as one of the largest growth areas in the coatings industry. Other forms of radiation, *e.g.* γ-radiation have been considered, but for reasons of safety or economy have not achieved wide interest. Electron beam cure has had a degree of commercial acceptance, yet has not been developed to anything like that seen for UV. Cost, and the need for complex equipment, have been suggested as reasons for the slow growth. Although relatively lower cost plant has recently become available, it still represents a substantial investment.

UV curing offers important benefits when compared with conventional thermal techniques. The "through-cure" is so rapid that film formation has been described as 'quasi-instantaneous'. Energy requirements are low and reactions can be carried out at ambient or sub-ambient temperature.

Despite its advantages, UV curing of activated olefins also brings certain limitations. It is clear from the curing mechanism that compositions with too much or too dark pigment, thick films, or work-pieces that are not sufficiently flat, will be difficult to process. Much work is in hand to address the issues, and sensitive photoinitiators that can facilitate the cure of thick, pigmented coatings by daylight have been claimed.

A further problem is air inhibition, where the intervention of diradical oxygen terminates chain growth prematurely and leaves a sticky surface. It is sometimes necessary to carry out the curing in an inert atmosphere, or otherwise to exclude oxygen. An alternative is to increase the radical flux to the point where it outstrips the effect of oxygen. Various diluents, usually containing allyl ether groups have shown themselves to be effective at reducing inhibition. The best known example is an oligomer of allyl-glycidyl ether.

Finally, UV-cured polyacrylate coatings often show disappointing weathering resistance, despite compositions that would be highly resistant to photooxidative degradation if otherwise assembled. Trapped radicals and the relatively high levels of photoinitiator residues have variously been blamed.

Attempts to develop visible light curing have shown considerable promise, and may address some of the problems noted above, but as yet the technology has not evolved to commercial exploitation in the coatings industry.

7.2.5.1 The Principles of Photochemistry. There are two basic laws in photochemistry that bear on its technological exploitation. They are the Grotthuss–Draper Law, that only radiation absorbed by the reacting system is effective in producing chemical change, and the Stark–Einstein Law that each molecule taking part in a chemical reaction absorbs one quantum of the radiation causing the reaction. Providing that the absorbing molecule reacts immediately, without further successive or side reactions, one molecule should react for every quantum absorbed.

For radiation of frequency v, the corresponding quantum is hv where h is the Plank constant, thus hv represents the amount of energy absorbed by each molecule reacting in accordance with the Stark–Einstein Law. For one mole of reacting substance the energy absorbed is Nhv, where N is the Avogadro number.

$$E = Nhv = Nhc/\lambda \qquad (7.9)$$

where c is the velocity of light and λ the wave length of the absorbed radiation. Using literature values for N, h and c, the energy is given in ergs. Division by 4.184×10^{10} gives the values in kcal. λ is in Angstrom units.

$$E = 2.86 \times 10^5/\lambda \; \text{kcal mol}^{-1} \qquad (7.10)$$

The quantity E is very often referred to as one einstein of radiation although the use is discouraged by the IUPAC nomenclature commission.

The intensity of the radiation transmitted through a layer of a solution is governed by the Beer–Lambert law (7.11)

$$\text{Log}(I/I_o) = elc \qquad (7.11)$$

where e is the molar absorption coefficient ($\text{L mol}^{-1} \text{cm}^{-1}$), c is the concentration (mol L^{-1}) and I is the thickness of the irradiated layer.

The efficiency of a photochemical process is a factor expressed as the quantum yield. Though composed of several considerations, for our purpose it can be defined as:

$$\Phi = \frac{\text{Number of moles converted or consumed}}{\text{Number of einsteins absorbed}} \qquad (7.12)$$

when $\Phi = 1$ – every quantum absorbed produces one chemical reaction,
  $\Phi < 1$ – other reactions compete with the main photochemical reaction,
  $\Phi > 1$ – a chain reaction is taking place.

We are concerned with the third case, where in some polymerisations quantum yields as great as 1×10^6 may be found.

There are three mechanisms that have been used for the UV curing of coatings: free radical, cationic and charge transfer. The first is the most generally used whilst the second is becoming increasingly attractive. The third is of recent origin and although not industrially developed offers certain environmental advantages.

7.2.5.2 Initiation of Free Radical Polymerisation. The direct formation of reactive species by the action of UV radiation on the monomer is not an efficient process, therefore the initiation step requires an initiator than can generate active species when irradiated. There are two mechanisms for the generation of free radicals:

1. Cleavage of the initiator by the energy of the radiation employed,

$$I \rightarrow 2A^{\cdot}$$

   Examples of such initiators are azobisisobutyronitrile (AZDN), hydrogen peroxide and 1,1-azidocyclohexanecarbonitrile.
2. 'Sensitised' reactions in which a sensitiser is used in conjunction with an initiator, the sensitiser absorbing light energy and attaining an excited state. From this state it may either transfer its energy to the initiator molecule through a hydrogen transfer mechanism (exiplex formation, Scheme 7.9) or transfer energy directly to the initiator in the ground state (Scheme 7.10).

$$S \longrightarrow S^*$$
$$S^* + I \longrightarrow (S\text{-}I)^*$$
$$(S\text{-}I)^* \longrightarrow S\text{-}H + I^{\cdot}$$

Scheme 7.9

$$S \longrightarrow S^*$$

$$S^* + I \longrightarrow S + I^*$$

$$I^* \longrightarrow 2A^{\bullet}$$

Scheme 7.10

It has been proposed that initiators which undergo intramolecular bond cleavage be designated Type P1 indicating that the reaction is unimolecular. Initiators that undergo intramolecular hydrogen abstraction from a hydrogen donor are designated P2 because the reaction is bimolecular. Examples of P1 initiators are benzoin ethers (**13**), acyloxime esters (**14**) and substituted acetophenone derivatives (**15**). P2 initiators are exemplified by benzophenone (**16**), Michler's ketone (**17**) and thioxanthones (**18**).

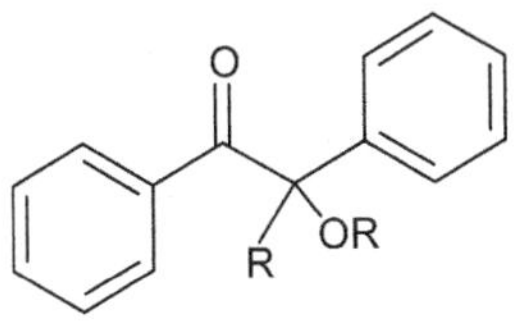

(**13**) 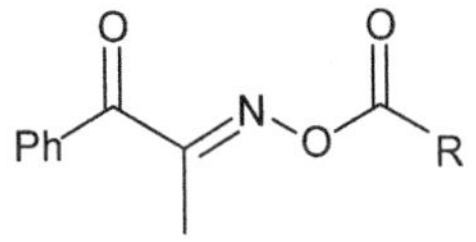   (**14**) 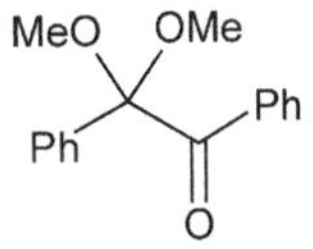   (**15**)

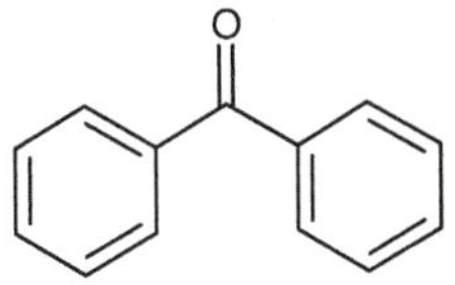

(**16**) 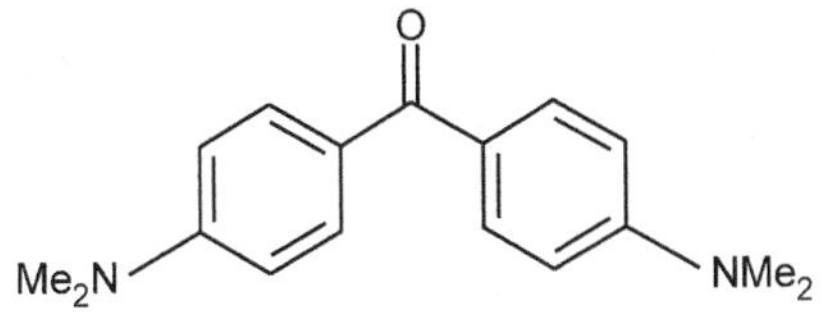   (**17**)

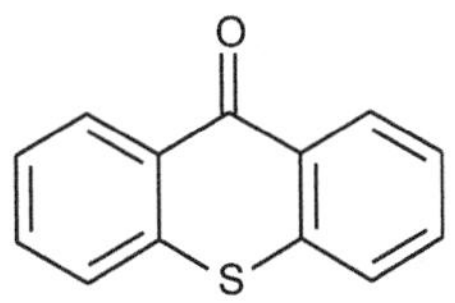

(**18**)

7.2.5.3 Acrylates and Other Activated Olefins for Photochemical Cure. Once radicals have been generated by the photochemical process, polymerisation can take place by a radical mechanism. It has been observed that monoacrylates often lead to thermoset films, when polymerised in the absence of solvent. Chain propagation outstrips termination as the viscosity increases, leading to accelerating reaction and eventual gelation (the 'Trommsdorff' or 'gel' effect). However, it is usual to incorporate di- or higher functional acrylates to guarantee a certain minimum level of crosslinking. Engagingly, the usual order of events is inverted: the crosslinks are factory made and sold to the applicator who builds polymer chains around them!

The crosslinking monomers can have a profound effect on the properties of the final network. Their level controls crosslink density, and consequently mechanical and other properties, whilst the structure and length of the spacer group is also very influential. Spacer groups derived from polyether, polyesters, epoxy resins or polyurethanes can constitute a significant fraction of the network, and import their own particular properties. The degree of reaction at the gelation point can commonly be 10–100 times larger than the calculated value, indicating a very large number of defects in the network.

The usual acrylic monomers are toxic, and relatively volatile, so considerable effort has been spent to develop replacements. One approach is to employ acrylate esters of relatively bulky alcohols to render them less volatile. Another is to use methacrylates, being generally less toxic than acrylates, but also less reactive. There is, however, increasing interest in using pairs of electron-donor, electron-acceptor monomers, that form alternating copolymers very readily although either alone may be reluctant to polymerise.

Examples of electron-rich monomers are styrene, allyl or vinyl ethers, allyl or vinyl formamide and α-olefins, although the cost of vinyl ethers has limited their use to specialised markets. Suitable electron-poor counterparts are maleic anhydride, maleimides (**19**), maleate or fumarate esters and benzylidenecyanoacetates (**20**).

(**19**)   (**20**)

Appropriate combinations can polymerise even more rapidly than acrylates and in certain cases, the charge transfer complex set up between the two types of monomer can take the place of the photoinitiator.

The polymerisation of charge transfer complexes is of further interest since the presence of a photoinitiator is not required, and problems associated with photoinitiator and the leaching of its residues from films are eliminated. However, as the mechanism involves the generation of free radicals, the presence of oxygen still needs to be controlled.

Fundamental studies have largely centred on complexes from maleimides or *N*-alkyl maleimides and acrylates, fumarates and various electron donors and on maleimides with vinyl ethers. The development of the XeCl lamp has played a significant role in this work, since it is more efficient than the commonly used medium pressure Hg lamps.

7.2.5.4 Initiation of Cationic Photopolymerisation. Although the free radical mechanism is very widely applied to UV curing of acrylates and similar compounds, it is not applicable to many important film formers such as epoxides. There has been increasing interest in cationic polymerisation that can be used with epoxides, vinyl ethers, lactones, sulfides and cyclic ethers. Unlike its free radical counterpart the ring-opening mechanism is not inhibited by oxygen, and lower film shrinkage has been claimed. Also unlike free radical curing, the polymerisation does not cease when the radiation is removed, but continues for an appreciable time. The development of very efficient initiators has also added to interest in the areas of coatings, inks and adhesives so we may look forward to strong growth in this area.

Photochemistry can be used to generate strong Brönsted acids that initiate cationic polymerisation. However, advances in its use for curing tended to be retarded by the ease with which chain termination and chain transfer reactions occurred at ambient temperature. It was not until *ca.*1972 that aryldiazonium salts were shown to initiate the polymerisation of epoxy-functional monomers and oligomers. A notable advance came a few years later when Crivello introduced triarylsulfonium salts, diaryliodonium salts and triaryl selenonium salts, all of which contained complex metal halide anions. Unfortunately all the low nucleophilicity anions, particularly the largest which gave the fastest cure rates, introduced an undesirable toxicity. Typically the anions were $^-BF_4$, $^-PF_6$, $^-AsF_6$ and $^-SbF_6$. More recently, the pentafluoroborate anion (**21**) has been promoted. It reduces the toxicity enormously (LD50 > 2000 mg kg^{-1} – rat) and is claimed to be very efficient.

(**21**)

Another recent introduction to the cationic initiation scene has been the use of iron arene salts having anions of low nucleophilicity. It has been claimed that members of the CpFe(η^6-arene)$^+$ family may be used as visible light sensitive initiators for a variety of monomers, *e.g.* epoxies and acrylates.

7.2.5.5 Epoxies and Related Materials for Photochemical Cure. 'Internal' epoxies are most reactive towards cationic cure since the generated carbenium ion is relatively stabilised by substitution. A range of cycloaliphatic epoxies such as (**22**)

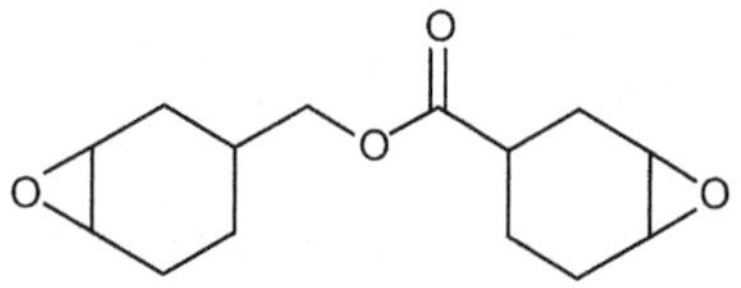

(**22**)

and epoxidised oils that have low reactivity with nucleophiles and have little other coatings use except as reactive diluents come into their own in photocuring technology (Scheme 7.11).

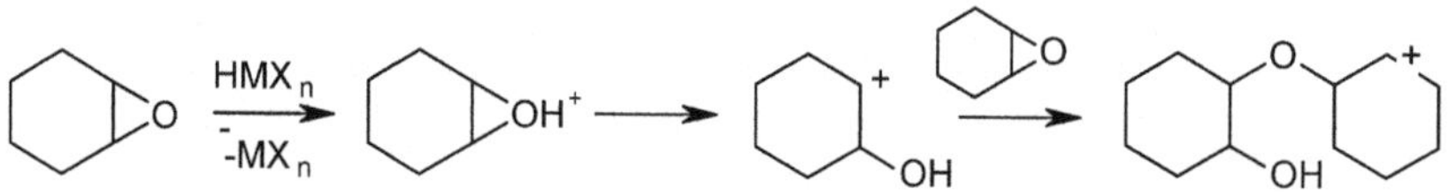

Scheme 7.11

Their relatively flexible structures are often well adapted to high solids applications. Nevertheless, the more familiar terminal epoxies derived from phenols and epichlorohydrin have also been used extensively.

Cationic polymerisation in general is susceptible to moisture and the presence of other nucleophiles, but photochemical epoxy cure can be carried out in the open air without undue precautions.

7.2.6 Chemically Initiated Chain-growth Crosslinking

We previously mentioned the incorporation of methacrylates and the like into auto-oxidising systems. Effectively, auto-oxidation is a redox process, and in that case was generating free radicals to initiate the polymerisation of the methacrylate.

More direct use is made of redox chemistry in unsaturated polyesters. The polyesters are made from diols and diacids, part or all of the diacid residues being derived from maleic anhydride. At the high processing temperatures used, most of the olefinic bonds convert to the *trans*-form, so that the resulting polymer contains largely fumarate residues. Happily, fumarate double bonds are more reactive then maleate in the subsequent chemistry. Traditionally, the polyester has been dissolved in styrene, but since the mutagenic potential of styrene was recognised, vinyltoluene has been preferred. It is also possible to use various methacrylates.

The preferred redox initiator is a combination of cobalt(II) octoate and methylethylketone hydroperoxide. The cobalt soap is usually blended with the polymer, and the hydroperoxide is added at the point of use.

Catalytic quantities of cobalt can decompose an unlimited amount of peroxide. Since oxygen centred radicals are not very effective at initiating polymerisation, they first abstract hydrogen from a convenient molecule to generate the initiating radical. It is common practice to incorporate a tertiary amine to provide a suitable source of abstractable hydrogen (Scheme 7.12).

$$Co^{2+} + ROOH \longrightarrow RO^{\cdot} + {}^{-}OH + Co^{3+}$$
$$Co^{3+} + ROOH \longrightarrow RO^{\cdot} + {}^{1}/_{2}O_2 + H^{+} + Co^{2+}$$
$$RO^{\cdot} + R^{1}H \longrightarrow R^{1\cdot} + ROH$$

Scheme 7.12

In an alternative process, a tertiary amine accelerates the decomposition of benzoyl peroxide leading directly to phenyl radicals.

Like UV-cured acrylates, unsaturated polyesters suffer from oxygen inhibition resulting in a sticky surface, unless steps are taken to exclude air. When they were fashionable as deep gloss coatings for wood, the practice was to add a small amount of wax to the formulation. It floated to the surface to exclude air and could be removed by polishing once the coating was cured. The principal modern use for unsaturated polyesters is as binders in glass reinforced composites, and the gel coats used to coat them. The gel coat is usually applied to the inside of a mould, so the stickiness occurs on what will become the interface with the structural layer after the glass fibre has been 'laid up'. It helps to promote adhesion, and the eventual outside surface is completely cured.

Other polymerisable systems initiated by redox processes have seen limited use. Notable amongst them are the acrylate esters of epoxy resins.

7.2.7 Step-growth Crosslinking

7.2.7.1 Monomeric Curing Agents. Crosslinkers with low molecular weight and consequently low viscosity are obviously very helpful, but may bring toxicological problems in their wake. For example, the use of monomeric amines to cure epoxies gives the best possibilities for low solids, but the materials are much more severely corrosive to human tissue than the polyamides or epoxy–amine adducts that would otherwise be used. They are also particularly susceptible to carbamation by atmospheric carbon dioxide (*cf.* Chapter 6). Some more subtle issues lie in the province of the formulator, for example, a crosslinker that is more viscous than the main film former should have as low an equivalent weight as possible, so that the minimum amount is used in the composition. Conversely, it may be desirable to maximise the amount of a low viscosity curing agent.

The isocyanate and amino curing agents for the various active hydrogen-functional film formers are prepared with as well-defined structures as possible. Typical are the isocyanurate trimer of hexamethylenediisocyanate, and hexamethoxymethylmelamine (HMMM). They can both be prepared in a fairly monomeric form. HMMM has a distinct advantage over the partially methylolated aminos which have a strong tendency to self polymerisation in manufacture and are therefore oligomeric, so unsuitable for high solids applications. Methoxymethylolated glycoluril, as noted in Chapter 6, is even more satisfactory since the precursor possesses only secondary amine sites and so can only yield fully methylolated systems.

7.2.7.2 Amines as Alternatives to Hydroxyl Groups. Hydroxyl functional materials are the common currency of much thermoset chemistry, being cured with isocyanates and various amino-formaldehyde prepolymers. They are unfortunately very strong H-bond donors and there is opportunity to reduce viscosity by derivatisation or replacing them with other materials of comparable reactivity. Trimethylsilyl derivatives, though costly, have been claimed to achieve a dramatic reduction in the viscosity of various polyols and hydrolyse readily on exposure to the atmosphere. Amines are markedly weaker H-bond donors but cannot be conveniently used with formaldehyde resins because their basic character negates the effect of the necessary acid catalyst. On the other hand, they are excessively reactive with isocyanates and must be blocked as aldimines or ketimines, that release aldehydes or ketones on exposure to ambient moisture. β-hydroxyamines can be blocked as oxazolidines, and regenerated on hydrolysis. Less frequently β-diamines can be blocked as imidazolidines (*cf.* Chapter 6).

Secondary amines, in their reaction with isocyanates, are generally of comparable reactivity to primary amines, but highly hindered versions have been proposed as having reactivity closer to that of alcohols. One successful innovation uses 'aspartate esters' (**23**), made by reacting a primary amine with a maleate ester. Their reactivity towards isocyanates is lowered by steric hindrance and H-bonding, though they are proposed primarily as reactive diluents (*v.i.*).

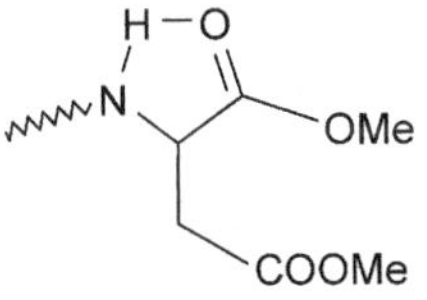

(23)

Successful co-reactants for formaldehyde resins have been based on amides or carbamates that also show reduced H-bonding.

7.2.7.3   Thiols as Alternatives to Amine and Hydroxyl Groups.   Thiols are interesting co-reactants for isocyanates and epoxies. (Amino resins are rarely cured with thiols, perhaps because the smell would be unacceptable in a stoving system.) They have only very weak H-bonding character, yet react with isocyanates to produce thiocarbamate linkages, that H-bond as strongly as urethanes (Scheme 7.13).

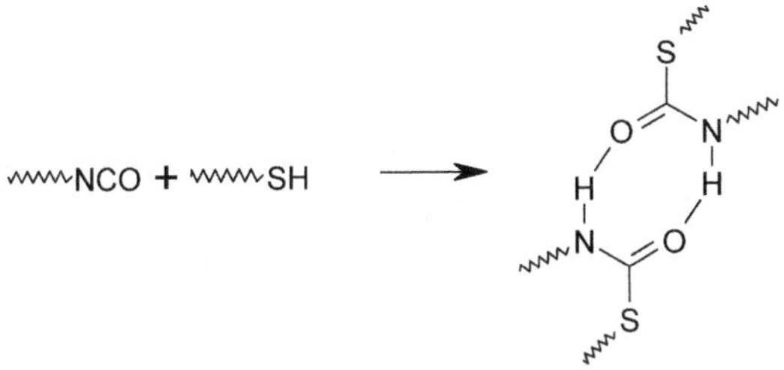

Scheme 7.13

7.2.7.4   Thiol–Ene Chemistry.   The easy radical reaction between thiols and various olefins is the basis of many proprietary technologies.

Electron-rich olefins such as vinyl ethers and aliphatic terminal olefins generally react satisfactorily with the electrophilic thiyl radical. Bicyclo-heptenes are even more highly reactive, perhaps because of ring strain, but also because the allylic bridgeheads resist H-abstraction since they cannot adopt a flat shape.

The thiol–ene reaction is not inhibited by oxygen, but a certain amount of oxidation to disulfide and β-hydroxysulfoxide occurs when the reaction takes place in air (Scheme 7.14).

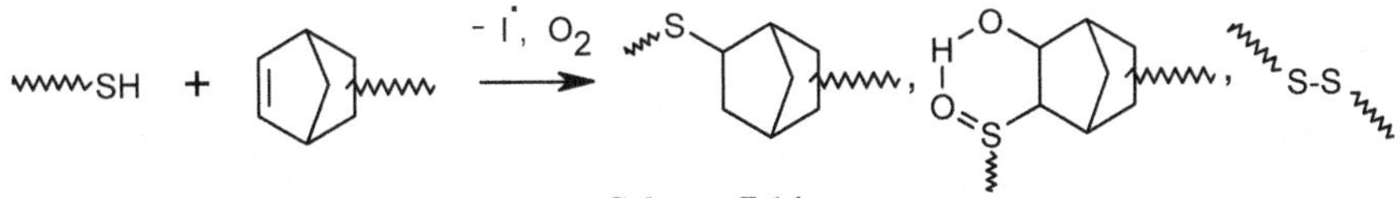

Scheme 7.14

The side reactions are not necessarily unhelpful, and a thick coating might be expected to show increasing polarity from the substrate to the top surface.

Radical thiol–ene chemistries have been used in adhesives, fibre optic coatings and electronic coatings. They are increasingly popular in an alternative UV-cured technology. Obvious attractions are the absence of air inhibition, low toxicity, ability to adhere strongly to various substrates and the chemical inertness to be expected from such coatings.

The Michael-type addition of thiols to activated olefins such as maleate or acrylate esters under basic catalysis is also of interest in high solids coatings, but should not be confused with the thiol–ene chemistry described above (*cf.* Chapter 6).

7.2.7.5   Acetoacetate Chemistry.   The synthetic versatility of acetoacetate esters is reflected in their increasing use in polymer chemistry, strongly promoted by the Eastman Company. They can react with isocyanates and amino resins through their activated methylene sites. Since only about 7% of the molecules exist in the enol form, H-bonding is minimised as compared with similar polyols.

They can also undergo several other interesting reactions that have been exploited in coatings systems. The Michael reaction with acrylate or maleate groups is of special value (Scheme 7.15).

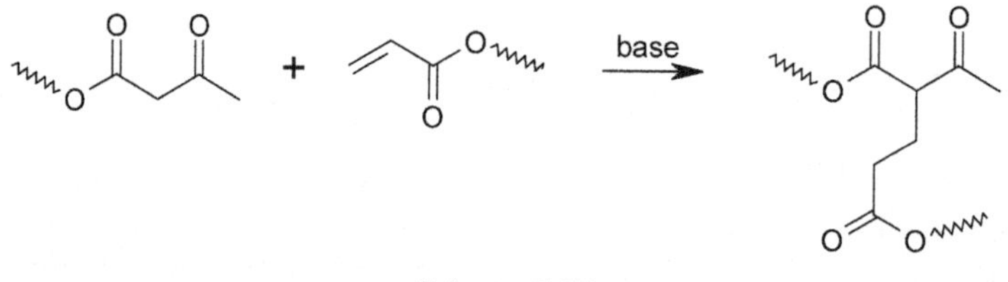

Scheme 7.15

Reaction with primary amines occurs very readily to form an imine that rearranges to an enamine structure, stabilised by H-bonding (Scheme 7.16).

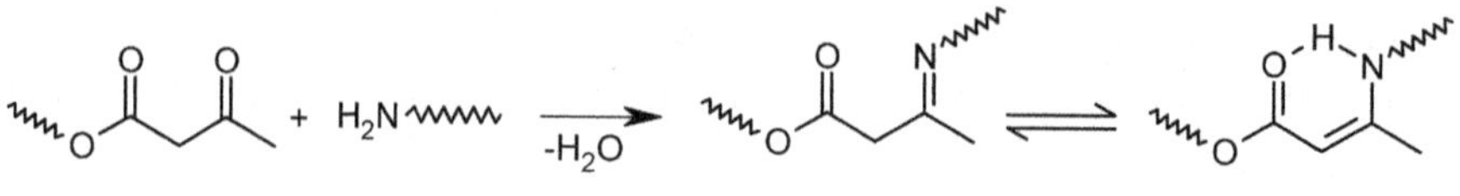

Scheme 7.16

It is common to present the amine as its aldimine or ketimine of an aliphatic ketone or aldehyde. The expected ambient hydrolysis takes place before reaction, or possibly a transimination reaction takes place *in situ* (Scheme 7.17).

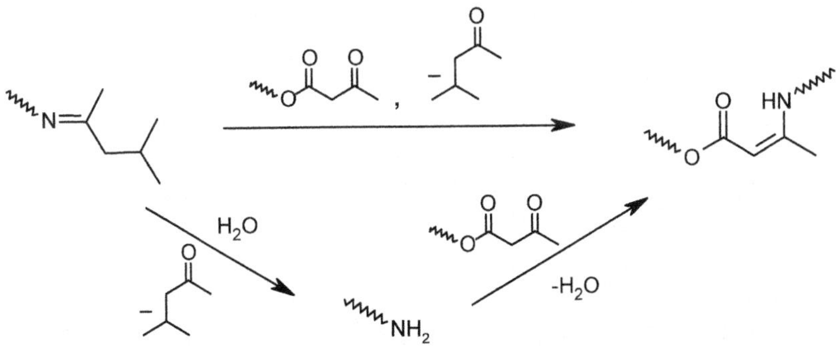

Scheme 7.17

The behaviour of acetoacetates with aromatic aldimines is completely different and unexpected. In the absence of catalyst it can proceed to the point when each Schiff's base group forms a linkage so that multifunctional Schiff's bases can be used to crosslink an acetoacetate functional polymer (Scheme 7.18). The reaction was discovered by chemists at Reichhold and immediately proposed as a curing chemistry for coatings. However, researchers at Rohm and Haas soon showed that the reaction could continue with elimination of the amine and formation of a crosslink between two acetoacetate groups by the aromatic aldehyde, which then underwent an internal condensation to a cyclic species (Scheme 7.18). The net result was the same as would be obtained by Knoevenagel reaction, and multifunctional acetoacetates could be cured with monofunctional Schiff's bases. The formation of a ring structure on curing satisfies one of the presumed needs of high solids cure chemistry – the development of auxiliary hardening features.

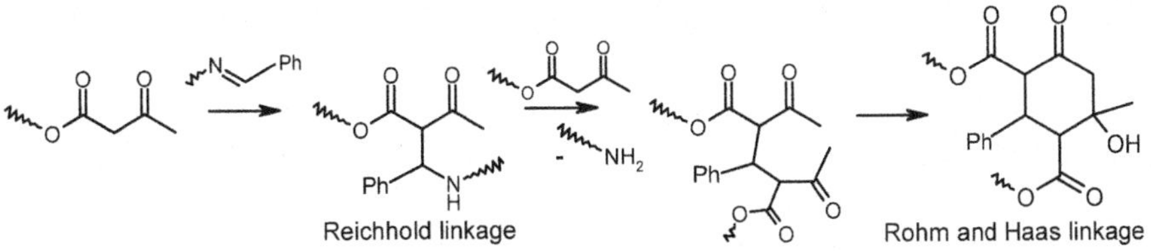

Scheme 7.18

7.2.8   Reactive Diluents

Diluents of all kinds are added to compositions to reduce their viscosity to levels suitable for application. They are for the most part fugitive, and their release into the atmosphere creates the problems that high solids technology sets out to address. Materials of lower volatility may remain in the film indefinitely and have a serious impact on its properties. The plasticisation effect may be tolerated or embraced, but is the proper province of the formulator and will be discussed in Chapters 10 and 11.

A third type of diluent remains in the film because it undergoes reaction and becomes a permanent part of the network.

If a reactive system contains principal components with functionalities A and B that react with each other, a reactive diluent might have A or B functionality. Alternatively it could have A$'$ or B$'$ functionality, where A$'$ is different from A but still reacts with B, and B$'$ is not B, but reacts with A. Again, it could have C functionality and react on its own or with different parts of the developing network. Reactive diluents with a single A, B, A$'$ or B$'$ group have given reactive diluents a rather bad name. Their effect is usually to use up the other type of function and degrade the network properties to a greater or lesser extent. However, multifunctional diluents, A–A for example, may not be detrimental and may actually enhance the properties of the composition. Diluents of the A–B type have the particular advantage of not disturbing the stoichiometric balance of film former and diluent.

Considerations that apply to film formers are equally valid for reactive diluents, so a low viscosity is achieved by low molecular weight, low T_g, and by minimising the interactions due to functional groups. With all their physical properties conspiring to deliver soft and unsatisfactory coatings, it is essential that the reactive moieties in the diluents work hard to redress the balance.

The highest linkage efficiency is necessary, and it is important for the chemical changes to do more for the network structure than simply attach 'passenger' molecules. The use of diluents that participate in the crosslinking process has been mentioned, and is hardly distinct from the selection of crosslinkers with an eye to their fluidity. It is also helpful if reaction leads to structures that interact strongly with each other or different parts of the molecule, for example, by hydrogen bonding, or if stiff ring-structures are formed. Lowering of the molecular weight must not result in significant volatility, particularly in stoving systems. The low molecular weight, reactive species used as reactive diluents are likely to be toxicologically active, and no less harmful to the environment than commonly used organic solvents. An indication of the viscosity reducing effect of a diluent can be gained by estimating the combined T_g of the mixture using the Fox equation, usually expressed as Equation (7.13), or more conveniently as Equation (7.14) for present purposes:

$$\frac{w}{T_g} = \sum \frac{w_i}{T_{gi}} \qquad (7.13)$$

where w is the weight of each component.

$$\frac{1}{T_g(\text{mix})} = \frac{\phi}{T_g(\text{film former})} + \frac{1-\phi}{T_g(\text{diluent})} \tag{7.14}$$

where ϕ is the mole fraction of each component. It follows that a low molecular weight has a double effect – both in lowering the diluent T_g and increasing the mole fraction represented by a given weight addition.

7.2.8.1 Reactive Diluents for Alkyds. Suitable reactive diluents for alkyds are materials that participate in the auto-oxidation process. Linseed oil, or other triglycerides well-furnished with allylic sites are the oldest and ultimately the best materials for the purpose, being reasonably free of unwanted side reactions, though they inevitably have a softening effect on the final film.

Other allylic materials have found widespread use, but allyl ethers are particularly reactive. An oligomer of allylglycidyl ether is dramatically effective. Unfortunately, allyl ethers have an unwelcome tendency to form the very toxic lachrymator, acrolein, under radical conditions and are no longer widely promoted.

Another approach is to incorporate a radically polymerisable monomer, as discussed in Chapter 6. Acrylates are regarded as too toxic, but methacrylates are equally effective. One proprietary material (**24**) made from the inexpensive dicyclopentadiene contains methacrylate and allylic moieties in a single molecule, but its true value lies more in its potential for improving the coating properties and drying, than its ability to reduce viscosity.

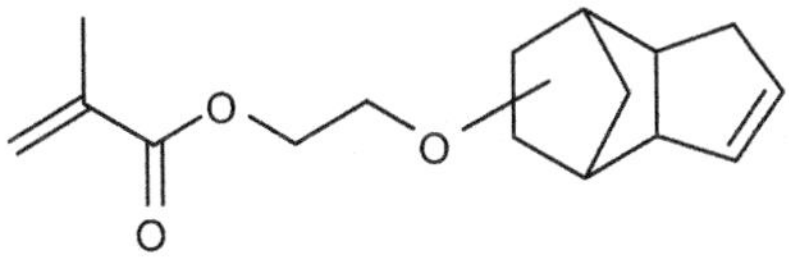

(**24**)

In compositions seething with the potential to create free radicals, it is difficult to ensure the stability of methacrylates. They are nevertheless quite widely used.

Alternatives that sometimes find favour are the esters of maleic or fumaric acid. Relatively stable alone, they are always ready to copolymerise with electron-rich olefins that might be present as part of the alkyd structure, or added separately.

7.2.8.2 Reactive Diluents for Epoxies. High performance epoxies being based on intrinsically viscous precursors, there has always been a strong demand for diluents, and an enormous number of candidates has appeared in the literature over the years.

Typical monofunctional diluents are glycidyl ethers of alkylated phenols and the glycidyl ester of neodecanoic acid. A–A types include diepoxy-octane (**25**), vinylcyclohexene dioxide, (**26**) assorted glycidyl ethers of diols and many others.

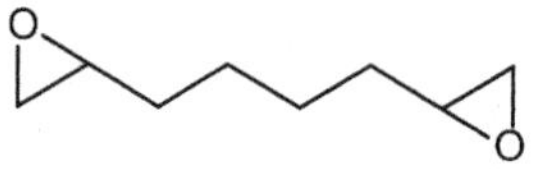

(**25**) 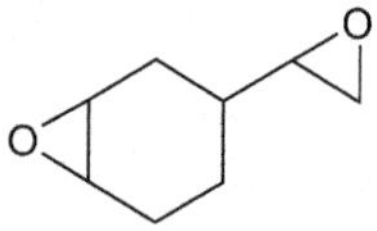   (**26**)

They are rarely pure materials and those made from epichlorohydrin often contain much more chloride ion than the aromatic film formers, with consequent potential to degrade the corrosion resistance of compositions in which they are used. Moreover, most low molecular weight aliphatic epoxies are toxicologically suspect. One interesting material is vernonia oil, a naturally epoxidised oil containing primarily the triglyceride of vernolic acid (**27**). It is obtained from *Vernonia galamensis*, a bushy plant native to Eastern Africa, and is capable of auto-oxidation as well as slow reaction with amines. There have been many proposals to farm it for the oil, but it has yet to become a significant commercial item.

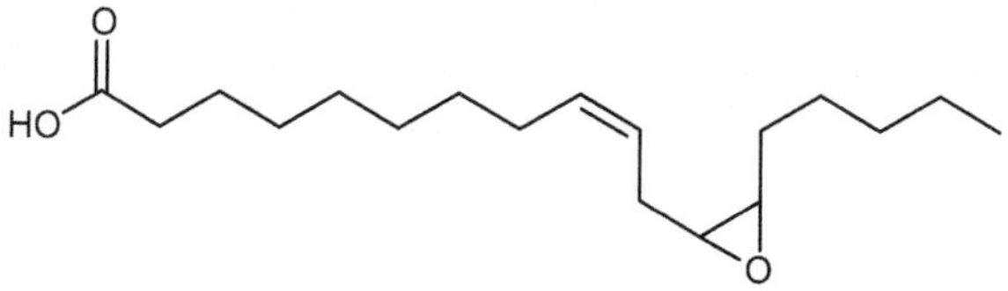

(**27**)

An alternative approach is to use acrylates, and occasionally fumarates, as diluents for epoxies that are to be cured with amines. Both species undergo rapid Michael type reactions with amines and addition of, for example, trimethylolpropane *tris*-acrylate to an epoxy–amine system will increase the cure rate as well as reducing the viscosity. Unfortunately, most suitable materials are esters and may introduce a degree of hydrolytic susceptibility.

Cyclic carbonates such as ethylene or propylene carbonate – both excellent solvents regardless of reactivity, and toxicologically benign – have also been used as diluents for epoxy amines. Reaction with amines leads to β-hydroxyurethane groups (*cf.* Chapter 6). Since urethane groups hydrogen bond to each other so strongly as to provide a 'secondary crosslinking' their introduction to an epoxy network may have a beneficial effect on its properties.

7.2.8.3   Reactive Diluents for Urethanes.   Urethanes are normally formed by the interaction of isocyanates with polyols. As noted above, the existence of urethane groups in a network improves its mechanical properties considerably. Unfortunately, hydroxyl groups hydrogen bond very strongly and increase the viscosity of the precursors.

Suitable amines have been used as diluents, as they have a much reduced propensity for H-bonding and tend to be more fluid than equivalent diols. At the same time, they produce urea structures that H-bond as strongly as urethanes (*cf.* Chapter 6). Unfortunately, most amines react with isocyanates with extreme rapidity, and would produce uncontrollable coatings systems. Several latent amine diluents have appeared, where the amine and hydroxyl functionality are protected as oxazolidines, ketimines or aldimines, *e.g.* (**28a,b**). The materials are very fluid and produce amine and hydroxyl groups for isocyanate reaction after hydrolysis on exposure to atmospheric moisture.

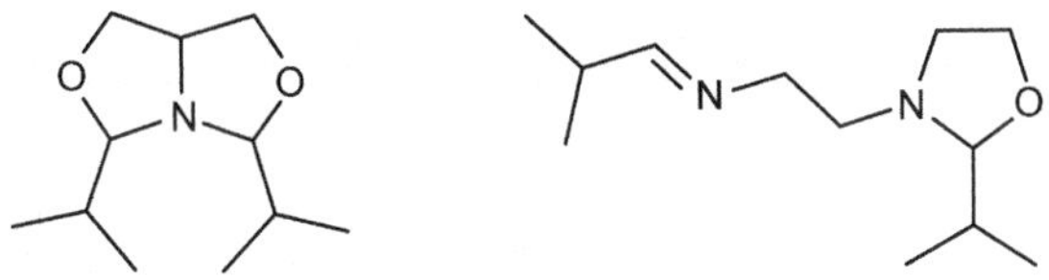

(**28a**)   (**28b**)

However, the blocking agent (isobutyraldehyde in the case of (**28a,b**)) makes its own contribution to the volatile organic emissions, and tends to offset the advantage of using a reactive diluent. A recent innovation uses molecules containing both isocyanate and a blocked amine as reactive diluents. Where equal numbers of reactive groups are produced, the diluent can be added to curing systems without upsetting their stoichiometry. An example is the allophanate (**29**), where the two isocyanate groups are equivalent to the hydroxyl and secondary amine generated on hydrolysis of the oxazolidine.

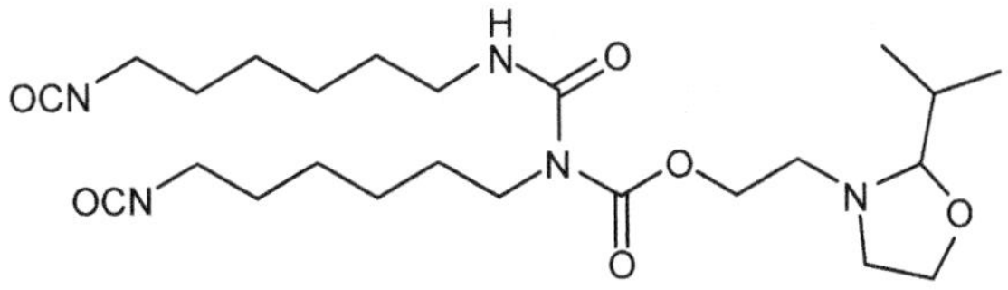

(**29**)

Another approach uses an amine highly hindered and deactivated by hydrogen bonding. The so-called aspartic esters are made very simply, by reacting a primary amine with a maleate ester. They react smoothly with isocyanate and can replace all or part of a polyol reactant.

7.2.8.4 *Reactive Diluents for Stoving Systems.* Caprolactone is a time-honoured diluent for a variety of stoved systems as it is a fluid material of good solvent powers and low toxicity. It is able to behave as a 'masked AB' system, reacting with hydroxyl groups to produce new hydroxyl groups. It is, however, rather too volatile to be entirely satisfactory. Another AB system (30) contained a conventional hydroxyl group and a methylol group derived from formaldehyde and a carbamate group. It could be added to amino formaldehyde systems cured with polyol.

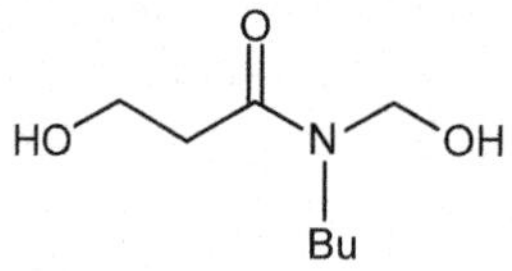

(30)

An interesting A–A' diluent (31) contained both phenolic and hydroxyl end groups. It was made from glycidyl neodecanoate and *p*-hydroxybenzoic acid. Both groups could react with the methylol groups in an amino crosslinker, but the thermoset produced showed substantially better properties than one using a comparable diol. It was believed that the phenolic site reacted with two methylol groups in the amino crosslinker, so as to introduce additional cyclic structures (32).

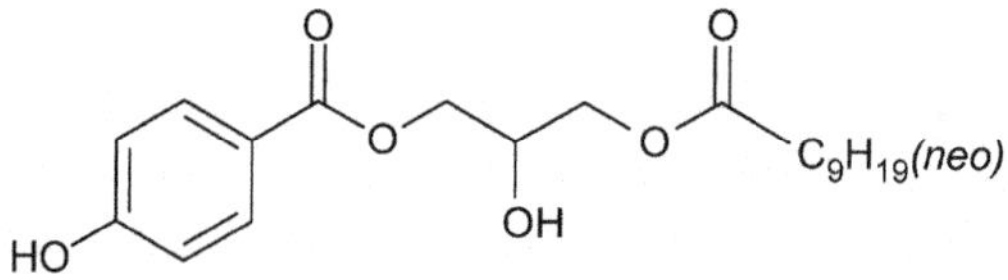

(31)

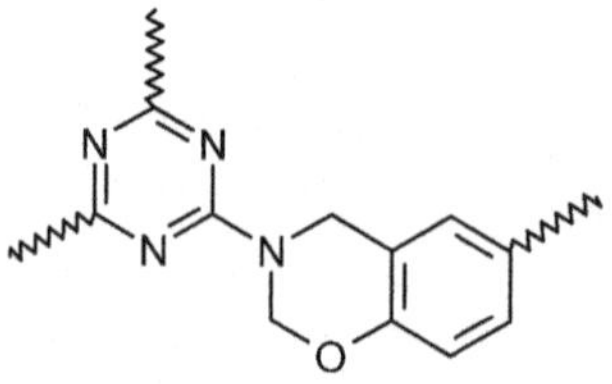

(32)

7.3 POWDER COATINGS

Powder coatings, in the ideal case, are amongst the most environmentally benign coatings. Systems with no significant volatile emissions not only avoid the problems directly associated with them but are also very economical in their use of energy. Because there is greatly reduced condensation in the stoving ovens, ventilation can be kept to a minimum and heat is not lost to the atmosphere.

They are made from polymers that are solid at room temperature, but can be compounded with pigments, curing agents and other components in a heated extruder. They are then ground to small particles. The powdered material is applied to articles by electrostatic spray, or by immersion in a fluidised bed. Particles adhere to the substrate by electrostatic attraction, and flow and coalesce when it is heated.

It is usually necessary for the manufactured powder to be completely stable under the hottest ambient conditions, yet flow and form smooth films rapidly on stoving. (Refrigerated storage is possible but expensive.) It is also required that any curing chemistry takes place during the brief period of stoving, yet heated extrusion can be carried out without any premature reaction.

The size, shape and dispersity of the particles, the ability of the surface chemistry to promote the development of electrostatic charge, and the crucial role of certain additives are all of considerable importance to the process of film formation, but are outside the scope of this chapter. The factors that can be addressed are the bulk properties of the binder compounds and the crosslinking chemistry. They also have a very strong impact on the quality of the final coating. The exacting requirements for coalescence of reactive films before gelation occurring in reactive systems were discussed in Chapter 4, and mean that the onset and duration of crosslinking have to be very carefully controlled.

7.3.1 Thermoplastic Powders

Thermoplastic powder coatings have a certain advantage in that film formation is not curtailed by crosslinking. They also tend to give more flexible films than thermoset systems. Nevertheless, suitable materials tend to be excessively viscous, for the reasons given earlier, and thick uneven films can be expected unless very high temperatures are used. Even so, molecular weights are kept as low as possible consistent with achieving the desired end properties.

Typical thermoplastic powder binders are low density polyethylene, polypropylene, vinyl chloride and polyamides (Nylon-11 and Nylon-12, *i.e.* the self condensation products of 11-aminoundecanoic acid and 12-aminododecanoic acid). Acid functional materials such as the copolymers of ethylene and acrylic or methacrylic acid have particular value since they bestow greatly improved adhesion on their compositions. Polyesters, modified epoxies and polyketone resins are also used in smaller quantities. Fluoropolymers such as poly(vinylidine fluoride) and copolymers of chlorotrifluoroethylene, are used where the need for weathering resistance, dirt shedding or chemical resistance justifies their high cost.

Thermoplastic powders, because of their tough character are difficult to grind to a small particle size, and are consequently unsuitable for electrostatic spraying. They are usually applied by fluidised bed and are stoved at temperatures as high as 200–300 °C.

There is also a significant industry applying protective coatings *in situ* to steel structures using flame spraying. Films as thick as 1.5 mm can be achieved. The usual media are employed, but there is also scope to apply engineering plastics for truly resistant coatings, and thermotropic liquid crystalline polyesters such as poly(phenylene terephthalate) have been examined.

Reversibly crosslinking coatings show some of the characteristics of thermoplastic and thermosetting technologies. One system employed a copolymer of ethylene and acrylic or methacrylic acid compounded with zinc or other divalent metal ions to form ionic crosslinks.

A range of 'reactive thermoplastics' technology has also been deployed in powder coatings. Frequently the functionality of binder and curing agent may fall short of that required to give a true network. Because the binder precursors selected have intrinsically high glass transition temperatures, a degree of molecular weight advancement may be sufficient to achieve the required properties.

There have also been proposals to form high molecular weight powder binders such as polyesters by ring-opening macrocyclic polyesters *in situ*.

7.3.2 Reactive Binders for Powder Coatings

7.3.2.1 Epoxies. The earliest powder coatings were based on solid bisphenol A–epichlorhydrin condensates of a few repeat units – the familiar 'epoxy resins'. They were of course furnished with terminal epoxy groups for curing. The high aromatic ring content of such materials ensured a sufficiently high T_g for them to form stable powders.

7.3.2.2 Polyesters. Polyesters were developed next and have remained the staple of the powder coatings industry. They could be predominantly acid or hydroxyl tipped but acid groups were preferred for most curing chemistries.

The requirement for binders that were stable as powders at up to 40 °C, yet became highly fluid at temperatures as low as 120 °C, indicated a high T_g and a low molecular weight with little or no branching.

The T_g of 60 °C or higher was achieved by using predominantly aromatic acids in combination with low molecular weight diols. Terephthalic and isophthalic acids have usually been preferred though both are rather intractable materials and require severe processing. They can also lend a degree of crystallinity to their polyesters that itself increases T_g and is helpful to powder stability. Isophthalic acid cannot exist in resonance forms with as long conjugation as terephthalic, so absorbs less strongly in the solar ultraviolet spectrum. Consequently, its polyesters are more weatherable than those of terephthalic acid. On the other hand terephthalic acid produces films with better mechanical properties, and the two isomers are often used in combination. Low levels of adipic acid are used to adjust the T_g and achieve the necessary high fluidity.

When functionality higher than about two is required, it may be more desirable to develop bifunctional tips with trimellitic anhydride than risk the increase in melt viscosity that the branching due to a multifunctional monomer would produce. A hydroxyl tipped polymer is first prepared, then reacted with the anhydride ring in trimellitic anhydride, under relatively mild conditions.

Considerable interest has attached to the use of cycloaliphatic diacids and diols (chiefly cyclohexane-1,4-dicarboxylic acid and cyclohexane-1,4-dimethanol) to prepare powder coatings of high weatherability. They are able to deliver high T_g polyesters because the increased rigidity of the cyclic diol compensates for the reduced stiffness of the non-aromatic diacid. Unfortunately, the rewards are not as great as the absence of conjugated chromophores would lead one to hope. The easy abstraction of hydrogen radicals from their tertiary carbons introduces a new susceptibility to degradation processes, and they are also rather less resistant to hydrolysis than the aromatic polymers.

7.3.2.3 Acrylics. Acrylic polymer technology satisfies some important *niche* markets in the powder coatings industry, for example as clear coats in parts of the automotive industry, but it has not rivalled polyesters in most applications. One important factor is that acrylic and polyester powder coatings find it difficult to co-exist in the same manufacturing or application plant. Very low levels of cross contamination by airborne particles can lead to serious defects in the stoved films because of the incompatibility of the two polymer systems. A change to a different binder system is therefore not undertaken lightly. On the other hand, many technologies involving blends or grafts of acrylic and polyester polymers have been perfectly satisfactory.

As with the high solids liquid technologies, conventional acrylic polymer technology has some inherent drawbacks for powder applications. Principal amongst them are the difficulty of achieving low molecular weights, and retaining sufficient functionality in a high proportion of the molecules when molecular weight is lowered. The incorporation of functional chain transfer agents to supplement the pendent functional groups has been applied to powder coatings with some success. More recently, the expanding range of controlled polymerisation techniques is beginning to be brought into play, as noted above, and the prospects of highly weatherable powder coatings based on acrylic technology are very encouraging.

A severely practical problem is that low molecular weight acrylic polymers for coatings are usually prepared in organic solvents that would need to be removed completely for powder application. Though straightforward, such distillation processes are time consuming and costly, and subject the polymers to thermal abuse that they would otherwise not encounter. Attempts to polymerise acrylic monomers in bulk usually lead to excessive molecular weight or gelation due to the Trommsdorff effect.

There have been several attempts to prepare acrylics in reactive solvents such as fumarates, and in supercritical carbon dioxide which can be removed easily. However, the most favoured solution is to carry out the process in aqueous suspension. It is very simple and may be considered as a water cooled bulk polymerisation of monomer which is dispersed in the form of small droplets in water. The polymer is formed as spherical particles having diameters ranging from about twenty microns to a few millimetres. It is difficult, but not impossible, to obtain particles in the 1.0–10.0 μm range desirable for powders due to the concomitant formation of sub-micron sized particles which give rise to latexes.

In a typical suspension polymerisation, a water insoluble, or very sparingly soluble monomer (or monomers) is mixed with a monomer-soluble free radical forming initiator, the solution being dispersed in water that contains a suspension stabiliser. The agitation is adjusted till the desired droplet size is attained, and polymerisation initiated. There are though many other variables that influence the particle size of the final polymer. They include the type of dispersant used, the viscosity of the system and the interfacial tension between the disperse and continuous phases. The design of the reactor and the agitator can also play a major role.

Many suspension stabilisers have been suggested for use in this polymerisation technique and they may be either organic or inorganic in nature. Inorganic compounds such as talc, colloidal silica, bentonites and freshly precipitated calcium phosphate have all been used with success. Organic polymers that have frequently been reported include poly(vinyl alcohol), poly(vinyl pyrrolidone), poly(acrylic acid) and carboxymethyl cellulose. The efficiency of poly(vinyl alcohol) can depend on its source, the blockier structures being superior in stabilisation.

An important parameter in this method is the 'particle identification point' which is defined as the percent of monomer conversion after which the polymer particles retain their identities for the remainder of the polymerisation. An indication of this point may be obtained even in an unsuccessful first run, and changes in initiator or temperature made to adjust the polymerisation rate. Suspension polymerisation has also been used to prepare step-growth polymers.

7.3.2.4   Other Binders.   Several other polymer types have been adapted to reactive powder coatings. Most notable are polyurethanes, or molecules containing some urethane character, the main backbone being usually acrylic or polyester, but with a surfeit of OH groups rather than acid.

7.3.3   Crosslinking Chemistry and Crosslinkers

As we have seen, the reactive binders available for powder coating usually present carboxyl, hydroxyl or epoxy groups for crosslinking. The crosslinking reaction is required to take place over a fairly short stoving schedule at temperatures in the range of 160–200 °C. Typical would be 180 °C for 10 min or 160 °C for 30 min. At the same time, the reaction must not occur in prolonged storage at temperatures as high as 40 °C. The fact that the binder is necessarily below its vitrification temperature in storage helps to suppress reaction. It can also be useful to contrive to disperse the curing agent as crystalline particles that melt or dissolve below the curing temperature.

7.3.3.1   Crosslinking of and by Epoxy Resins.   One of the earliest crosslinkers for powder coatings was dicyandiamide (**33**).

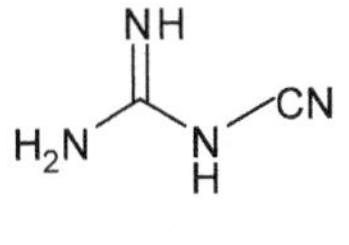

(**33**)

The mechanism of epoxy–dicyandiamide cure is complex and still not fully understood, but all four active-hydrogen sites can react with epoxy though the process is accompanied by the elimination of ammonia. The amidine group can catalyse epoxy homopolymerisation, and it is also understood that the activated nitrile residue will react with hydroxyls under certain conditions to produce urea linkages (Scheme 7.19).

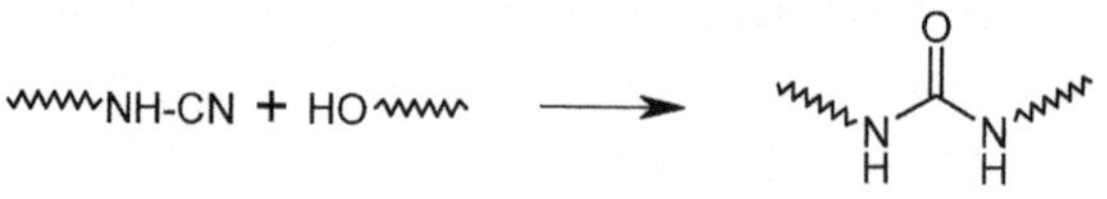

Scheme 7.19

Dicyandiamide systems required high stoving temperatures, in the range of 200 °C or higher, because of its high melting point and insolubility in epoxies. Adducts of dicyandiamide with monofunctional epoxy, and alternatives such as (**34**) have been offered as giving lower stoving temperatures, and are used in epoxy 'rebar' coatings, in view of their outstanding corrosion protection.

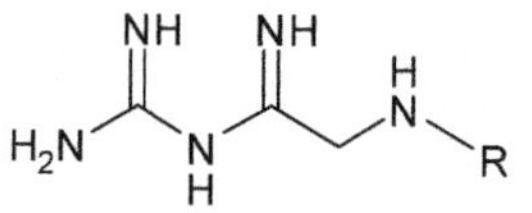

(**34**)

Other amines have been used from time to time, including aromatic amines such as (**35**) but their toxicity has prevented widespread acceptance. Dihydrazides of difunctional acids such as adipic have enjoyed a certain popularity in view of their suitable melting points.

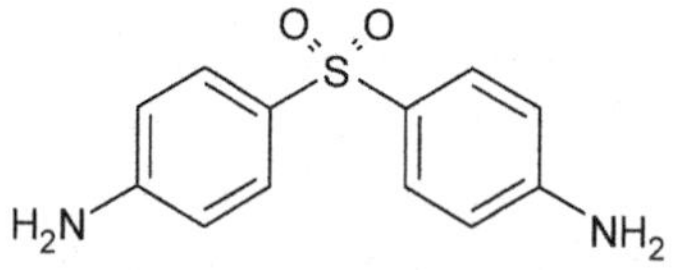

(**35**)

One of the most popular epoxy systems uses a phenolic polymer to cure the epoxy film former. Novolacs are most frequently used (*cf.* Chapter 6). They react at 150 °C and form coatings of outstanding chemical resistance.

Polymeric acids have been widely used to cure epoxies, leading to the 'hybrid' systems that contained comparable amounts of acid functional polyester and epoxy resin. Similarly, acrylic copolymers of glycidyl methacrylate have been cured with polyesters or simple crystalline diacids such as decanedioic or dodecanedioic.

Mono and difunctional cyclic anhydrides have also been exploited and the anhydride elements undergo a chain-growth reaction with the epoxides as described in Chapter 6. Typical anhydrides are phthalic (**36**), trimellitic (**37**), and related molecules (**38–40**).

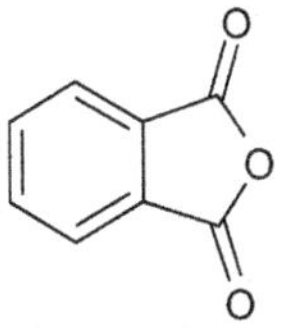

(36) 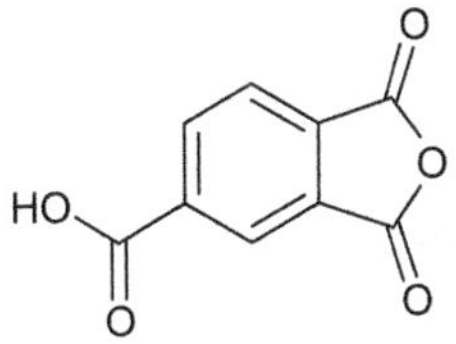   (37) 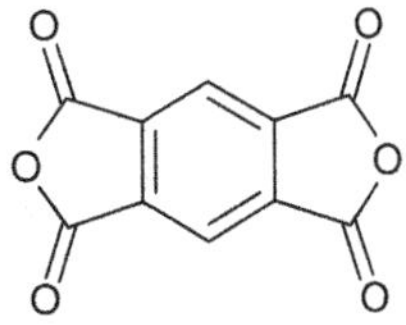   (38)

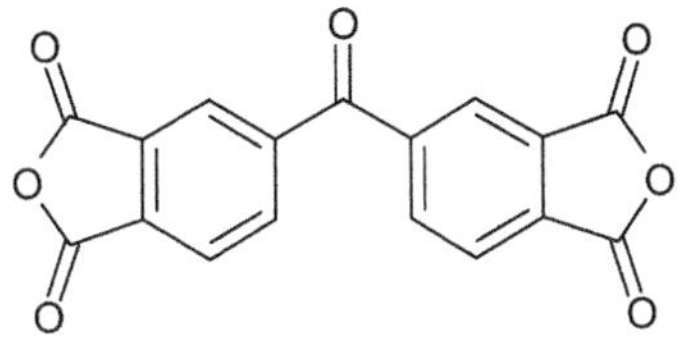

(39) 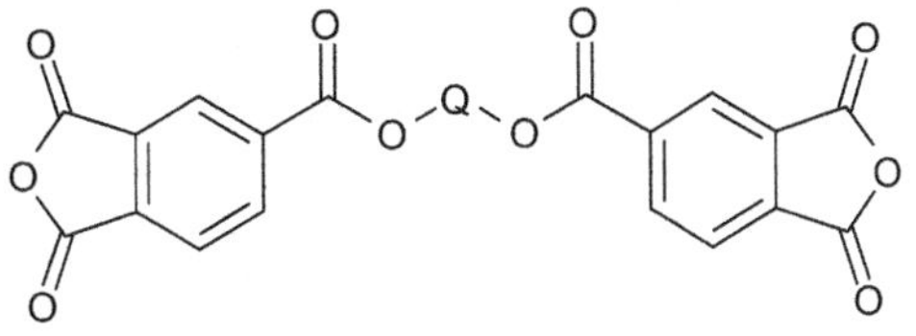   (40)

7.3.3.2 Crosslinking Acid functional Binders. The curing agent that came to set the standard for polyester powders was triglycidylisocyanurate (**41**). TGIC is highly resistant to weathering, unlike some of the other epoxy systems mentioned. It is also a crystalline solid and trifunctional, so can be used to cure the essentially difunctional linear polyesters.

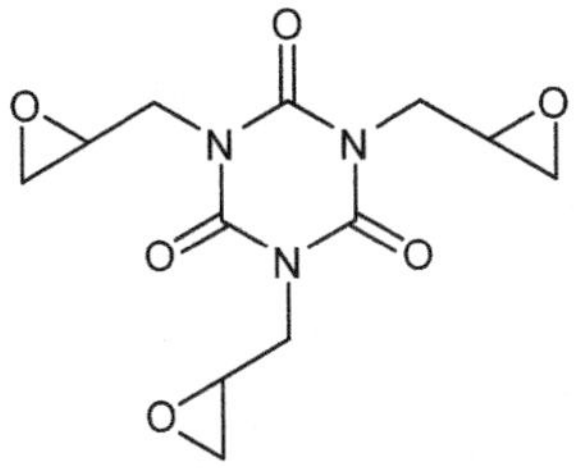

(41)

Like the other epoxies, its reaction with acid groups is catalysed by tertiary amines and phosphines, triphenylphosphine being particularly preferred. The use of TGIC has been inhibited in recent years by reports that the crosslinker may act as a human mutagen. The extent of the risk to end users who only encounter the material when it is encapsulated in polymer is controversial, nevertheless the pursuit of alternative cure chemistry has been given a considerable impetus.

Oxazolines react with organic acids without catalysis under typical powder stoving conditions to form ester–amide linkages, the so-called Fry reaction (Scheme 7.20). Difunctional materials such as (**42**) are crystalline solids, and have been advanced as powder crosslinkers for many years.

Their penetration of the market has been rather limited; however, perhaps because of their tendency to yellow.

Scheme 7.20

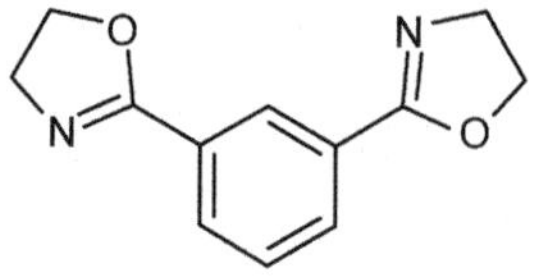

(**42**)

Chemistries involving residues due to the amide of diethanolamine have become widely used. The preferred product is the bisamide of adipic acid (**43**), a crystalline solid, although a wide range of low molecular weight and polymeric variants has been envisaged.

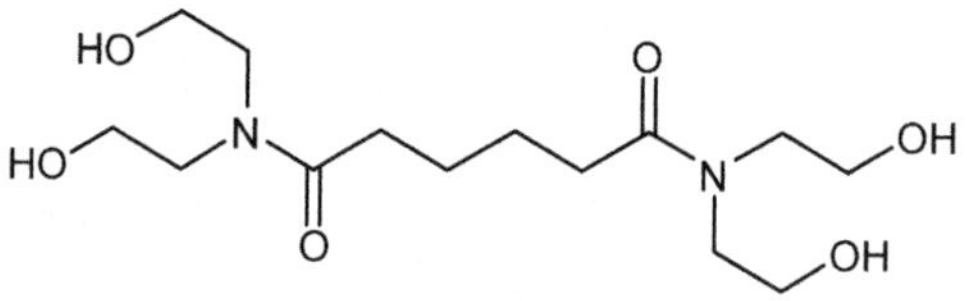

(**43**)

The enthanolamide hydroxyls are activated to condensation with organic acid groups, reflecting a tendency of the hydroxyamide to interchange with aminoester *via* an oxazoline-like intermediate. Interestingly β-hydroxy-amides prove to be more reactive with aromatic than aliphatic acids, the reverse of the behaviour of ordinary alcohols. They are also indifferent to the usual esterification catalysts.

Ethanolamide-acid condensation of course releases water, so is not regarded as an ideal powder curing chemistry. Indeed, pinhole formulation due to escaping water can be a problem in thick films. Nevertheless, the low toxicity, low cost and general robustness of the system are bringing it increasingly into favour.

7.3.3.3 Crosslinking Hydroxyl Functional Binders. There is substantial technology dealing with curing agents for hydroxyl functional binders (usually polyesters). The multifunctional anhydrides mentioned in the epoxy binder section have been historically important, but of course react with hydroxyl groups in a step-growth fashion to produce half esters.

Azlactones react with hydroxyl groups in a way analogous to the acid–oxazoline (*cf.* Scheme 7.21), and bisazlactones have been claimed for low temperature cures.

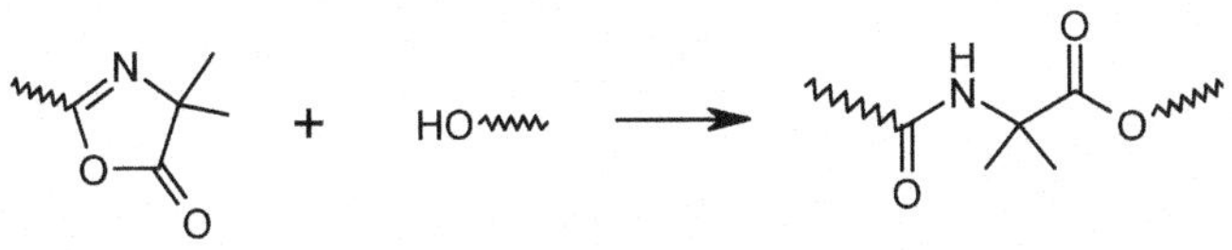

Scheme 7.21

Unfortunately, azlactones tend to be rather costly.

Successful systems have also used solid amino-formaldehyde cross-linkers, despite their unwelcome tendency to emit organic volatiles. Glycoluril systems, in particular, have been appreciated for the reasons given in Chapter 6.

However, the most important curing agents for hydroxyl functional powder binders are blocked isocyanates. Because of the need for solid materials, and because of its good weatherability, isophoronediisocyanate, its adducts with various polyols, and particularly its isocyanurate trimers have featured in most 'urethane' powders. Trimers of hexamethylene-diisocyanate have also been used. Caprolactam has been a favourite blocking agent, because of its reactivity and propensity to form solid adducts (**44**).

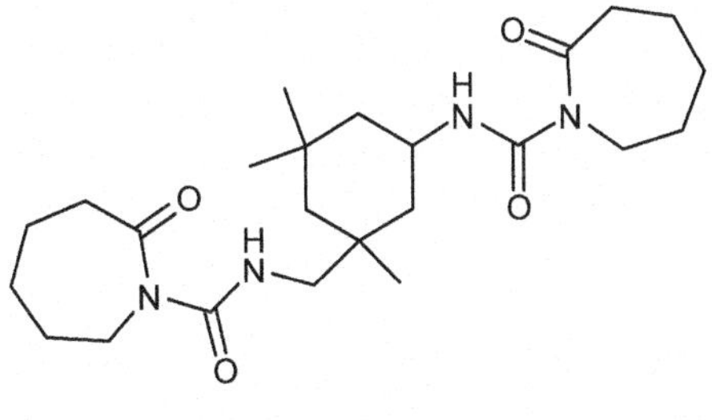

(**44**)

It tends not to be completely eliminated from the film, somewhat mitigating the problem of condensation within the stoving oven. (The possibility that some of it could polymerise to Nylon 6 has been suggested, but not substantiated. Certain amides also tend to accelerate the curing reaction.)

A still more satisfactory solution is to prepare uretidinediones from diisocyanates. A low molecular weight polymer (**45**) is developed and compounded with the hydroxyl functional binder.

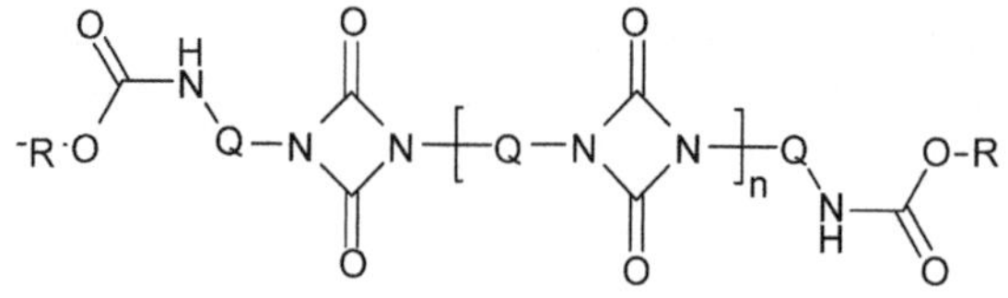

(**45**)

On stoving it either dissociates into the diisocyanate precursors that are available for crosslinking, or it first undergoes nucleophilic attack (*cf.* Chapter 6). The terminal isocyanate groups can be pre-reacted with hydroxyl compounds if necessary. A related chemistry contains uretoneimine groups (**46**) that can dissociate into isocyanates and carbodiimides. Oxadiazacyclohexatriones (**47**) have been prepared by reacting diisocyanates with carbon dioxide and proposed as powder curing agents. The evolution of CO_2 is probably not a helpful factor.

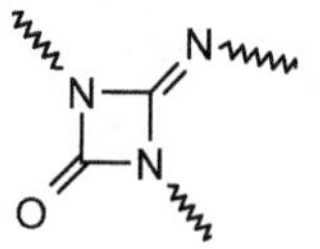

(**46**) 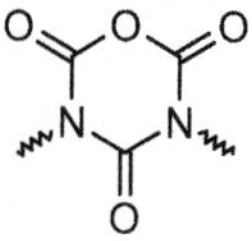   (**47**)

Again isocyanates of reduced reactivity such as *m*-xylylenediisocyanate have been used in formulations without any blocking chemistry.

In any event, the usual catalysts such as amines and tin salts are used to promote cure, and the mechanical properties of the final material benefit from the 'secondary crosslinking' due to hydrogen bonds.

7.3.3.4 UV Cure for Powders.

Radiation curing offers the possibility of decoupling curing from melt and flow in powder coatings with considerable advantage to the applicator. He can design a stoving schedule to achieve the optimum film formation, then cure the coating at will.

Typical technologies involve the attachment of acrylate residues to polymer spacer groups with the usual high T_g backbones. Epoxies, urethanes and polyesters are favourite candidates as in the liquid systems. They can readily be made to react with, respectively, acrylic acid, hydroxyethyl (meth)acrylate or glycidyl methacrylate.

It is even known to include crystalline monomers such as diacetoneacrylamide (**48**) as reactive diluents.

(48)

Alternatively, a vinyl ether tipped urethane can be made to copolymerise with an unsaturated polyester containing fumarate residues.

The usual problems associated with UV curing, such as oxygen inhibition and the difficulty of curing thick or pigmented films, apply equally to powder coatings. The networks are highly heterogeneous, and there is also a risk that premature reaction might take place as the coatings are being fused. Perhaps for those reasons, radiation cure has not penetrated the powder coatings market as rapidly as might have been expected. Nevertheless, interest is increasing as the possibilities of powder coating wood products are investigated.

7.4 NON-AQUEOUS DISPERSIONS

Conventional emulsion polymerisation of hydrophobic monomers in non-aqueous media was first suggested in a patent of 1937, and was further studied in the late 1940s. However, little success was achieved and the dispersion of polymers in non-aqueous media was confined to processes in which a monomer was dissolved in an organic solvent that was a non-solvent for the polymer. In the presence of certain polymeric additives the precipitated polymer could be obtained as a colloidal dispersion in the solvent. A great deal of work was carried out, by companies such as ICI and PPG, to develop the technique, and coatings applications evolved. Non-aqueous dispersions were also found to have an important use in the production of waterborne coatings (*cf.* Chapter 8).

The continuous phase employed was frequently an aliphatic hydrocarbon, so that ionic stabilisation by electric charges on the polymer particles was not available to prevent flocculation. Instead, steric stabilisation was used to achieve stability. It involved the use of block or graft copolymers, one part of which adhered strongly to the polymer particles whilst the other part extended into the continuous phase. The arrangement prevented particles from approaching each other closely enough for van der Waals attractive forces to promote coalescence. Steric stabilisation is considered further in Chapter 8, since it is equally applicable to the use of non-ionic and polymeric surfactants in waterborne binders.

Although the use of non-aqueous dispersions in coatings has diminished due to the failure of standard types to comply with VOC regulations, they may yet have value in high solids systems. It is possible to achieve polymer contents above 80% by weight in bimodal distributions of spherical particles when the ratio of the diameters is greater than 12:1 and 70% of the total volume is occupied by the larger particles. It has been claimed that packing fractions up to 95% of the theoretical density can be achieved with a tetramodal distribution of spherical particles (British Patent 1,157,630). It should also be noted that the aliphatic hydrocarbons used for non-aqueous dispersions can be selected so as to be toxicologically benign by comparison with most paint solvents.

Dispersion polymerisation as described here is very useful in the preparation of polymer particles in the size range 1.0–15.0 μm, sizes that are not readily achieved by suspension or emulsion polymerisation techniques.

To a very large extent, non-aqueous polymerisation has been practised using the free radical mechanism, but it should be noted that condensation polymerisation is also possible although the range of reactions is limited. Anionic polymerisation has also been conducted using block polymer stabilisers.

7.4.1 Dispersion Polymerisation

Since the early 1990s there has been a considerable body of work in the use of supercritical carbon dioxide as a solvent or as the continuous phase for the polymerisation of vinylic and acrylic monomers. The results have led to efforts in the coatings industry to utilise these methods for the production of low to zero VOC coatings. However, progress has been slow, probably due to the costs involved in handling supercritical carbon dioxide. In the context of dispersion polymers we should mention that the principle of stabilisation is that relevant to non-aqueous dispersion. For example, a block polymer of styrene and 1,1-dihydroperfluorooctyl acrylate acts as an efficient stabiliser for the dispersion polymerisation of styrene, the styrene component acting as the anchor whilst the fluorinated polymer extends into the continuous carbon dioxide to give a well solvated stabilising chain.

The polymeric stabiliser that is an essential component of dispersion polymerisation recipes may be a block or graft co-polymer in which one part has a strong affinity for the polymer colloid whilst the other has a strong affinity for the continuous phase. Such polymers are said to be amphipathic and dispersions containing them are said to be sterically, or entropically, stabilised. The term entropic more accurately describes the mechanism, which will be outlined later. A particularly recommended graft structure is

(**49**) one in which the grafts (m) are equidistant along the polymer backbone (M) (comb graft polymers).

```
wwww MMMMMMMMMMMMMMMMMMMM wwww
       m         m         m         m
       m         m         m         m
       m         m         m         m
       m         m         m         m
       m         m         m         m
       ≷         ≷         ≷         ≷
```

(**49**)

In these structures we may look upon the M component as the part that associates with the polymer particle, whilst the m component extends into the continuous phase and is well solvated by it. The M associated with the polymer particle is known as the anchor component, and may be connected to the polymer by physical attraction, covalent bonding or by incorporation into the particle. In a qualitative way we may say that stabilisation results from the repulsive forces due to the loss of configurational entropy when the particles approach one another, and the solvated chains of the stabiliser begin to compress or overlap. This loss of entropy is sufficient to overcome van der Waals–London forces of attraction so that the colloidal polymer is stabilised against flocculation, the steric barrier formed by the well extended and solvated part of the stabiliser preventing close contact.

7.4.2 Plastisols

A particular type of dispersion that has found wide use in coating applications consists of poly(vinyl chloride) and/or co-polymers of vinyl chloride along with, *e.g.* diethyl malonate for flexibility and compatible plasticisers, *e.g.* tributyl phosphate. Such dispersions are known as plastisols, a term which should be reserved for resin–plasticiser systems. In practice these dispersions are too viscous for coating, and adjustment is made by addition of volatile solvent. The thinned product is known as an organosol.

PVC plastisols have been used for over 50 years, the name plastisol having been first used in 1946. The partial crystallinity of poly(vinyl chloride) allows the polymer, which is in contact with a plasticiser, to be diluted with solvent without excessive swelling of the dispersed polymer. On stoving, the PVC particles melt and dissolve in the plasticiser to afford the plasticised coating material, used for example as seals in screw-topped food jars.

Some attempts to extend the technique to acrylic polymers have been made but technical difficulties have impeded progress. Small sized acrylic particles are required that can be modified or placed in an environment where swelling on storage could be eliminated or very much reduced.

An interesting development that is somewhat akin to plastisol technology was pioneered by Asahi Glass Co. using poly(vinylidene fluoride). Fluorine-containing polymers are well known for their excellent weather resistance but the high temperatures required for acceptable application have limited the use of many thermoplastic fluoropolymers. By using acrylic resin solutions along with poly(vinylidene fluoride), melt flow properties are improved and the blended system has found use in coil coatings. Thin, pinhole-free films are achieved whilst the poor adhesion of the fluorine-containing polymer is overcome.

Poly(vinyl fluoride) also has a crystalline structure that does not swell to any great extent when solvent is added, and smooth flowing compositions are formed at elevated temperatures, resulting in the formation of defect-free coatings. As is the case for poly(vinylidene fluoride), the solvents used should be looked upon as latent since they are only effective when the resin–solvent mixture is hot.

The advantage of the plastisol approach is that it overcomes the viscosity constraints regarding polymer concentration that are inherent in solutions. In the present time when environmental considerations seek to reduce the use of organic solvents the applications of plastisols and organosols has diminished, although there are still applications for them in specialist applications. However, if some means could be found whereby resins of different types could be formulated into very high ($>85\%$) solids the plastisol technique might still be utilised. Unfortunately, the effects due to the crystallinity of poly(vinyl chloride) and the fluorinated polymers does not appear to be duplicated by conventional crosslinking of non-crystalline polymers.

7.5 CONCLUDING REMARKS

We have seen that the potential of organic solvents to cause environmental damage, as outlined in Chapter 2, has been addressed by the Coatings Industry in many innovative ways. The physics of polymers, and the chemistry of crosslinking have been exploited to create 'high solids' systems to the point where some successful liquid coatings contain no solvent at all. Meanwhile, powder coatings routinely confer outstanding finishes on stoved items, and are competing strongly with liquid paints in the industrial arena.

7.6 BIBLIOGRAPHY

1. W.J. Blank, The Slow and Winding Road to 'Zero' VOC, *Proceedings of 28th International Waterborne, High Solids, and Powder Symposium*, February 21–23, 2001, pp. 1–16.

2. C. Decker, The use of UV irradiation in polymerisation, *Polym. Int.*, 1998, **45**, 133–141.

3. J.P. Fouanier, *Photo-initiation, Photo-polymerisation, Photocuring*, Hanser, Munich, 1995.

4. J. Hubrechts and K. Dusek, Star oligomers for low VOC polyurethane coatings, *Surf. Coat. Int.*, 1998, (3), 117–127; 1998, (4), 172–239.

5. K. Matyjaszewski and T.P. Denis (eds), *Handbook of Radical Polymerisation*, Wiley, New York, 2002.

6. T.A. Misev, *Chemistry and Technology of Powder Coatings*, Wiley, Chichester, 1991.

7. D. H. Napper, *Polymer Stabilisation of Colloidal Dispersions*, Academic Press, London, 1983.

8. R. Parvani and M.C. Shukla, Recent developments in resin systems for high solids, coatings, *Paintindia*, 1991, **41**(5), 21–29.

9. S. Paul, *Surface Coatings: Science and Technology*, Wiley Interscience Publishers, New York, 1995.

10. Z.W. Wicks Jr. and L.W. Hill, Design considerations for high solids coatings, *Prog. Org. Coat.*, 1982, **10**, 55.

11. Z.W. Wicks Jr., G.F. Jacobs, I-Chyang Liu, E.H. Urruti, L.G. Fitzgerald, Viscosity of oligomer solutions, *J. Coat. Technol.*, 1985, **57**, 725.

CHAPTER 8

Binders for Water-Borne Coatings

W.A.E. DUNK

8.1 INTRODUCTION

A fully formulated paint is a complex system which comprises pigments and a variety of additives that vary with the final demands of the coating. These components are dispersed in a fluid continuous phase that may be aqueous or non-aqueous, and which contains a polymer either as a solute or as a colloidal dispersion. This continuous phase is known as the binder and largely determines the character of the coating. It should, however, be emphasised that although solvent-based coatings are still used, the increasing attention to environmental preservation by health and safety agencies has led to an important growth in the development and adoption of water-based resins (*cf.* Chapter 2). As a consequence of the shift in emphasis, this chapter will give preferential attention to those techniques that are solvent free, or confine their use to a minimum. Thus non-aqueous dispersions are considered briefly in view of theoretical background concerning colloid stabilisation that can be applied to aqueous environments. More emphasis will be placed on emulsion polymerisation and its ability to exercise morphological and particle shape control of film-forming polymers.

8.2 COLLOIDAL AND NON-COLLOIDAL SYSTEMS

A colloid may be defined as any material that contains a physical boundary dimension in the range of 1–1000 nm. When the upper bound is exceeded, a suspension is formed that quickly settles under gravity: a colloidal dispersion is not so affected. When considering colloidal dispersions, it is important to differentiate between those in which the disperse phase is solvent compatible (lyophilic) and those in which it is incompatible with the solvent (lyophobic). When the continuous phase is water, the terms hydrophilic and hydrophobic are used. In the case of a solvent or water-soluble polymer, we have a lyophilic colloidal dispersion, whereas when the polymer exists as finely divided solid dispersed in a solvent or water, we have a lyophobic colloid dispersion.

In the coatings industry there has been a degree of ambiguity and obscurity in the definition of colloidal dispersions, and this is most commonly seen in the use of the terms emulsion, latex, dispersion, and suspension which have sometimes been used synonymously. Table 8.1 defines these terms as they are understood here. It should be noted that latexes as defined in the table are all dispersions, but all dispersions are not latexes. Similarly, the definition of an emulsion is sufficient to preclude its use in describing a latex as an emulsion polymer, a common misuse frequently seen in trade literature. Presumably this comes about because the process by which the binder is made commences with the emulsification of liquid monomers that are then polymerised. The misnomer has very recently drawn comment in the *Handbook of Radical Polymerisation* (2002, p. 302). On another point of nomenclature the terms 'nano-particle' and 'nano-latex' have appeared in the literature, but they seem to be linked to the recent boom in 'nano-technology'. However, in the present context the terms will be confined to particles and latexes containing polymer particles that have a diameter of <100 nm. Thus, a nano-latex would be one that had a mean particle size <100 nm.

In this chapter the subject of emulsion polymerisation will take priority, not only because it has great importance in the coatings industry and its environmental obligations, but also for its overall importance in the industrial synthesis of polymers (Fitch, 1997). Water reducible systems, plastisols and micro-gels along with their applications will also be introduced. Non-aqueous dispersions, although scientifically interesting, will receive brief attention since their utilisation for coatings has diminished as VOC legislation has become more severe. Suspension polymerisation for

Table 8.1 *Definitions of terms encountered in disperse phase polymer science*

Dispersion	A distribution of finely divided solid particles in a liquid phase to give a system of very high solid–liquid interfacial area
Emulsion	A heterogeneous system consisting of at least one immiscible liquid dispersed in the form of droplets in another liquid
Latex	A colloidal dispersion of a solid polymer in a liquid continuous phase. Such a system should be termed a latex only when it has been prepared by emulsion polymerisation
Suspension	A non-colloidal dispersion of a solid in a liquid. In such systems the solid–liquid interfacial area is low and size is >1 μm

hydrophobic monomers is a water-based technique but it is introduced in Chapter 7 since there may be an application in the synthesis of polymers for powder coatings.

Before embarking on a descriptive tour of the various techniques, it is important to draw attention to the advantage of polymer dispersions over polymer solutions in terms of viscosity–molecular weight relationships. In Chapter 4 it was shown that film strength increases as molecular weight increases; thus, high molecular weights are generally desirable. However, polymers having such high molecular weights would need to be present at uneconomically low concentrations in solution for acceptable application viscosities to be attained. This is reflected to some extent in water reducible systems where the pseudo-dispersion nature of these types allows polymers having molecular weights in the range 20,000–50,000 to be used at acceptable concentration levels. At the other extreme, latexes may have their disperse phase polymer with a molecular weight of >1,000,000 yet exhibit viscosities that are applicable at economically viable concentrations. Viscosity in a polymer dispersion is independent of molecular weight, the factors affecting it being the polymer particle shape, the packing factor and the volume fraction of the dispersed polymer. The relationships may be expressed mathematically by the Mooney equation (Equation (8.1)):

$$\ln \eta_{rel} = \frac{k\phi}{(1-s)\phi} \tag{8.1}$$

where η_{rel} is the relative viscosity, ϕ is the volume fraction of disperse phase, s is the self-crowding (packing) factor and k is the Einstein coefficient, 2.5.

Other colloidal dispersions are somewhere in between the water reducibles and the latexes with respect to viscosity–molecular weight dependence.

8.3 EMULSION POLYMERISATION

The importance of the technique of emulsion polymerisation in industry cannot be over-emphasised. It has been claimed that over 50% of all polymers synthesised by free radical polymerisation are produced in this way (Fitch, 1997). Fifteen years earlier George Ham had commented that there were more published studies in the field of emulsion polymerisation than in any other area of polymer chemistry, and this is probably as true today as it was in 1982. In the coatings industry where there is an ever-increasing demand for reducing solvent emissions, the need for water-based polymeric binders leans heavily on emulsion polymerisation technology.

Moreover, the ability to prepare polymers having well-defined morphologies and particle shape is a further advantage for enhancing film performance and paint rheology.

Perhaps the greatest advantage that emulsion polymerisation brings to industry is the unique mechanism which allows the rate of polymerisation to be increased without loss of molecular weight. This important factor is not available in bulk, solution or suspension systems. In this chapter we shall review at an introductory level some techniques for the preparation of polymer dispersions, but in view of its importance emphasis will be placed on oil-in-water emulsion polymerisation. In passing it should be noted that water-in-oil emulsions of aqueous solutions of monomers may be polymerised to give latexes. However, these have little application in coatings, although the method has been used to prepare hydrophobically modified water-soluble polymers that find use as associative thickeners in water-borne coatings (*cf.* Chapter 4).

Commencing with a brief review of the evolution of the procedure, we shall then consider some of the attempts that have been made to formulate qualitative and quantitative theories which elucidate the experimental observations. The practice of emulsion polymerisation will then be considered in a general way, and some of the variations to give desired properties to the polymers, will be introduced. Latex properties and application to coatings will precede a description of a procedure which, although not a polymerisation reaction, has had some importance to the coatings industry. This concerns the preparation of pseudo-latexes for polymers that cannot be synthesised by emulsion polymerisation.

At this point it is pertinent to recall a comment of the late John Vanderhoff regarding emulsion polymerisation: 'Although its practice can be based on the kinetic principles of free radical initiated addition polymerisation, the complex interactions of monomers, initiators, surfactants and other ingredients leave the quality of the final latex very much in the hands of the operator'.

8.3.1 Historical Background

The first approach to the preparation of a latex may be traced back to the work of Hoffmann and co-workers at Bayer in the first decade of the 20th century. The objective was to imitate the physiological conditions under which rubber is formed in the tree, then to prepare synthetic rubber under these conditions from isoprene or butadiene. These monomers were mixed with water in the presence of albumin or gelatine that acted as protective colloids, then heated for days and even weeks. Polymerisation occurred, probably through peroxides formed randomly by reaction of atmospheric oxygen across the double bonds in the monomers. These results were published in patents but there seems to have been little further published work until after World War I.

Somewhat unsystematic studies were made, chiefly in Germany, during the 1920s. It was observed that the addition of soaps such as oleates, alkylarylsulfonates and abietates to mixtures such as had been used by Hoffmann, gave stable emulsions that led to stable latexes after polymerisation. At about the same time, Luther showed that by introducing certain oxidising compounds, *e.g.* persulfates, perborates, hydrogen peroxide to the mixture, the polymerisation time could be reduced from days to hours. However, these techniques were essentially what is now known as suspension polymerisation, and it was not until 1929 that the first real emulsion polymerisation based on a rational recipe was disclosed (R.P. Dinsmore). This involved the synthesis of a rubber from dimethylbutadiene. Subsequently, the development of emulsion polymerisation was rapid, urged on by the need for synthetic rubber, although the first technical application of a latex appears to have been an acrylic for leather finishing (I.G. Farben, 1931). Numerous other latexes were produced over the next two decades, finding uses in adhesives, coatings and synthetic leather to name but a few. Table 8.2 gives a few examples of monomers that were successfully polymerised to latexes during this period, whilst the interested reader is directed to a review of this era's technology and some of the bizarre recipes used (G. Whitley, *Ind. Eng. Chem.*, 1933, **25**, 1204 & 1388).

The complexity of a typical emulsion polymerisation recipe has been alluded to earlier in this chapter, and the formulation of an industrial process using this technique can be a formidable task. In addition to the continuous water phase, monomer(s), surfactant and initiator, there may be a buffer and a chain transfer agent. In some recipes there can be additional surfactants and initiators, thus the understanding of such systems is far from complete,

Table 8.2   *Some polymers prepared by emulsion polymerisation*

Monomer	Year	Patent reference
Acrylic ether	1930	Ger 654989
Vinyl ether	1930	Ger 634408
Vinyl chloride	1932	US 2068424
Vinyl esters	1934	Ger 727955
Ethylene	1938	Ger 737960
Vinylidene chloride	1951	*Makromol. Chem.*, 1951, **6**, 39

and the comment of Vanderhoff on the role of the operator becomes even more apt. We shall now proceed to a closer look at the theoretical and practical aspects of the subject.

8.3.2   Qualitative Theories

The growth of emulsion polymerisation through the 1930s was rapid, but although an extensive patent literature accumulated there were very few scientific studies published. An investigation of the polymerisation of butadiene under what amounted to emulsion polymerisation conditions was published in Russia in 1936, followed a couple of years later by an important paper on the locus of polymerisation by Fikentscher. Only after 1939 several researchers independently published qualitative pictures of the process, culminating with the 'classical' summary of Harkins in 1945.

The particular concern of workers in this field was with the locus of particle nucleation and the growth of polymer particles. Staudinger in 1935 had proposed that the locus of nucleation was at the surface of the monomer droplets, but this was subsequently shown to be erroneous. In 1938 Fikentscher suggested that polymerisation commenced in the aqueous phase, monomer dissolved in the water being initiated by the free radicals generated from the water-soluble initiator. The monomer concentration would be maintained by dissolution of more monomer from emulsified monomer droplets. In the presence of a micelle generator, monomer would be solubilised in micelles, and these monomer-swollen micelles were even then believed to be the principle locus of polymerisation.

The proposed picture of Harkins was of a compartmentalised system of monomer(s), water, emulsifier and initiator.

The monomer has only a low water solubility and the free radical generator is water soluble. Figure 8.1 depicts this situation, and it should be noted that the monomer(s) may be in three domains:

- dissolved in the continuous water phase
- contained in surfactant micelles
- as emulsified monomer droplets

Similarly, the surfactant may be found in each of these domains, but once initiation of polymerisation has commenced, it will also be found at the surface of monomer-swollen polymer particles.

Conceptually similar to polymerisation in micelles is the Medvedev mechanism of initiation in the adsorbed emulsifier layer. This may be true for some systems but, as Napper has pointed out, it is not possible to differentiate the mechanisms by kinetic studies.

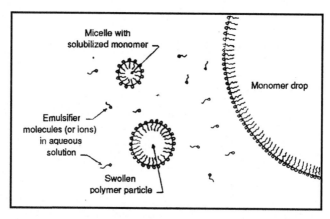

Figure 8.1 *The Harkins picture of emulsion polymerisation (water-soluble initiator and dissolved monomer not shown)*
(From J.W. Vanderhoff, *et al.*, *Adv. Chem.*, 1962, **34**, 32)

Once free radicals have been generated in the aqueous phase, the initiation of the monomer(s) can occur by two distinct mechanisms:

- micellar nucleation where a free radical enters a monomer-swollen micelle and creates a monomer radical
- a homogeneous nucleation where free radicals generate monomer radicals from monomer dissolved in the water

In the second case, known as the Hansen, Ugelstad, Fitch, Tsai theory, the oligomeric polymer radicals may either enter monomer-swollen micelles, or they may propagate to a point where they reach water insolubility, further growing by monomer adsorption to become primary particles stabilised by emulsifier adsorption. The viability of this mechanism is upheld by the fact that monomers having very low water solubility, *e.g.* stearyl methacrylate, *p-t*-butyl styrene and most fluorinated monomers, cannot be polymerised under standard conventional conditions of emulsion polymerisation (see Section 8.5).

When free radicals enter monomer-swollen micelles, the mechanistic picture is readily envisaged in terms of the difference in size of the monomer domains. Micelles consisting of typically *ca.* 100 emulsifier molecules are present in an emulsion polymerisation at some 10^{18} units mL^{-1} and have a size of *ca.* 5 nm when swollen with monomer. By comparison, stabilised monomer droplets are present in far smaller number (*ca.* 10^{10} units mL^{-1}), but are much larger at *ca.* 1000 nm. Not only does the preponderance of monomer-swollen micelles lead to their being the principle target for initiation, but also their much larger surface area enhances the capture of free radicals.

Once polymerisation has commenced in the micelles, monomer is replenished by diffusion through the aqueous phase from the monomer droplets. Micellar swelling continues through polymer growth until the size reaches *ca.* 100 nm diameter, whilst their number reaches *ca.* 10^{14} cm^3. As the monomer droplets are replaced by polymer particles, the average particle diameter of the system is reduced, leading to an increase in total surface area. As more of the emulsifier is adsorbed, a point is reached where the micelles disappear completely and the surface tension of the system becomes constant.

At the point where the micelles disappear, the monomer conversion is between 10 and 20% and the reaction rate becomes constant, *i.e.* zero order with respect to monomer. Since no new latex particles can now be generated, the monomer concentration is kept constant by the monomer droplet reservoir. When all the monomer is in the latex particles, the reaction becomes first order with respect to monomer. The stages described here are conventionally referred to as intervals and only give qualitative data on emulsion polymerisation kinetics. The progress of an emulsion polymerisation in the terms of surface tension, rate of polymerisation and conversion is illustrated in Figure 8.2.

It should be noted that the mechanisms outlined here represent limiting cases. However, there is a third theory of nucleation due to Napper and co-workers at Sydney (1983–1991). This mechanism involves the formation of primary particles by either of the previously described processes, followed by their limited aggregation in a two-step process.

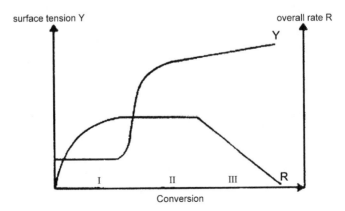

Figure 8.2   *The progress of an emulsion polymerisation in terms of surface tension, overall rate of polymerisation, and conversion*

8.3.3 Quantitative Theories

The complete modelling of emulsion polymerisation is extremely complex and even today efforts are hampered by the numerous factors involved (van Herk *et al.*, 1998). However, to appreciate the problem, it is well to be aware of the pioneering efforts that were made over 50 years ago when the Harkins qualitative picture was presented. The names of Montrol, Corrin and Haward are associated with worthy contributions, but it is to Smith and Ewart that we owe what is regarded as the canonical theory of the process. Although idealised and not explaining all the observed phenomena, it has served as a reliable base for modification and refinement. Here we give an outline of what Smith and Ewart proposed, but for a clear and comprehensive account of the micellar mechanism the treatise of Blackley should be consulted. A rigorous and thorough account of the current theories is to be found in the monograph of Gilbert (see Bibliography).

Smith and Ewart considered the number of particles formed in Interval 1 and the number of free radicals per particle present in Interval 2 (see Figure 8.2). Although monomer conversion in Interval 1 is only 10–20%, most of the latex particles are generated here and this number may significantly influence the course of the subsequent intervals. In the period of zero-order reaction where the overall reaction rate is constant, monomer concentration [M], particle number N and the number of latex particles per centimetre with n active radicals, N_n will be constant. When this stationary state is fulfilled, the three reactions leading to formation of latex particles with n free radicals are:

- entry of a radical in a latex particle with $(n-1)$ radicals
- exit of a radical from a latex particle with $(n+1)$ radicals
- termination between two radicals in a particle with $(n+2)$ radicals

These reactions will be equal to those reactions which cause their disappearance, *i.e.* entry, exit or termination, the overall situation being expressed in the recursion equation (Equation (8.2)) that was derived by Smith and Ewart, and which bears their names:

$$N_{n-1}\left(\frac{C'}{N}\right) + N_{n+1}\left(\frac{K_sS}{N_A}\right)\left(\frac{[n+1]}{V_s}\right) + N_{n+2}\left(\frac{K_t}{N_A}\right)\left(\frac{[n+2][n+1]}{V_s}\right)$$
$$= N_n\left\{\left(\frac{C'}{N}\right) + \left(\frac{K_sS}{N_A}\right)\left[\frac{n}{V_s}\right] + \left(\frac{K_t}{N_A}\right)\left[\frac{n(n-1)}{V_s}\right]\right\} \tag{8.2}$$

where N_A is Avogadro's number, K_s is the radical exit rate constant, S is the surface area of swollen latex particle, K_t is the radical mutual termination rate constant, V_s is the volume of swollen polymer particle and C' is the rate of entry of free radicals into latex particles.

The intractable nature of this equation led its formulators to consider just three limiting cases for which explicit solutions were obtained. A few years later Stockmeyer presented a general analytical solution, modified later by O'Toole.

The three limiting cases defined by Smith and Ewart were:

Case 1 where $n \ll 1$. Here the rate of exit of free radical species is very much greater than their rate of entry. This behaviour is not often encountered, vinyl acetate being the best known example.

Case 2 where $n = 0.5$. Here the termination rate is greater than the free radical entry rate, and the exit of free radical species is negligible. Styrene is the classical example of such behaviour.

Case 3 where $n \gg 1$. In this case the rate of radical entry is greater than the rate of termination. Polymers having crystalline structures at the polymerisation temperature exhibit this behaviour; poly(vinylidene chloride) is a good example. Large polymer particles also give rise to this situation.

The theory also allows the prediction of the effect of certain variables, *e.g.* temperature and initiator concentration, on the reaction rate (R_p) and degree of polymerisation (DP). The following expressions (Equations (8.3) and (8.4)) for these two factors are easily derived:

$$R_p = K_p[M]\bar{n}N \qquad (8.3)$$

$$DP = \frac{K_p[M]\bar{n}N}{R_i} \qquad (8.4)$$

where K_p is the rate constant for propagation, N is the number of particles, n is the average number of radicals per particle, $[M]$ is the monomer concentration and R_i is the rate of radical generation.

Such a kinetic condition is attained when the radicals are segregated and the number of loci available for segregation is within a few orders of magnitude of the number of free radicals present. This situation is in direct contrast to the inverse variation of R_p and DP which is observed in free radical bulk, suspension and solution polymerisations; thus, it is easy to see the attraction to industry of a process that permits the rate of polymerisation and degree of polymerisation to increase simultaneously.

8.4 SOME POST-HARKINS, SMITH–EWART DEVELOPMENTS

With some understanding of the early rationalisation of the emulsion polymerisation technique, we may look briefly at some developments that have subsequently appeared. For nearly 20 years the micellar theory was so entrenched that to question it could lead to rejection of papers by referees. However, attitudes softened and progress was made.

8.4.1 Mini-emulsion Polymerisation

The first non-micellar system to achieve success, and which currently finds application in coatings, was the mini-emulsion technique that was developed between 1972–1980 by Ugelstad and collaborators.

Initially, it had been demonstrated that under certain conditions, monomer droplets could be the main loci for initiation in an emulsion polymerisation. By adding a long chain alkane or alcohol to an emulsion of water, monomer and surfactant, it was possible to form relatively stable droplets in the size range 50–500 nm. In this range the total surface area of the droplets is comparable to that of monomer-swollen micelles and initiation can occur in them. This type of emulsion has been given the name mini-emulsion, hence mini-emulsion polymerisation.

In a well-prepared mini-emulsion the droplets will give latex particles of very nearly the same size, indicating that coalescence of droplets and Ostwald ripening are suppressed. To achieve a narrow droplet size distribution, high intensity sonification is required and the surfactant/co-surfactant pair should have the same long chain component, *e.g.* sodium hexadecyl sulfate is paired with either hexadecane or hexadecyl alcohol.

Mini-emulsion polymerisation is claimed to give higher system stability as compared to the conventional method, whilst the slower reaction rate allows improved control in industrial scale polymerisations. It is also easier to control particle size by varying the concentration of co-surfactant. The procedure is also claimed to improve the preparation of composite polymer particles and inter-penetrating polymer networks, both of which command interest in coatings technology.

8.4.2 Surfactant-free Emulsion Polymerisation

The presence of surfactant in latexes used in coatings can have undesirable effects on performance. As long ago as 1949, similar observations were made in the synthetic rubber industry, and attempts to prepare latexes in the absence of emulsifiers were investigated. Stabilisation was achieved by using a persulfate initiator that provided ionic stabilising groups in the polymer; the addition of monosodium phosphate as a mild colloid stabiliser reinforced the stability. Subsequently it was shown that mono-disperse poly(styrene) could be prepared, albeit at low concentrations, that found use in colloid research.

For this introduction it will suffice to mention the reactive components that impart colloid stability, leaving kinetic and mechanistic aspects, which may be found in references given in the Bibliography (Fitch). The components may be one or more of the following three types:

- initiators that provide charged end groups, *e.g.* potassium persulfate, 2,2-azobis(2-amidinopropane)dihydrochloride
- hydrophilic co-monomers, *e.g.* acrylic acid, acrylamide and derivatives
- ionic co-monomers, *e.g.* sodium styrene sulfonate, quaternised 4-vinyl pyridine

Polymerisation rate is increased by the addition of solvents that are solvents for the monomer but not for the polymer thus increasing mass transfer to growing polymer. The same effect can be achieved by the use of phase transfer catalysts, *e.g.* crown ethers which are also claimed to provide higher polymer contents in the latexes. In terms of polymer content it has very recently disclosed (2001) that 60% has been achieved using a polymerisable ionic monomer and a persulfate initiator. Such a system might find application as a binder for water-borne coatings.

8.4.3   Composite Polymer Latexes

The desirable properties of latex polymers used in coatings are inevitably achieved by the co-polymerisation of two or more monomers. A typical composition is a co-polymer of methyl methacrylate ($T_g = 100\,^{\circ}$C) and *n*-butyl acrylate ($T_g = -50\,^{\circ}$C) for a polymer giving film formation at ambient temperatures, the filming property being governed by the shell polymer (see Chapter 5). For example, it has been found that enhanced performance can be attained if polymers have a core–shell morphology, where the core is a hard polymer and the shell is relatively soft, or *vice versa*. Polymers having this type of morphology are found to have lower minimum film-forming temperatures, good early block resistance and high temperature hardness.

An interesting variation of the core–shell structure is that of gradient control whereby linear compositional gradient is imposed from the core to the periphery of the particle. This is achieved by application of a technique known as 'power feed', developed by Union Carbide Corp., and involves the continuous change of co-monomer composition as it enters the polymerisation reactor. For example, a feed tank containing a 10:90% mixture of methyl methacrylate and *n*-butyl acrylate is attached to a second

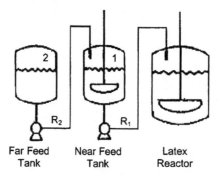

Figure 8.3 *The arrangement for continuously changing the monomer composition entering the reactor*

tank, which contains a 70 : 30% mixture of the same monomers. Each tank contains half of the total batch to be prepared. Whilst feeding the 70 : 30 mixture to the reactor, the 10 : 90 mixture is fed to the 70 : 30 tank at the same rate as that mixture is passing to the reactor. Figure 8.3 illustrates the set-up.

The resulting co-polymer is found to possess a linear compositional gradient ranging from the 70 : 30 composition to the 10 : 90 composition. Compared to a co-polymer of the same composition prepared by uniform feed, the gradient co-polymer is found to possess a broader glass transition, a lower minimum film-forming temperature and an enhanced high temperature block resistance.

8.4.4 Miscellaneous Variations

Although latex particles are usually uniformly spherical, it is possible to prepare lobed modifications that have been claimed to give enhanced rheological performance in water-borne coatings. It is also possible to prepare latexes in which the polymer particles are rod-like, even to the extent of being present as extended filaments. Such latexes are claimed to impart increased resistance to mud-cracking in polymer films when pigment loading is high.

Increasing the polymer content of latexes is a challenge that has generated much interest and there is a substantial patent literature devoted to the subject. One approach is to adjust the proportions of particle size domains to give a multi-modal distribution in the latex. Concentrations as high as 75% have been claimed by application of this technique. Viscosities in the range 500–1000 mPa s^{-1} at shear rates of 20 s^{-1} for polymer contents >70% have been achieved. Strategies for producing such concentrated latexes have been discussed in detail in a series of papers, by Schneider and co-workers (see Bibliography).

8.5   PRACTICAL ASPECTS OF EMULSION POLYMERISATION

With a feeling for the theoretical aspects of emulsion polymerisation, we shall turn our attention to the practice of the method. Let us first consider the four principle components in a polymerisation recipe:

- the continuous aqueous phase
- the monomer or monomer mixture
- the free radical source, *i.e.* the initiator
- the emulsifying agent

In addition to these components it may be necessary to include a buffer, *e.g.* sodium hydrogencarbonate, since some initiators are sensitive to pH changes. Similarly, monomers may need protection from changes in acidity or alkalinity. A chain transfer agent, *e.g.* dodecyl mercaptan may also be added for molecular weight control.

Initial trials must be carried out to determine the best way in which a stable emulsion of monomer(s) can be prepared, and subsequently maintain the stability of the polymer dispersion. It is usual to conduct these experiments under batch conditions, although once a suitable system has been developed, it may later be used in a continuous or semi-batch procedure. The traditional method used for polymerisation trials has been to seal the monomer emulsion and the initiator in a bottle and rotate it in a water bath at the desired temperature until polymerisation is complete. By using several bottles, one can terminate the reaction at intervals, analyse the contents and prepare time/conversion profiles, which should give sufficient information regarding system stability and kinetics to formulate a preliminary working recipe. This formulation may now be used in a polymerisation in a multi-necked glass reactor of 0.75–2.0 L capacity fitted with a mechanically driven stirrer. An inert gas inlet and a thermometer are fitted in other necks whilst one should be left free for sampling. Provision of further inlets will be necessary if continuous monomer and initiator addition are desired.

A typical latex composition would consist of:

Monomer(s)	40–60 vol%
Water	40–60 vol%
Initiator	0.1–1.0 wt% (on monomer weight)
Surfactant	1.0–5.0 wt% (on monomer weight)

Monomer selection will depend on the application of the latex and there will be variations to be made depending on the monomer(s). However, some generalisations may be made in surfactant and initiator selection.

8.5.1   Surfactant Selection

The surfactant plays a critical role since it not only has the primary function of generating the desired number of particles, but must also provide a stable monomer emulsion at the start of the polymerisation. This stabilisation must extend to the polymer particles as they are formed. Surfactants may be ionic or non-ionic, the most commonly used being anionic where the hydrophilic part is an anion, *e.g.* a sulfate or carboxylate. Cationic surfactants are less commonly utilised in latexes destined for coatings application and may be exemplified by the compounds in the group coconut-alkyldimethylbenzyl-ammonium chloride. Non-ionic surfactants find uses in coatings when relative insensitivity to pH change is required. They are also used in conjunction with anionic surfactants to improve colloid stability. Ethoxy-lated long chain phenols, *e.g.* nonyl phenol, or acids, *e.g.* stearic acid, are commonly encountered examples. Polymeric surfactants of all three types have been widely studied but the only ones to have reached large-scale industrial importance are the non-ionic ABA block polymers of ethylene oxide and propylene oxide. However, in certain specialised applications polymeric surfactants may be synthesised 'in house'.

A fairly recent development has been the introduction of polymerisable surfactants that have the advantage of reducing transport of surfactant through the film thus improving water resistance. It is also claimed that shear resistance of the latex is enhanced. As examples we may cite methacryloyl tipped polyglycols and sodium allyl lauryl sulfosuccinate.

The generation of micelles lies at the heart of conventional emulsion polymerisation and an awareness of the concept of critical micelle concentration (CMC) is important. A micelle may be described as an aggregate of surfactant molecules in which the hydrophobic parts are at the centre with the hydrophilic groups surrounding them. The CMC is defined as the concentration of surfactant at which micelles form. It is readily determined by measuring the surface tension of surfactant solutions having a range of concentrations. The CMC is the point where the surface tension becomes constant (Figure 8.4).

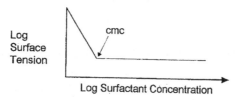

Figure 8.4   *The critical micelle concentration (the concentration at which aggregation of surfactant molecules occurs)*

It is also important to know that there is a minimum temperature for each surfactant below which micelles do not form. This is known as the Krafft temperature.

When a micellar process of polymerisation is desired, one should always work at the minimum concentration that combines micelle formation with system stability.

8.5.2 Initiator Selection

The temperature at which it is desired to conduct the polymerisation plays an important role in deciding which initiator to use. For a thermally cleaved initiator it is a useful rule of thumb to select an initiator having a half-life approximately one-third of the time required for complete polymerisation. Thus, for a polymerisation time of 3 h one should choose an initiator having a half-life of 1 h at the desired temperature. Alkaline metal persulfates are commonly used but require the presence of a buffer since the half-life is pH dependent. Azo initiators such as 4,4'-azobis(4-cyanopentanoic acid) (**1**) and 2,2'-azobis(2-amidinopropane) dihydrochloride (**2**) are often used.

(**1**)   (**2**)

When polymerisation is to be conducted at temperatures below *ca.* 35 °C, it is convenient to use one electron transfer redox couples as the source of initiating radicals. Persulfates with bisulfites are commonly encountered and provide ionic end groups. When non-ionic groups are required, the couple consisting of hydrogen peroxide–ascorbic acid couple is often used.

8.5.3 Monomer Considerations

Latexes used in the coatings industry are invariably based on co-polymers, the composition of which will be dictated by the final demands of the coating. In an earlier part of this chapter, it was noted that for emulsion polymerisation to succeed monomers must have at least a low solubility in water, but for some applications the presence of a very insoluble monomer may be required. Lauryl and stearyl methacrylate as well as many fluorine-containing monomers are examples of such monomers. Transport of the monomer to the polymerisation site may be achieved by the addition of a solvent for the monomer, but in recent times it has been shown that phase transfer catalysts such as cyclodextrins may be used with advantage.

8.5.4 Techniques

We have seen that both the degree and rate of an emulsion polymerisation may be increased simply by increasing the number of particles in the system. It is useful to know, when designing a polymerisation, that the following approximations are valid:

- Increasing the emulsifier concentration increases the number of particles, N, proportionally to the emulsifier concentration to the power 3/5.
- The overall reaction rate is proportional to N, hence to the emulsifier concentration to the power 3/5.
- Increasing the initiator concentration results in both N and the overall reaction rate increasing with the initiator concentration to the power 2/5.

The addition of monomer and initiator continuously throughout the polymerisation reaction can lead to a broad, and sometimes bimodal, distribution. The distribution may be narrowed considerably if part of the monomer is emulsified and polymerised after which the remaining monomer is added in the form of an emulsion.

At industrial scale levels the number of particles nucleated in Interval 1 is difficult to reproduce, but this can be overcome by the use of a previously prepared seed latex having the desired particle size. This is used as the medium for the polymerisation of a new batch of monomer, the emulsifier and initiator concentrations being controlled to prevent the generation of a new crop of particles. With the polymerisation commencing at the particle growth stage, the difficultly controlled nucleation step is avoided.

When monomer addition is continuous through the polymerisation, it is common to find the terms 'starved' or 'flooded' conditions. These terms simply indicate that the addition is slower than the rate of polymerisation or faster, respectively. Further details may be found in references given in the Bibliography.

8.6 OUTLOOK FOR LATEX-BASED COATINGS

The application of latexes as binders in coatings having very low or zero volatile organic compound content is seen to be one of continuing growth. Higher polymer contents should be attainable, as has been noted earlier in this chapter, and it is likely that the increasing application of polymerisable surfactants may provide a viable route.

'Controlled' free radical methods have very recently been applied to emulsion polymerisation. Although extensive research has been conducted into bulk and solution controlled free radical polymerisation, there have been very few reported studies of heterogeneous systems such as emulsion and mini-emulsion polymerisation. With an ever-increasing demand for water-borne binders, the ability to use techniques that allow the synthesis of polymers having controlled microstructure in latex form should be of considerable interest to the coatings industry. The currently much studied methods of stable free radical and atom transfer polymerisation, as well as the reversible addition-fragmentation transfer chemistry all appear to be applied to heterogeneous systems (see Chapter 7 for a description of the control methods). In addition to the control of the polydispersity of the polymers, it should also be possible to prepare block polymers and other architectural variations. The current status of this field has recently been reviewed by Cunningham (see bibliography).

8.6.1 Applications of Latexes to Coatings

The advantages that may be gained in using latexes as binders in coatings have been noted, particularly in connection with the property whereby high molecular weight polymers may be used at high concentrations without the penalty of unmanageably high viscosities that would be expected in conventional polymer solutions. However, this desirable property is not entirely without problems since flow and levelling in water-based coatings tend to be less satisfactory than for solvent-based compositions. In order to introduce the thixotropic viscosity necessary for acceptable levels of application performance, the addition of rheology modifiers is needed, and there is a continuing search to optimise the performance of the associative thickeners that can control this factor, *cf.* Chapter 3.

Latex-based coatings for architectural use have been available for about 60 years and at the present time a very large proportion of both interior and exterior trade sales paints are water borne. On the other hand, the expansion of water-based binders into the industrial coatings sector has been slower than was envisaged at the time of this book's 1st edition. The search for a high performance ambient cure has not yet revealed a technology that might accelerate progress, but whilst the search continues, some acceptance of water-borne coatings into maintenance and heavy-duty industrial coatings markets has been achieved. Similarly, the need for very high gloss coatings in the aerospace industry proves to be a hurdle that water-borne coatings are slowly overcoming with some promising urethane-based binders. In other areas such as can coating, water-based coatings are increasingly used, mainly in the form of water reducible systems.

The latexes used in coating application are based on styrene–butadiene co-polymers, acrylic co-polymers, and vinyl acetate or vinyl acetate co-polymers. Various other monomers may be incorporated in small quantities to improve such properties as adhesion and corrosion inhibition. Poly(vinyl acetate) is susceptible to hydrolysis and this may be reduced by the inclusion of small amounts of certain vinyl esters of C_8–C_{12} acids in the polymer. Vinyl monomers from highly branched acids with 9, 10 or 11 carbon atoms can play a similar role. Such monomers are commercially available under the trademark VeoVa®. In addition to providing a resistance to hydrolysis of the vinyl acetate, these esters impart improved alkali and water resistance.

Research into latexes for coatings applications is a very active field as a glance at Chemical Abstracts or World Surface Coatings Abstracts will show. A considerable proportion of the work published is in the patent literature, since to a large extent technology lies in the expertise and experience of a few specialist companies, *e.g.* Rohm & Haas, S.C. Johnson.

8.7 PSEUDO-LATEXES

The utility of latexes prepared by emulsion polymerisation has been highlighted, but not all polymers can be prepared in this fashion. Since such polymers have important applications, it would be of considerable value to have them in latex form, *i.e.* as colloidal dispersions. An interesting and useful advance in achieving this goal was made by Vanderhoff and co-workers who disclosed a method which allows the preparation of sub-micron-sized polymer particles from polymers not available by emulsion polymerisation. The aqueous dispersions are known as pseudo-latexes.

The method is based on the same principles as those described under mini-emulsion polymerisation except that in the pseudo-latex case, the monomer is replaced by a solution of the polymer in a water immiscible solvent. The emulsified polymer solution is subjected to homogenisation until the droplet size of the disperse phase is between 100 and 500 nm. The solvent is now stripped out by distillation to leave a polymer latex with water as the continuous phase.

A typical procedure, taken from the patent disclosure will illustrate the method. Following the emulsification, homogenisation and solvent removal, a pseudo-latex containing 45% epoxy resin having a particle size <300 nm is obtained. The recipe given is:

Epoxy resin, *e.g.* Epon 1001	31.25 g
Toluene–methyl isobutyl ketone (1:1)	93.75 g
Hexadecyltrimethylammonium bromide	0.78 g
Cetyl alcohol	1.73 g
Water	375.00 g

In the past decade progress has been made in the preparation of colloidally dispersed resins, *e.g.* epoxies in water without the use of solvents, thus conforming to demands for the reduction or elimination of volatile organic compounds. Using a technique known as flow induced phase inversion (FIPI) (Akay, 1995), it was shown that subjection of water-in-oil emulsions of resins to very high deformation rate led to oil-in-water emulsions having very high disperse phase content and narrow size dispersity. To achieve the required deformation rate, the W/O emulsions are fed through a small volume mixer in which the inversion takes place. Resins with viscosities of 1000 Pa s could be accommodated by this technique using non-ionic–anionic surfactant blends. Subsequently, it was found that molecular surfactants that emulsified liquid resins would not perform for solid resins in the melt. However, hydrophobically modified water-soluble polymers were shown to give satisfactory results, high disperse phase dispersions with narrow size dispersities being formed.

A very recent advance has been the application of FIPI in the dispersion of poly(ethylene-*co*-vinyl acetate). Not only can high concentration dispersions be prepared, but also by controlling the conditions an intermediate 'powdery emulsion' may be formed (Akay *et al.*, 2002). This technology should be of interest in the powder coating sector.

8.8   DISPERSION POLYMERISATION

In considering emulsion polymerisation we have dealt with a technique of polymerisation that occurs in a compartmentalised system of monomer-swollen micelles and initiating free radicals generated in an aqueous continuous phase. The result is a colloidal dispersion known as a latex. Colloidal dispersions of hydrophobic polymers may also be prepared directly from monomers dissolved in organic solvents, and in the past such dispersions were important in the coatings industry. ICI in particular played a strong role in the development of this procedure and used such non-aqueous dispersions as a step to water-borne coatings (Aquabase® technology). Although it is possible to prepare non-aqueous dispersions having 85% polymer content, the complexity of the means required to achieve the multi-modal size distribution necessary is a barrier to industrial exploitation. Furthermore, environmental restrictions on the use of volatile organic solvents have seriously reduced interest in non-aqueous dispersions for coatings. However, there may still be application to high solids coatings and this aspect was considered in Chapter 7.

It is also possible to prepare colloidal dispersions of crosslinked hydrophilic polymers from mixtures of water-soluble monomers and crosslinking agents in the presence of a polymeric stabiliser. Application of this technique to micro-gel preparation will be described in a later section of this chapter. Aqueous dispersions of polymer colloids formed by condensation polymerisation have been reported, *e.g.* modified glyoxal–melamine polymer for use as a coagulant in water treatment.

8.8.1 Steric Stabilisation of Dispersions

Polymeric stabilisers are essential components in dispersion polymerisations in which the monomer is dissolved in a solvent before initiation. They are most often block or graft co-polymers (Chapter 7) in which one part has a strong affinity for the polymer colloid whilst the other has a strong affinity for the continuous phase. Such polymers are said to be amphipathic and dispersions containing these polymers are said to be sterically, or entropically, stabilised. It may be remarked that the term entropic more accurately describes the mechanism that will be briefly outlined later. A particularly recommended graft structure is that of comb graft polymers, in which grafts (of **m**) are spaced at uniform distances along the polymer backbone (poly**M**). This form of stabilisation is not only applicable to non-aqueous dispersions as the mechanism applies to non-ionic surfactants, polymeric surfactants and polymerisable surfactants in water.

In these structures we may look upon the **M** component as the part that associates with the polymer particle, whilst the **m** component extends into the continuous phase and is well solvated by it. The **M** associated to the polymer particle is known as the anchor component, and may be connected to the polymer by physical attraction, covalent bonding or by incorporation into the particle. In a qualitative way we may say that stabilisation results from the repulsive forces which occur from loss of configurational entropy when the particles approach one another, and the solvated chains of the stabiliser begin to compress or overlap. This loss of entropy is sufficient to overcome van der Waals–London forces of attraction so that the colloidal polymer is stabilised against flocculation, the steric barrier formed by the well-extended and solvated part of the stabiliser preventing close contact.

Steric stabilisation of colloidal dispersions has some advantages over electrostatic stabilisation:

1. It can be used with equal efficiency, and at both high and low solids concentrations, in aqueous or non-aqueous systems.
2. Sterically stabilised dispersions are quite insensitive to added electrolytes.
3. Flocculation in sterically stabilised dispersions is reversible.

8.9 WATER REDUCIBLE SYSTEMS

Although very frequently referred to as water-soluble polymer systems, the water reducible polymer system is not in fact a true solution, but a well-solvated collection of aggregates. The principle of introducing water solubility is very simple, merely involving the modification of the polymer chain to give anionic, cationic, non-ionic or zwitterionic functionality. Typically the polymers, *e.g.* alkyd, epoxy or polyurethane may be chemically modified, usually in an organic solvent, to impart acidic or basic character. This most often involves carboxylic or amine functions which are neutralised to give water-soluble salts, thus allowing dilution of the original organic solvent solution with water. The polymer aggregates, swollen with the water–solvent continuous phase, are usually present at concentrations of *ca.* 20%. The particle size is in the 10 nm range; hence the systems are transparent or translucent. A limitation of these 'pseudo-solutions' is that the viscosity is dependent on the molecular weight of the original polymer.

The presence of organic solvent in water reducible systems as described has caused attention to be focused on methods of preparation that would avoid their use. Similarly, increasing demands to reduce VOC have led to a move away from the use of amines. The use of non-ionic groups, *e.g.* polyols or polyethers, has been proposed since they may be easily introduced into polymer structures. However, this strategy is not without certain disadvantages, *e.g.* poor water resistance and adhesion to substrates, although it is possible to optimise performance to acceptable levels.

To avoid the use of amines for the neutralisation of acid function, the use of zwitterionic co-polymers in which the carboxylic and amine functions are covalently bonded to the resin to be dispersed has been proposed. The zwitterions undergo self-crosslinking on baking and this has led to further investigation for coating applications.

Clearly, there are many chemical routes available for introducing water solubility to resins and we shall mention just four:

1. maleated epoxy resin with dimethylaminoethanol
2. styrene maleic anhydride with dimethylaminoethanol
3. aryl cyclic sulfonium zwitterionic monomers
4. epoxy resins esterified with phosphoric acid and neutralised with dimethylaminoethanol

When using amines, the structure of the amine can have significant effects on final film performance, for example, the use of dimethylaminoethanol is advantageous since, being a tertiary amine, it does not form amides that might affect water resistance.

An interesting use of water reducible resins is as stabilisers in emulsion polymerisation procedures where they may be used at relatively high concentrations and become part of the final polymer film. Their use in this connection is noted in the later section on micro-gels.

The two most important applications of water reducible dispersion are in the packaging coatings area and electrocoating in the automotive industry. An epoxy resin may be reacted with *ortho*-phosphoric acid and the product advanced in molecular weight. Neutralisation of acid function formed by hydrolysis of the relatively susceptible di- and tri-phosphate esters leads to a product that is commonly used in the can coating area.

8.9.1   Electropaints

The importance of water reducible polymer dispersions as electrocoatings in the automotive industry, particularly as primers, is sufficient to devote a short section to the subject. The principle of the method lies in the susceptibility of lyophilic colloids to coagulation in the presence of electrically generated ionic species. This event leads to the deposition of the polymer on to the work-piece, which may be made the anode (in anaphoretic, or anodic, systems), or the cathode (in cataphoretic, or cathodic, systems). The typical reactions involved are depicted in Scheme 8.1.

ANODE

$$2H_2O - 4e^- \longrightarrow 4H^+ + O_2$$

$$Polymer\text{---}COO^- + H^+ \longrightarrow Polymer\text{---}COOH$$

CATHODE

$$H_2O + 2e^- \longrightarrow 2OH^- + H_2$$

$$Polymer\text{---}N^+R_2H + OH^- \longrightarrow Polymer\text{----}NR_2 + H_2O$$

Scheme 8.1

Equivalent reactions at the opposite electrode lead to regeneration of the neutralising species that may be low molecular weight amine or acid, respectively. These can be segregated from the bulk by semi-permeable membranes or allowed to mix with the bath and removed at intervals by dialysis or ultra-filtration. Ultra-filtration has further value in that it provides a supply of 'permeate' that can be used to rinse the loosely adhering 'cream coat' from the coated articles and returned to the bath. The efficient utilisation of the electropaint is thereby greatly increased.

Anodic electropaints are very often polymeric carboxylic acids neutralised with tertiary amines, although other anionic groups and inorganic bases have been used. Backbone structures that have found favour include acrylics and maleinised fatty acids, epoxy esters of fatty acids, and polybutadienes. Although the performance of anaphoretic paints can be very good, they do suffer from reduced corrosion resistance when compared to cataphoretic paints. This is due to the participation of the steel substrate in an electro-chemical reaction that is outlined in Scheme 8.2.

$$Fe - 2e^- \longrightarrow Fe^{2+}$$

$$2Polymer-COO^- + Fe^{2+} \longrightarrow (Polymer-COO^-)_2 Fe^{2+}$$

Scheme 8.2

The outcome of this reaction is a disruption of the substrate surface and dissolution of iron that may promote corrosion.

Cataphoretic paints are not prone to the problems connected with the anaphoretic variety, and consequently, they are now very widely used. Aminated epoxy resins are frequently utilised, and are neutralised with acetic or phosphoric acid. Blocked isocyanates are the most widely used crosslinking agents. Along with pigments and other additives, a stable dispersion is formed in the presence of the neutralised film former.

Depending on the voltage used and a balance of the resistance of the liquid paint with that of the coated parts of the work piece, an electrocoat is 'self-limiting' in thickness and can 'throw' into enclosed spaces such as box sections and car doors. These properties result in unparalleled corrosion protection for objects of complex shape.

8.10 HYDROSOLS

In Chapter 7 polymer dispersions known as plastisols were introduced and it was seen that polymers required a degree of crystallinity to avoid excessive swelling. In some cases viscous systems could be diluted with organic solvents to give what are termed organosols. Here we briefly consider very small polymer particles of un-gelled polymer dispersed in water: they are known as hydrosols. In the next section we shall introduce similarly sized crosslinked polymer particles (micro-gel).

A hydrosol may be defined as a colloidal dispersion of a solid in water, the dispersion exhibiting fluidity. Du Pont gave the name to aqueous dispersions of acrylic co-polymer that differed from conventional acrylic latexes chiefly in the extremely fine particle size, typically 10–100 nm. In present day parlance such systems would be called nano-latexes.

The original technique developed for the preparation of these latexes was to co-polymerise 3–6% of a carboxyl-containing monomer, *e.g.* acrylic, or itaconic acid with methyl methacrylate and ethyl acrylate, or other acrylate, to give a latex. After neutralisation of the acid with ammonium hydroxide and swelling, the polymer particles were subjected to powerful mechanical shearing that split the particles to sub-100 nm size. The product was a fluid dispersion containing about 30% polymer with a molecular weight of 20,000–50,000. Use of these fine-sized polymer latexes in coatings was described in *J. Paint Technol.*, 1968, **40** (521), 263–270 by Beardsley and Selby. Since that time numerous modifications have been disclosed, mainly in the patent literature. Emulsifier free acrylic hydrosols have been claimed, with particle sizes in the 20–100 nm range. Poly(vinyl alcohol) having a degree of polymerisation >500 was used as a protective colloid, but the solids contents of the dispersions were relatively low.

Conventional methods of emulsion polymerisation limit the size of the polymer particles that may be prepared to around 75 nm. Micro-emulsion technology should be applicable to the synthesis of linear polymer dispersions in which the particle size is <50 nm, but at the present time high surfactant levels and low solids content are problems that have yet to be overcome. Typically the polymer/surfactant ratio in micro-emulsion systems has been <4, but recent disclosures have indicated that ratios of 7 or higher are feasible when monomer is slowly and continuously added to a polymerising micro-emulsion. It has further been claimed that this technique can lead to disperse phase contents in the range 10–40% with particle sizes <20 nm and polydispersities <1.5 (Ming *et al.*, 1998, see Bibliography).

8.11 MICRO-GEL

A form of polymer that is readily formed as a dispersion in non-aqueous or aqueous media is known as micro-gel. The dispersions are of intra-molecularly crosslinked polymer having particle sizes *ca.* <75 nm and as such may be considered as true molecules, which allows the dispersions to be treated by the thermodynamic equations that are applicable to polymer solutions. Although the term micro-gel has long been in use, these small particles might currently be termed nano-gels or nano-spheres.

Material of this type was first prepared as long ago as 1934, but its adventitious presence in certain alkyds was not recognised until 1949 when the name micro-gel was first introduced. Subsequently, structural studies elucidated their nature, and they found applications in the coatings industry where their film strength enhancement and contributions to rheology control were strongly championed by W. Funke. In the world of industry, Nippon Paint and ICI made important advances in synthesis and applications. Pittsburgh Plate Glass were also active in this field and both they and ICI successfully developed coatings for the automotive industry using the technology. In more recent times there have been developments in the application of micro-gels as catalysts and as phase transfer agents which have led to increases in commercial production.

Micro-gels may be prepared by the crosslinking of linear polymers, but in order to minimise inter-molecular bond formation, it is necessary to carry out the reaction in dilute solution. Such conditions are not economically acceptable and the preferred technique is to carry out the polymerisation of crosslinkable monomer mixtures in a controlled volume, in the same way as is achieved in emulsion polymerisation or non-aqueous dispersion procedures.

Careful selection of surfactant is required for successful micro-gel synthesis, the key requirement being the need for all the monomer to reside in the micelles. Conventional surfactants, *e.g.* sodium dodecylbenzene sulfonate, may be used but great improvements have been achieved when using polymeric surfactants. Pioneering studies by Nippon Paint showed that zwitterionic structures obtained from the modification of polyester or epoxy resins tended to give the best results. Using an epoxy-based surfactant containing the zwitterionic group (**3**), micro-gels from acrylic or vinylic monomers could be obtained in the particle size range 20–50 nm. Another useful surfactant, which is easily prepared by the reaction of a polyester with *N,N*-bis-2-(2-hydroxyethyl)-2-aminosulfonic acid (**4**), was found to give similar results.

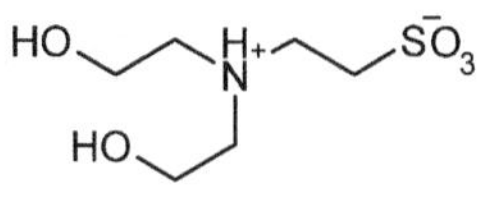

(**3**)

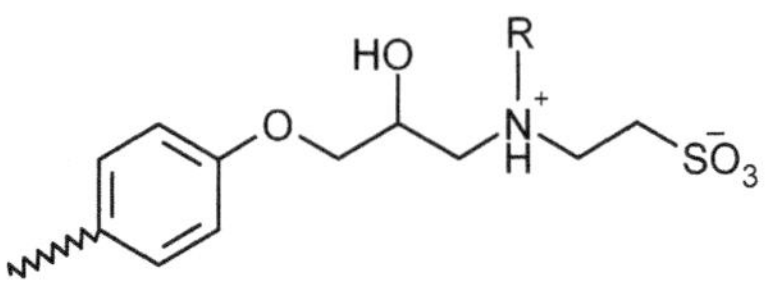

(**4**)

In recent times Antonietti and co-workers have developed metallo surfactants that can be used to prepare polymer particles in the 6–8 nm size range whilst preserving a narrow dispersity. However, such results are only attained by the use of high surfactant concentrations (30–50%) to prepare the micro-emulsions of monomer, but there seems to be a likelihood that an appropriate modification of the surfactant structure would enable the concentrations to be reduced substantially.

Specific applications for micro-gels in coatings lie mainly in the automotive industry where the incorporation of these polymers has been found to control metal flake alignment in metallic coatings. ICI's Aquabase® system was based on micro-gel technology and enjoyed considerable success in this area. In water-borne coatings, the presence of micro-gel has been shown to facilitate water release, whilst the low viscosities of micro-gel solutions, as compared to their linear or branched counterparts make their use attractive in high solids coatings. With solids contents of >80% in such coatings, the need for rheology modifiers that lead to the required pseudo-plasticity can be fulfilled by using micro-gels.

Suitably functionalised micro-gels have been considered as latent crosslinking agents. For example, a micro-gel having carboxyl functionality might be dispersed in carboxylated acrylic latex and crosslinking achieved by the introduction of metal ions such as Zn^{2+}, Zr^{2+} or Ca^{2+}.

It has long been known that the presence of micro-gel in alkyd resin films increased their mechanical strength and this observation has often been reported for other resin films. Speculatively there might be some application for sub-10 nm particles in corrosion inhibition by the application of the quantum size effects and electron transfer phenomena found as size is reduced.

8.12 CROSSLINKING OF WATER-BORNE COATINGS

Thermoplastic latex technology holds a predominant position in the water-borne coatings industry, because it is possible to manufacture and handle polymers of high T_g and very high molecular weight in water.

It is even possible to apply latices that are lightly crosslinked in advance. Nevertheless, potential advantages have been identified for reactive latices, and include enhanced block resistance, and the possibility of exploiting lower T_g, lower molecular weight polymers that coalesce more readily. It was remarked in the section on latexes in coatings that the absence of a truly effective crosslinking mechanism is one of the barriers to their exploitation in still wider markets.

There also remains a wide range of pseudo-latex and other water-based systems that are just as dependent on crosslinking as solvent-borne coatings. In some cases their dependence is higher as the curing process needs to overcome the intrinsic hydrophilicity of the system to deliver a water-resistant coating.

A feature of water-based crosslinking chemistry is that the reactive functions must generally exist in water for a considerable time before crosslinking takes place, so that hydrolytically unstable groups such as

isocyanate or certain esters are not usually acceptable. However, there
are successful technologies exploiting components such as isocyanate
held in a water miscible solvent and only introduced at the last moment.
Reactive latices pose a particular problem in that reaction is only likely
at the surface of the latex particles. It is necessary to ensure that the reactive
groups are mainly located at the surface, if most of them are not to be
wasted. If the crosslinker is a separate species, it may be another latex or
dispersion, or a water-soluble material dissolved in the continuous phase.

Hydrophilic groups such as –COOH and –OH tend to be concentrated on
the particle surfaces and are frequently exploited directly for crosslinking,
or modified with alternative species. There is also considerable interest in
crosslinking latex particles through the stabiliser residues necessarily
present at the water interface.

8.12.1 Crosslinking by Condensation Processes

A number of important curing chemistries for water-based systems involve
systems that form linkages by condensation. Where the condensation
process evolves water, the presence of the aqueous medium may be helpful
to storage stability.

8.12.1.1 Amino Resins. Amino resins have been very widely used in
stoving water-borne systems, drawing on their track record in conventional
coatings. Fully methylolated aminos are generally preferred, necessitating
the use of strong acid catalysts.

It is also desirable to incorporate a certain amount of alcoholic solvent to
ensure storage stability. The main film former of course must contain the
requisite number of suitability placed hydroxyl groups. Melamine, urea and
glycoluril systems have all been exploited, but acrylamide–formaldehyde
derivatives are particular interest in acrylic latex systems since they can be
incorporated directly into the polymer backbone. The monomers (**5**) and (**6**)
are examples of those commercially available.

(**5**)   (**6**)

Alternatively, acrylamide residues can be introduced into the polymer,
and subsequently converted with formaldehyde and optionally, alcohols.
They can either condense with hydroxyl groups or with each other.

Latexes containing glycoluril–formaldehyde or methylolacrylamide have reportedly been cured at ambient temperature, provided that steps were taken to remove all the cationic species normally present.

8.12.1.2 Tethered Aldehyde Technology. In an attempt to avoid the formaldehyde emissions associated with the amino resins, and to achieve a reduced stoving temperature, workers at Air Products developed amide monomers or crosslinkers with tethered aldehyde present as their acetals. They could undergo essentially the same chemistry as the aminos (Scheme 8.3).

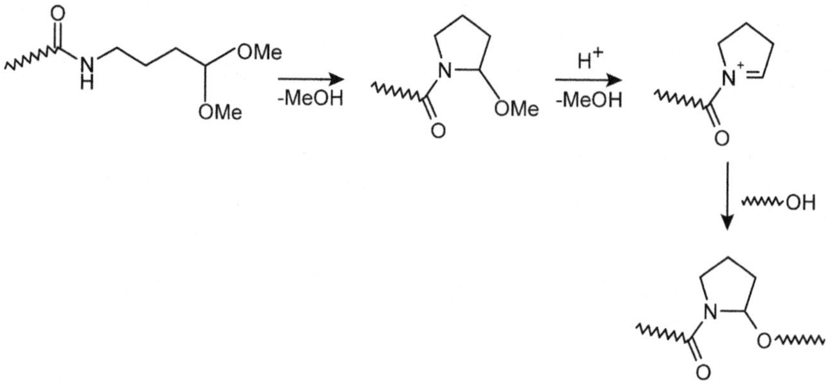

Scheme 8.3

8.12.1.3 Blocked Isocyanates. Blocked isocyanates have been used to cure cationic electropaints, where the basic environment obtaining after phase inversion would inhibit the cure of amino-formaldehyde systems.

Glycol ethers have proved to be effective blocking agents, and can be present in large excess to delay the hydrolysis of the urethane groups, or any premature reaction with amine groups present. At the same time, they provide useful coupling agents. Amine and hydroxyl groups are usually present in the film former.

8.12.1.4 Carbonyl-amine Condensations. Condensations between various amines and carboxyl groups have been exploited extensively in water-borne coatings technology. The best established is the so-called keto-hydrazide reaction to form a hydrazone derivative (Scheme 8.4)

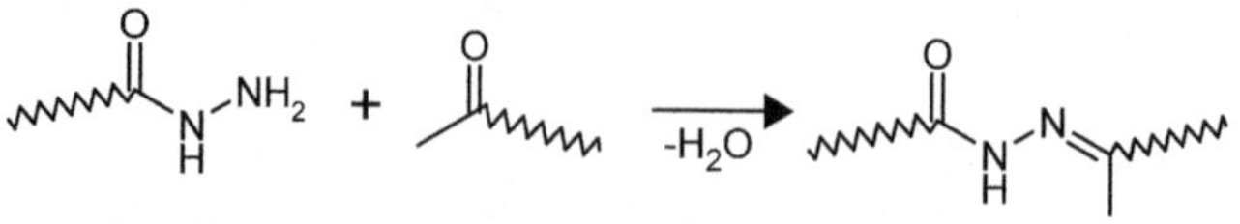

Scheme 8.4

It occurs spontaneously at room temperature and gives a stable linkage. Typically, a water-soluble dihydrazide would cure a latex furnished with ketonic groups derived from diacetone acrylamide (**7**). An attractive feature of the chemistry is that it is almost completely inhibited in the presence of water, so that stable compositions crosslink rapidly as water is lost from the films.

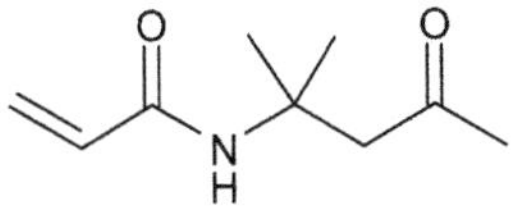

(**7**)

Numerous variations on the theme have been described, including polymeric hydrazides, and carbonyl components derived from acrolein, acetoacetates or occasionally vinylmethyl ketone.

Both aldehydes and acetoacetates also form fairly stable condensates with primary amines. Polymeric aldehydes can be prepared by copolymerising acrolein with acrylic monomers. Condensation with amines then yields aldimine linkages (Scheme 8.5).

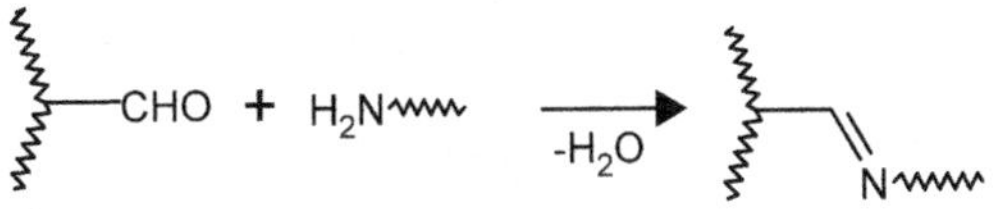

Scheme 8.5

Aromatic aldehydes give the still more stable Schiff's bases and are accessible by reacting aldehydic phenols such as salicylaldehyde with epoxy resins.

Acetoacetate groups can be introduced into acrylic systems using aceto-acetoxyethyl methacrylate (**8**), or by transesterification of hydroxyacrylics. Reaction with primary amines then produces enamine linkages (Scheme 8.6).

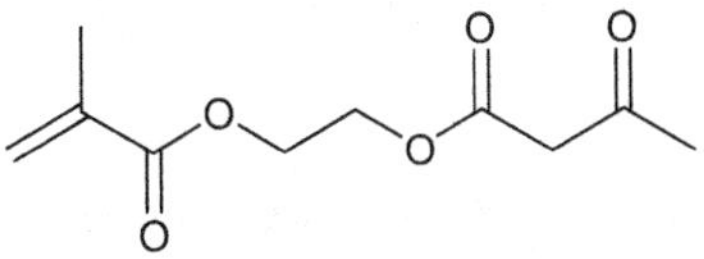

(**8**)

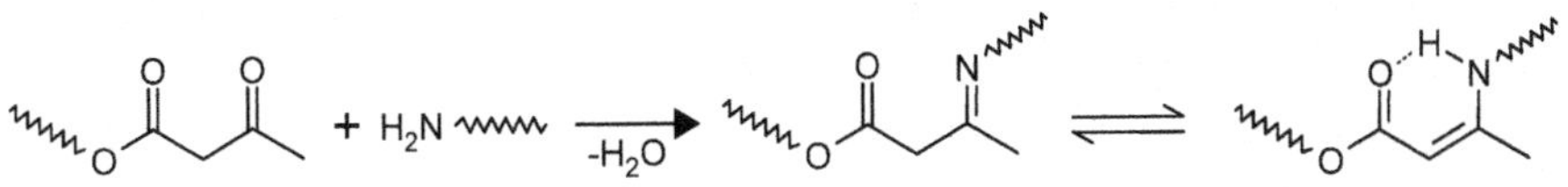

Scheme 8.6

In some cases the acetoacetoxy group has proved susceptible to hydrolysis, and a stable system has been created by adding ammonia to form the protective enamine (**9**). On drying in the presence of a diamine, the ammonia is lost and stable crosslinks are formed.

(**9**)

8.12.1.5 Diethanolamide–Acid Condensation. The adipoyl diethanolamide condensation chemistry described in Chapter 7 as a mainstay of powder coatings technology was originally developed for water-based systems, for which its water solubility would seem an ideal qualification. Similar positive and negative features apply, but it has made more headway in powders than in the stoved water-borne market.

8.12.2   Crosslinking by Addition Processes

Most of the electrophilic components exploited in addition cure for water-borne systems are strained rings or activated olefins, susceptible to a greater or lesser extent to premature hydrolysis. With two notable exceptions (aziridines and carbodiimides), they are better known in conventional, high solids, or particularly, powder coatings.

8.12.2.1 Epoxy–Amine Cure. Acrylic polymers including latices have frequently included glycidyl methacrylate to provide sites crosslinkable with water-soluble amines of various types, and amine dispersions made, for example, by reacting the surface –COOH groups in a latex with ethyleneimine. The glycidyl groups can prove insufficiently stable to hydrolysis, depending on pH, and have sometimes been replaced by cycloaliphatic epoxy groups.

In the growing area of aqueous epoxy dispersions, conventional epoxy resins such as liquid bisphenol A–epichlorohydrin condensates are dispersed in water, sometimes using the amine curing agents as stabilisers. Polyamidoamines, aminoamides and epoxy adducts possess the required ambiphilic properties, especially when they are partially neutralised with volatile acids. The acids also serve the purpose of extending pot life that is otherwise liable to be rather short in view of the catalytic effect of water. The inclusion of coupling solvents helps dispersion stability, but is discouraged by the drive to lower organic emissions.

8.12.2.2 Aziridine–Acid Crosslinking. Ethyleneimines or aziridines react readily with nucleophiles, especially carboxylic acid (Scheme 8.7). R can be H, alkyl or an electron-withdrawing group such as acyl. (The last named enhances the electrophilicty of the ring.)

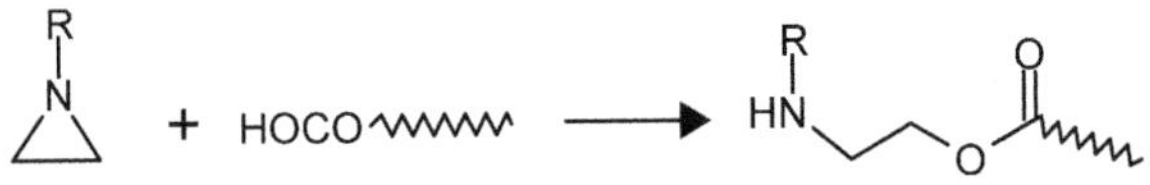

Scheme 8.7

Ethyleneimine itself is useful for introducing the aminoethyl group to acid functional polymers, and for making multifunctional aziridine cross-linkers by, for example, Michael type addition to multifunctional acrylates (Scheme 8.8).

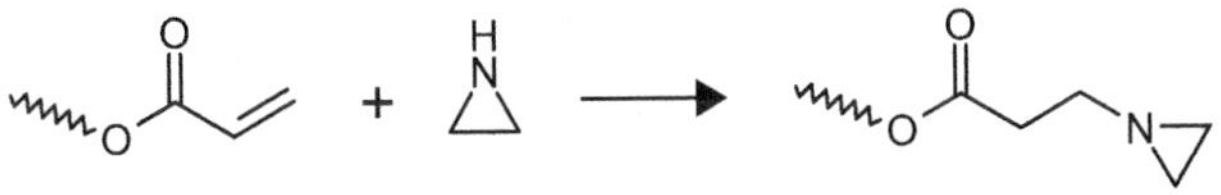

Scheme 8.8

Crosslinking of acid functional polymers with such materials occurs efficiently at ambient temperature and the technology has assumed some importance in the paper coatings industry. Unfortunately, low molecular weight aziridines have proved to be carcinogenic, and their use in many markets has been discouraged. Copolymers of the reaction product of N-hydroxyethylethyleneimine with m-isopropenyldimethylbenzyliso-cyanate (**10**) were recently claimed to have reduced toxicity.

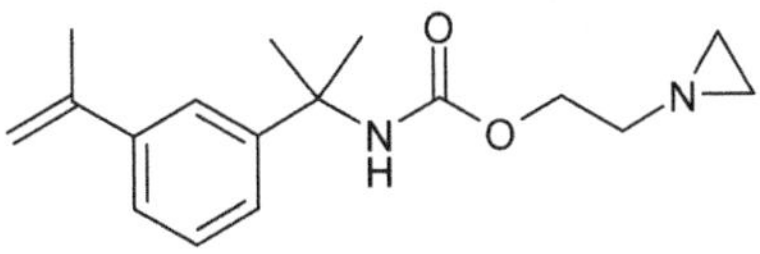

(**10**)

8.13.2.3 Oxazoline–Acid Crosslinking. The use of multifunctional oxa-zolines as crosslinkers, and oxazoline functional polymers to form films on stoving with carboxylic acid co-reactants, attracted considerable attention in the 1980s. Like several of the systems mentioned, they have achieved limited acceptance in water-based and powder coatings systems (*cf.* Chapter 7).

It has been pointed out that they may not be hydrolytically stable unless substituted in the 2-position with an aromatic ring (**11**).

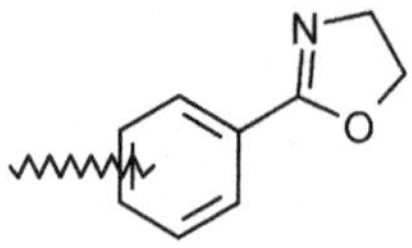

(11)

8.12.2.4 Azlactone-Hydroxyl and Amine Crosslinking. Azlactone-hydroxyl reaction requires stoving and again is of most interest in powder coatings (*cf.* Chapter 7). The reaction with amines takes place at ambient temperature to form a peptide-like linkage (Scheme 8.9) and has the potential to be useful in water-borne systems.

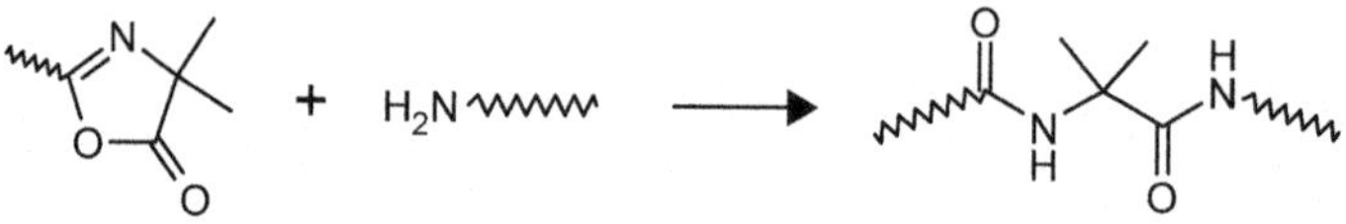

Scheme 8.9

Unfortunately, the expensive vinylazlactone is the only commercially available precursor, and its copolymers again appear to be of marginal hydrolytic stability.

8.12.2.5 Carbodiimide–Acid Crosslinking. Isocyanates can be catalytically converted to carbodiimides. A typical curing agent is prepared by condensing a trifunctional isocyanate with monoisocyanates, and reacts with acid groups to form acylurea linkages (Scheme 8.10).

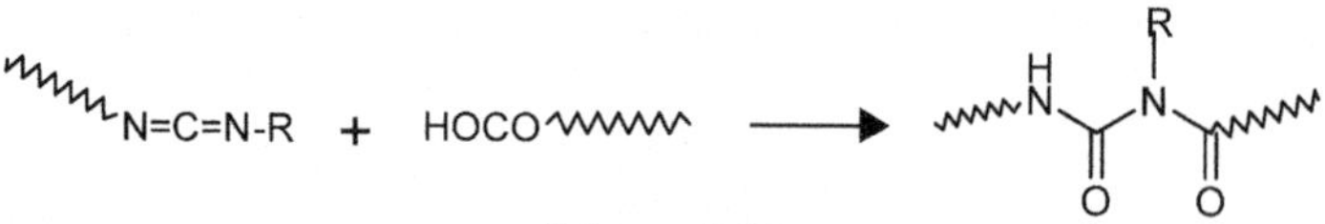

Scheme 8.10

An alternative crosslinker is based on copolymers of the carbodiimide derived from TMI and a monofunctional isocyanate (**12**).

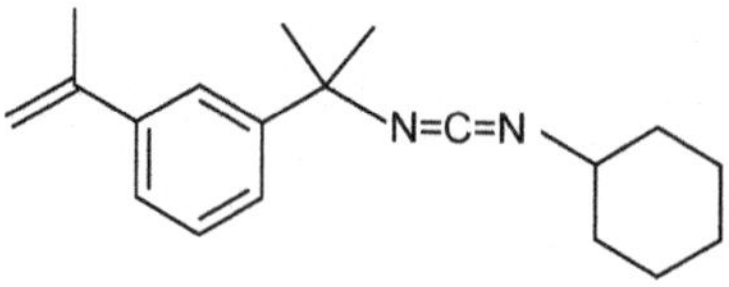

(12)

Neutralised acid groups are stable in the presence of carbodiimides, but crosslinking takes place when the neutralising amine or ammonia evaporates from the film and the pH moves towards neutral. Unfortunately, carbodiimides are, like azlactones, rather expensive.

8.12.3 Crosslinking by Radical Processes

8.12.3.1 Autooxidative Cure.   There is extensive technology and growing interest in water-borne alkyds. They are essentially polyester systems with pendent unsaturated fatty acid groups that can indulge in autooxidative cure as described in Chapter 6.

Water solubility is achieved by ensuring that sufficient carboxylic acid groups are present on the backbone and neutralising all or part of them with ammonia or a volatile amine. An 'acid value' of 30 is the minimum that will achieve water dispersibility in the absence of other hydrophilic groups (*i.e.* 30 mg of potassium hydroxide are required to neutralise 1 g of solid resin).

Water-borne alkyds are potentially susceptible to hydrolysis in storage. The tendency is countered by introducing as few additional ester linkages as possible. Additional acid groups are preferably introduced by maleinisation (*v.i.*). A coarse emulsion is maintained to minimise contact with water; particle size is controlled by the number of neutralised acid groups present.

Before the advent of cathodic systems, the best anticorrosive performance was obtained from anionic electropaints derived from maleinised polybutadiene. 'Maleinisation' is an 'ene' addition of maleic anhydride to an allylic group (Scheme 8.11). It introduces carboxylic acid groups for water solubility and shifts some of the double bonds in the backbone to the 'methylene interrupted' form that is most conducive to autooxidation. Since cure took place under stoving conditions, siccatives were not required.

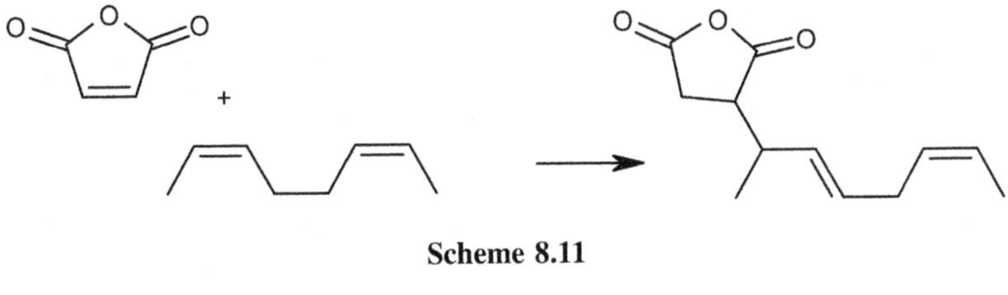

Scheme 8.11

There have been many attempts to introduce oxidative crosslinking to acrylic latices. Typical examples employed copolymers containing allyl methacrylate, methacrylates of other unsaturated alcohols or, for example, the species (**13**).

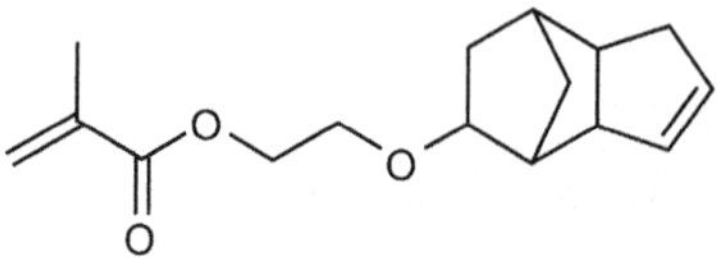

(**13**)

Enamines of acetoacetate have also shown a tendency to air-dry in the presence of drying oil fatty acids and metal driers.

8.12.3.2 Radiation Cure. UV cure of latices is a somewhat specialised area, but successful attempts have been made to furnish latex particles with acryloyl or methacryloyl groups and crosslink the products by radiation. Typical techniques used reactive monomers such as glycidyl methacrylate to attach unsaturated groups to the carboxyl residues present in the surfaces of preformed latex particles.

It might seem that the very rapid film formation achievable with UV cure is not quite compatible with the leisurely coalescence and water loss of a typical latex. Indeed, it has usually proved necessary to conduct the cure at an elevated temperature after an initial 'flash off' period.

8.12.4 Cure by Metal Salts

The simple process of adding the acetate or oxide of a divalent metal such as zinc to an acid-functional latex can bring about a useful ionic crosslink, as mentioned in Section 8.11. Zirconium has been used in the same way. The cations are usually introduced as an oxide or water-soluble salt such as an acetate.

An alternative approach is to decorate the polymer with acetoacetate groups that will coordinate strongly with certain metals such as titanium (**14**).

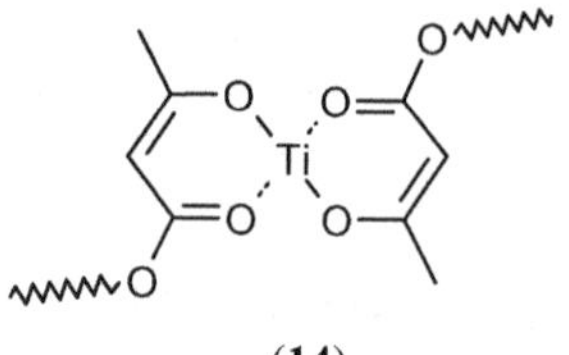

(**14**)

8.13   BIBLIOGRAPHY

1. P.C. Hiemenz and R. Rajagopalan, *Principles of Colloid and Surface Chemistry*, 3rd edn, Marcel Dekker, New York, 1997 (Chapter 1 supplements the introduction to colloids presented here).
2. Emulsion polymerisation receives varying degrees of coverage in textbooks of polymer chemistry. Here we give works devoted to the subject: D.C. Blackley, *Emulsion Polymerisation: Theory and Practice*, Applied Science Publishers, London, 1975 (now rather old but still a clear presentation of the micellar mechanism; useful chapters on the components used in this method); R.M. Fitch, *Polymer Colloids: A Comprehensive Introduction*, Academic Press, San Diego, 1997 (an in-depth presentation by the pioneer of the homogeneous nucleation mechanism); R.G. Gilbert, *Emulsion Polymerisation: A Mechanistic Approach*, Academic Press, San Diego, 1995 (a critical and thorough account of current theories for the specialist); M.S. El-Aasser and P.A. Lovell (eds), *Emulsion Polymers and Emulsion Polymerisation* (a multi-authored work that covers a wide range of topics that updates Blackley).
3. Practical polymerisation, which includes examples relevant to the material covered in this chapter, may be found in: D. Braun, H. Cherdron and H. Ritter, *Polymer Synthesis: Theory and Practice*, 3rd edn, Springer, Berlin, 2001 (a very useful introduction for newcomers to polymerisation procedures).
4. Free radical polymerisation is covered in detail, and experimental procedures for the latest techniques of "controlled" polymerisation are to be found in: K. Matyjaszewski and T.P. Davis (eds), *Handbook of Radical Polymerisation*, Wiley, New York, 2002.
5. "Controlled" radical polymerisation in disperse phase systems is reviewed by M.F. Cunningham in *Prog. Polym. Sci.*, 2002, **27**, 1039–1067.
6. High solids latexes and approaches to their preparation are dealt with by M. Schneider, *et al.*, *J. Appl. Polym. Sci.*, 2002, **84**(10), 1878–1948; E. Aramendia has reported the application of non-ionic polymerisable surfactants to the task in *J. Polym. Sci., Part A: Polym. Chem.*, 2002, **40**(10), 1552–1559.
7. G. Akay presents the application of flow induced phase inversion to dispersion preparation in *Chem. Eng. Sci.*, 1998, **53**(2), 203–223; L. Tong and G. Akay, *J. Mater. Sci.*, 2002, **37**, 4985–4992 give examples of the use in epoxy dispersion.
8. W. Ming, *et al.*, *Macromol. Chem. Phys.*, 1998, **199**, 1075; *Polym. Bull.*, 1998, **40**, 749.
9. D.H. Napper, *Polymer Stabilisation of Colloidal Dispersions*, Academic Press, London, 1983 (a thorough and comprehensive presentation of the steric stabilisation mechanism).

CHAPTER 9

Inorganic and Hybrid Coatings

C. CAMERON

9.1 INTRODUCTION

It is reckoned that paints have been used by humans for at least 15,000 years, most notable early examples being at Altamira in Spain and Lascaux in France. The impulses which drove Palaeolithic man in those first early days of consciousness to express his dreams, needs, ritual and religion in art and paintings is a fascinating story, but one which is beyond the scope of this book. There have been many milestones since in the history of coatings, demonstrating a remarkable sequence of technological innovation. By far and away the bulk of paints used during these 15 millennia could be classified as organic coatings. It is believed that new coatings technologies now coming to the fore and based on inorganic materials will be recognised by future generations as another key milestone. It is the purpose of this chapter to give the reader a flavour of the technologies used in inorganic coatings, their applications and the diverse possibilities for future new technologies.

Most of the polymer materials used in coatings during the past 40 or 50 years are organic, derived from petroleum products or from living things. Low cost starting materials and the enormous scope of organic chemistry to convert the available precursors into useful materials, and the properties of the resultant products are probably the most important reasons for this situation. Organic polymers are light in weight, most are excellent electrical insulators, resistant to corrosion and are relatively easily fabricated at moderate temperatures. Good mechanical properties such as impact resistance or flexibility are either an intrinsic feature of the polymer or else can be engineered by suitable modification. These properties often derive from the long chain structure of the polymer, as a result of entanglements and crosslinks.

Carbon-based polymers do have limitations though. Few organic substances can be heated for long periods above 150–200 °C without degradation, most commonly caused by reaction with oxygen (see Chapter 5). Fluorocarbon polymers or some types of aromatic polymers are perhaps exceptions here. Most organic polymers will also burn. Organic polymers very commonly dissolve in a variety of solvents or else undergo considerable solvent-induced swelling when crosslinked. This gives limitations in comparison to inorganic materials in advanced engineering applications where exposure to oils or hydraulic fluids is common. (Polysulfide polymers, commonly used to give fuel resistance as aircraft sealants, may be regarded as an inorganic–organic hybrid polymer.) Other applications, where flame retardancy, prolonged exposure to ultraviolet radiation, welding or maintenance of properties over very wide temperature ranges are required, have encouraged chemists to seek solutions by exploring the replacement of carbon atoms in polymers with other elements. These alternative elements are common components of two other main classes of engineering material – ceramics and metals.

From an engineering point of view, ceramic materials overcome many of the disadvantages of organic polymers – they are heat and oxidation resistant, hard and resistant to chemical attack. However, their manufacture is normally by high temperature processing which takes large amounts of energy. Ceramics are often brittle, lacking the flexibility of many organic polymers.

The other common class of engineering material – metals and their alloys tend to be tough, ductile materials with reasonable high temperature behaviour. They are, however, heavy and prone to corrosion.

Each of these three types of engineering materials clearly has advantages and disadvantages not shared by the other two. A common approach to the development of new materials is to modify the properties of organic polymer systems by the formation of hybrid systems, combining the desirable aspects of organic polymers with those of inorganic solids. Such an approach should provide useful polymer materials for coatings as long as the modification minimises the less desirable aspects of the components.

This chapter describes some important inorganic-based coatings which are currently used in the industry, and tries to give some idea of their importance for the future. It is emphasised that the vast majority of inorganic polymers used in coatings are indeed hybrids of inorganic elements and organic systems. A brief description of the main inorganic polymers follows, noting if and where they have some application in coatings technology. This is followed by a more detailed discussion on some key coating performance areas where inorganic polymers provide outstanding and in some cases unique attributes.

9.2   INORGANIC POLYMERS AND THEIR USE IN COATINGS TECHNOLOGY

There are very few purely inorganic polymers. Most systems are based on a mixture of organic groupings and inorganic elements. The inorganic atoms provide heat, fire and radiation resistance, biological inertness and electrical conductivity; the organic portions provide solubility, functionality (*e.g.* for crosslinking), hydrophobicity and other surface properties and many other characteristics.

Polymers based on a backbone comprising inorganic atoms with organic or organometallic side chains are conventionally described as inorganic–organic polymers (often just abbreviated to inorganic polymers). Organic–inorganic systems represent the opposite situation with an organic backbone and inorganic elements in side chains. Most inorganic polymers are hybrids of one kind or the other, and this is the classification we will use. Indeed, with one possible exception, there are no wholly inorganic polymers used in coatings – they are all organically modified in one way or another.

It should come as no surprise to the reader that silicon is by far the most important inorganic element in polymers next to oxygen. Indeed, polymers containing silicon and oxygen – polysiloxanes – are very well established. However, a glance at the periodic table should serve to convince us that many inorganic elements are capable of incorporation into polymers. Phosphorus, nitrogen, halogens, sulfur, boron, aluminium, germanium, tin, titanium, zirconium and a variety of other elements have all been used and incorporated in polymers. Polymers containing metals, either by direct bonding or by co-ordination is a rapidly increasing area of academic and practical interest. Polymers containing only carbon, hydrogen, oxygen and nitrogen will not be considered further as they really represent mainstream organic polymers.

9.2.1   Polymers with Inorganic Side Chains

The range of polymers with organic backbones with side chains containing inorganic elements is virtually inexhaustible. It would therefore be fruitless to attempt a comprehensive catalogue and only a few examples will be given.

Vinyl systems perhaps represent the most common polymers with inorganic side chains. By the very flexible and versatile methods of vinyl polymerisation, the amount of inorganic modification can be easily tailored to yield a polymer of the desired characteristics, such as T_g, solubility, reactivity or co-ordinating ability. Some examples are given below (Scheme 9.1), some of which exemplify the 'grey' area between describing the polymer as organic or inorganic.

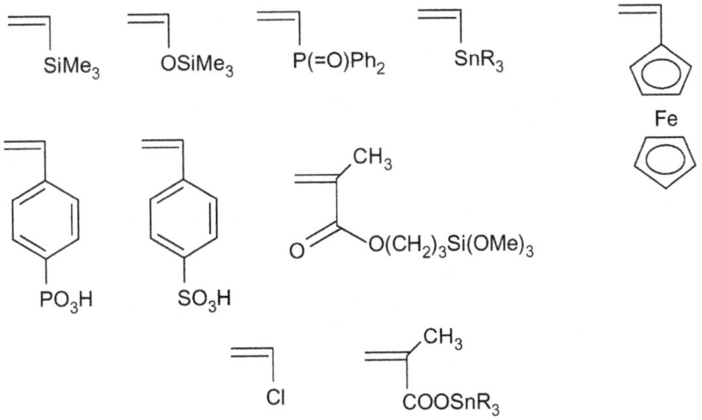

Scheme 9.1 *Examples of vinyl monomers used to prepare organic–inorganic polymers*

Polyvinyl chloride (PVC), for instance, is very important in some coatings technologies – coil coatings ('plastisol') and in beverage can coatings (where it is used as a sealant in 'easy-open' ends, where inexplicably it is called 'compound'). PVC, in principle, could be classed as an organic polymer with inorganic side chains, but in practical terms all polymer chemists accept it as an organic polymer.

Scheme 9.1 also contains an example of particular relevance to anti-fouling coatings. An acrylic copolymer backbone with tributyl tin ester side chains has, for many years, represented the state of the art in fouling prevention. In seawater, the hydrolysis of the tin ester bond (Scheme 9.2) provides the double benefit that the fouling organisms are killed by the tributyltin oxide, and the nascent carboxyl groups, once their concentration has exceeded a critical value, render the polymer soluble in the mildly alkaline seawater. This provides a very important enhancement of the economic performance of the vessel. The polymer dissolution provides a polishing action, resulting in a smoother outer hull, resulting in very significant fuel savings.

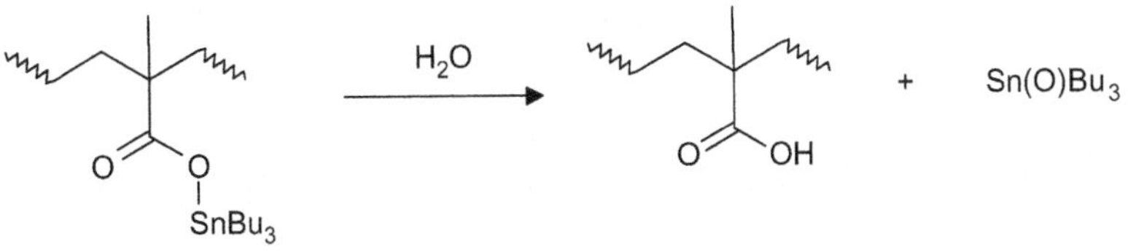

Scheme 9.2 *Hydrolysis of acrylic tributyltin ester groups*

A non-biocidal variant on this theme is to substitute the tin ester with a trimethylsilyl ester.

The need for these anti-fouling polymers to be soluble implies they are thermoplastics. Such non-crosslinked systems suffer from the disadvantage that high molecular weights are required to generate sufficiently robust mechanical properties. This increases the solvent requirements in order to deliver the right application viscosity. Coatings with a high solvent content are the subject of intense environmental pressures. However, it is the toxic tributyl tin oxide by-product which has led to a worldwide ban on tin-based anti-foulings. The mechanical properties of some otherwise weak anti-fouling coatings can also be enhanced by the incorporation of mineral silicate microfibres, but this does not on its own really count as a valid inorganic modification – it could simply be regarded as a special type of pigmentation.

Functional groups which form strong complexes with inorganic elements can also provide a good strategy for attaching desired inorganic groups on polymers. For example, poly(vinyl pyridine) – not an organic–inorganic polymer itself, according to our definition – possesses a strong tendency to co-ordinate to many metals.

A number of polymerisable phosphorus-based monomers are available and these, in particular, can be used to provide coatings with improved flame retardancy, improved adhesion or improved pigment dispersion characteristics. Vinyl polymers with cyclic phosphazene (discussed later) or borazine substituents are known. Indeed, it is even possible to attach inorganic rings to inorganic backbone polymers, giving a variety of new materials with interesting properties (*e.g.* a borazine ring pendent to a polysilazane chain), although only so far exploited in ceramics.

Brominated epoxy resins, for example, at one time represented a very important class of flame retardants, but their use is sharply diminishing, as a result of some very justified environmental concerns. Phosphated epoxy resins are often used in beer and beverage coatings to provide excellent adhesion and chemical resistance, as noted in Chapter 8.

Polymers with fluorine atoms in side chains are increasingly promoted as providing excellent solutions to adhesion and other problems encountered with service in extreme environments. Fluorinated epoxy resins (**1**) represent a good example and have been promoted for adhesive end-uses.

(**1**)

There are many other examples of polymers with inorganic side chain modification.

9.2.2   Main Chain Inorganic Polymers

Backbones can also incorporate inorganic elements. A simple method which results in only a small modification is to attach the desired species in a chain transfer agent. For example, the use of phosphorus-based chain transfer agents can have a dramatic effect on the pigment wetting ability of certain acrylic polymers. However, we are mainly concerned with polymers whose backbones themselves mainly comprise inorganic elements. Some of the more important systems will be briefly described, and their importance for coatings outlined.

9.2.2.1   Poly(organosiloxanes) or Silicones.   At the extreme, with no orga-nic modification, silica itself, SiO_2, is used in coatings polymer technology, although strictly speaking in a form that is itself inorganically modified. Silica glasses have very high ultimate T_g and a 3D network of silica may only be achieved by very severe calcination, requiring some 1800 °C. Under condi-tions of cure at ambient temperature, such a difference between $T_g(\infty)$ and the temperature of cure would almost certainly mean that, given ideal random step-growth reactions in homogeneous solution described in Chapter 4, no substantial network would be formed at all. Yet networks and gels are known. This implies a radically different evolution of the network structure, in part, due to the very low solubility of even oligomeric silica in water. However, it is also true that the process of random step-growth network formation is not always a very good model for what occurs in siloxane chemistry. This gives us an inkling that perhaps the simple 'rules' coatings scientists have developed and applied so successfully to organic systems may, in some cases, have to be radically altered for some important inorganic systems.

Polysiloxane chains, as found in glass and other mineral silicates, are often used as pigments and fillers in coatings. Without modification, such highly functional polysiloxane chains with very few atoms between the network junction points tend to be very brittle (**2**). Replacement of some of the siloxane groups with organic groups allows the flexibility and other properties to be dramatically improved.

         (2)   (3)

Polydimethylsiloxane (**3**), normally abbreviated as PDMS, represents the best known example and is used in stop-cock grease, car polishes and many other applications, including biomedical devices (low tendency to cause irritation or blood clotting). In particular, silicones are used as flow agents and anti-foam agents in coatings but care must be taken to optimise the amount present as an excess can easily lead to dewetting and the appearance of other defects in the film. Silicone polymers also have importance as release agents, providing a very low surface energy substrate from which it is very easy to detach any film applied on top of it. Such low energy silicone coatings are widely used by technologists involved in the characterisation of coatings – many useful techniques require a coating to be separated from its substrate in order to study its properties. Many adhesive technologies rely on silicone-coated paper to allow easy release of the adhesive tape. In the marine coatings industry, polydimethylsiloxane and related systems form the basis of non-biocidal anti-fouling coatings. Fouling organisms faced with such a low energy surface either cannot adhere or if they do, it is very easy to clean them off. These systems are discussed in more detail later.

The methyl-substituted siloxane polymers are often made from the catalysed ring opening polymerisation of cyclic species, such as octamethyl-cyclotetrasiloxane (**4**), which itself is prepared from dichlorodimethylsilane.

(**4**)

Other cycles and indeed linear polymers are also possible depending on the reaction conditions. The competition between linear polymerisation and cyclisation is subject to the usual considerations of thermodynamics and kinetics. Ring strain in siloxanes tends to somewhat favour the eight-membered ring, the probability of two unreacted ends meeting to react decreases as the ring size increases. It is possible to substitute the silicon with different alkyl groups, leading to a whole range of different polymers.

The cycles polymerise to linear polymers by heating in the presence of an acid or base catalyst. Low molecular weight systems ($<25,000$ Da) are fluids, and higher molecular weight systems are gums. Elastomers result if the materials are lightly crosslinked. Functional resins are increasingly important, and depending on the substitution, may be liquid or solid. Resins can also be made by hydrolytic degradation of specific mineral silicates such as olivine in the presence of trimethylsilylating agents.

It should be noted that poly(organosiloxanes) do have a tendency to depolymerise (not unlike sulfur) and form cycles at temperatures above 250 °C (the thermodynamic 'ceiling' temperature'). This temperature can be increased by phenyl substitution (5), crosslinking and by other chemical modifications beyond the scope of this chapter:

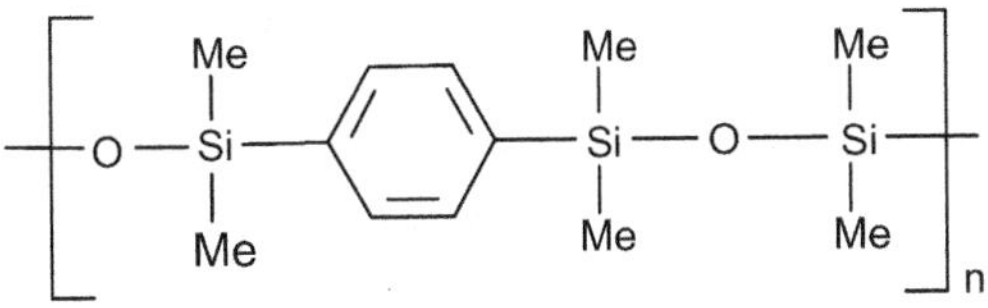

(5)

Crosslinking is a very important tool in increasing the maximum use temperature of siloxanes. If it is assumed that the networks degrade by the reverse of the crosslinking process and undergo a random bond scission, more densely crosslinked systems will be able to tolerate a greater degree of degradation (loss of crosslinks) before the integrity of the film is unacceptably compromised. Modelling of crosslinking is a very well trodden path: scientists working in the field of polymer degradation have only relatively recently turned their attention to modelling the degradation process. Expectations that design criteria for optimising systems for thermal resistance and other properties are high.

Compounds containing Si–OR bonds are known as silicon esters, or alkoxy silanes. Tetraethyl orthosilicate (6), TEOS, is a very common example, and this material has considerable importance in coatings, described later.

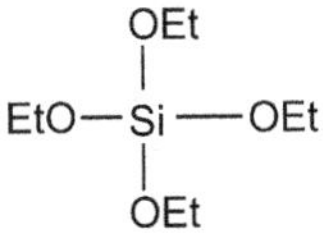

(6)

A very large number of silicon esters are available, covering all degrees of substitution with both alkyl and aryl groups, giving a wide range of physical properties. Thus, there is much scope to prepare an enormous number of variants, depending on particular requirements. A simple nomenclature (Scheme 9.3) has been devised for describing complex siloxane systems and is based on the degree of organic substitution. Compounds described as Q possess four Si–O bonds per silicon atom, T possesses three Si–O and one Si–C bond per silicon, D has two Si–O and two Si–C bonds per silicon and M has one Si–O and three Si–C bonds per silicon atom.

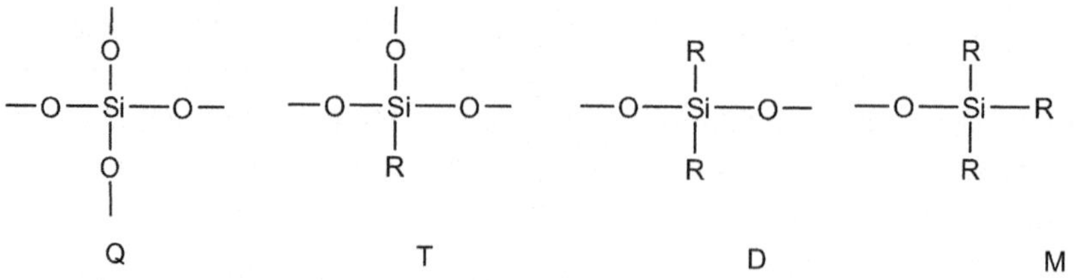

Scheme 9.3

Thus, hexamethyldisiloxane becomes MM, end-capped linear polymers are MD_nM and octamethylcyclotetrasiloxane is D_4. Many systems, made by mixing Q, T, D and M species in a desired proportion, are only partially hydrolysed and condensed, leading to unreacted alkoxy functionality, which is used in later reactions. DT, MQ and many other resins are commercially available. High levels of Q or T species lead to increased hardness and brittleness. The nature of the R group also has a significant effect on reactivity.

For crosslinked systems, Q and/or T species are essential. Crosslinking and other reactions may also be effected through the R group. By suitable functionalisation, the whole range of organic chemistry can be brought to bear on siloxane technology by producing reactive substituents. Materials are available where R is as shown in Scheme 9.4.

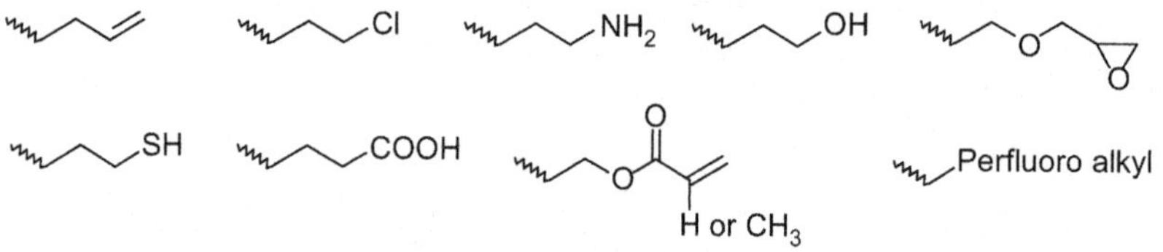

Scheme 9.4

This list is not exhaustive. Non-functional organic substituents may also be readily obtained. Many of the monomeric functional T materials are used as so-called adhesion promoters and are widely thought (and often assumed) to provide covalent bonding between a metallic substrate and an organic polymer, ideally as (**7**):

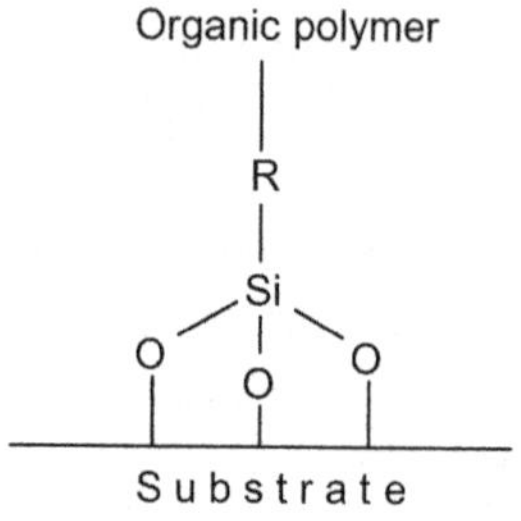

(**7**)

Many coatings industries use such adhesion promoters as additives in their formulations. Aminopropyltriethoxysilane (APTES) is very commonly used, particularly with epoxy systems. Although less reactive than the corresponding methoxy derivative, it is often preferred due to the toxicity of methanol. Aerospace sealants and coatings are, in fact, very rarely formulated without an 'adhesion package', and are often 'optimised' through trial and error, or in a statistical experimental design where the response measured is the adhesion. Various spectroscopic methods have been used to prove the presence of the silane at the substrate coating interface. The Si–O–metal bonds are thought to be much more resistant to hydrolysis than direct polymer–metal interactions. However, often no specific characterisation is undertaken to ensure the alkoxy silane, in fact, reaches the intended location at the substrate – it is rather assumed that any improvement in adhesion under the conditions of use is due to covalent bond formation as exemplified by (7). The use of silane coupling agents can appear to be a bit of an art – when used as a very thin pre-coat, the concentration of the solution, the time between application of the silane pre-coat and the overcoating, the pH, the temperature and even the direction of brushing the silane coating(!) have been claimed to be important. Non-functional silanes, in some cases, have been demonstrated to be more effective than the functional materials. Nevertheless, when formulated and used correctly, it is clear that such materials can provide significant improvements in adhesion and corrosion protection in many coatings technologies. More success has been obtained with aluminium as a substrate than with steel. Titanium, zirconium, zircoaluminate and other metal-based adhesion promoters are also available.

If, instead of an alkyl substitution on silicon, a hydrogen is present, further possibilities result, including hydrosilylation as a crosslinking reaction (Scheme 9.5).

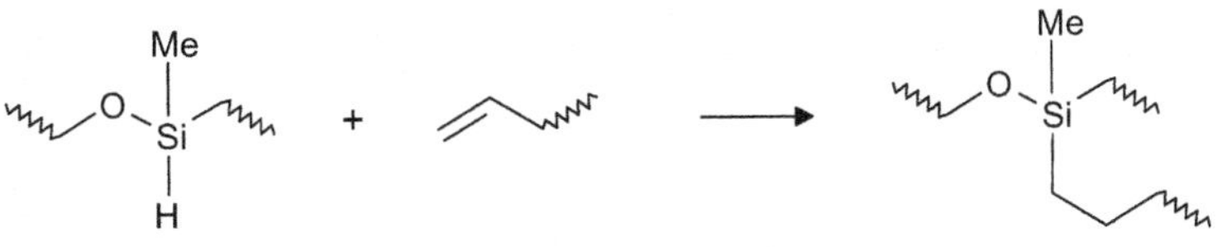

Scheme 9.5

A platinum catalyst is usually required for this reaction to proceed at a reasonable rate. Model siloxane networks of varying structure and functionality using the hydrosilylation reaction have been used to provide insight into many of the relationships between structure and properties in crosslinked silicones. Indeed, it was shown that mixing high molecular weight chains with low molecular weight chains (so-called bimodal networks) could give mechanical properties much improved over the properties of the individual mono-modal networks.

Siloxane systems have many desirable properties, depending on the particular system used. In terms of general coatings properties, systems can be designed for performance benefits in many applications. Some examples are given in Table 9.1.

Table 9.1   *Range of application of siloxane materials*

Property	Possible application
Low surface energy	Easy clean surfaces
Thermal stability	Heat resistant coatings
UV stability	Durable coatings
Interfacial activity	Adhesion promotion
	Corrosion protection
Low VOC	Compliant coatings

In addition, the mechanical properties can be very easily manipulated to provide for high flexibility (low T_g), or for high hardness. In inorganic–organic hybrid systems, the properties depend strongly on the morphology of the system. Very high chemical resistance can be achieved, by developing a combination of high crosslink density with as big a mismatch as possible between the solubility parameter of the inorganic polymer with the organic or other liquids with which it may be in contact. An exception, however, is resistance to alkali. The Si–O–Si bond is not resistant to hydrolysis – this makes it possible to generate low molecular weight polymers by the hydrolytic degradation of mineral silicates.

Many technologies make use of the reactivity of the silicon ester or alkoxy silane bond. In the presence of water such esters are hydrolytically unstable and convert to the silanol group and release an alcohol, depending on the nature of the alkyl ester (Scheme 9.6).

$$\text{—Si—OR} + H_2O \rightleftharpoons \text{—Si—OH} + ROH$$

Scheme 9.6   *Hydrolysis of alkoxy silanes*

The reaction is very heavily influenced by catalysis. Once the silanol has formed, several possibilities may occur (Scheme 9.7).

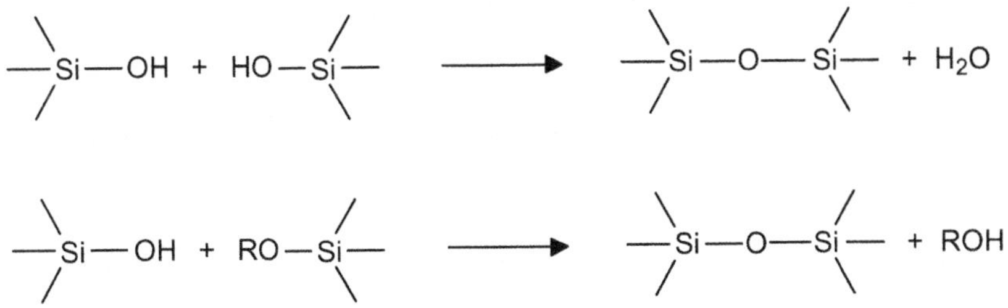

$$\underset{/}{\overset{\backslash}{-}}Si-OH \; + \; HO-\underset{\backslash}{\overset{/}{Si}}- \quad \longrightarrow \quad \underset{/}{\overset{\backslash}{-}}Si-O-\underset{\backslash}{\overset{/}{Si}}- \; + \; H_2O$$

$$\underset{/}{\overset{\backslash}{-}}Si-OH \; + \; RO-\underset{\backslash}{\overset{/}{Si}}- \quad \longrightarrow \quad \underset{/}{\overset{\backslash}{-}}Si-O-\underset{\backslash}{\overset{/}{Si}}- \; + \; ROH$$

Scheme 9.7 *Condensation reactions of silanols and alkoxy silanes*

Thus, two silanol groups may self-condense to form a siloxane bond and eliminate water, or a silanol may react with unhydrolysed ester to form the siloxane and release alcohol. Silanol groups will also condense readily with the groups shown in Scheme 9.8.

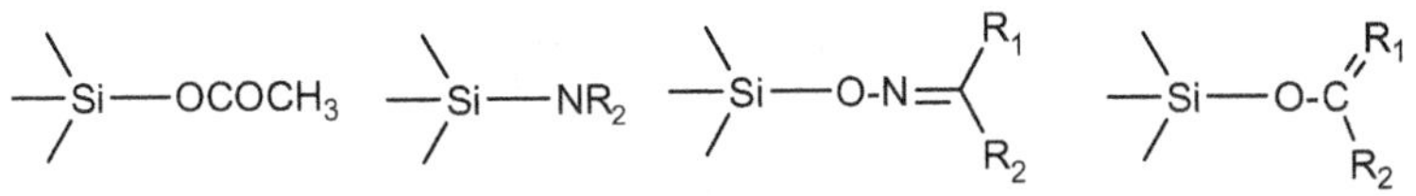

$$\underset{/}{\overset{\backslash}{-}}Si-OCOCH_3 \qquad \underset{/}{\overset{\backslash}{-}}Si-NR_2 \qquad \underset{/}{\overset{\backslash}{-}}Si-O-N{=}\!\!<^{R_1}_{R_2} \qquad \underset{/}{\overset{\backslash}{-}}Si-O\text{-}C{\overset{R_1}{\underset{R_2}{<}}}$$

Scheme 9.8 *Further reaction possibilities with silanols*

The reaction of alkoxy silanes is the most important in coatings and occurs readily at room temperature and above. It is the basis of an ever increasing number of coatings technologies, as evidenced by the explosion in publications concerning alkoxy silanes in the polymer and coatings journals and in the patent literature. The need for hydrolysis of the alkoxy silane to yield the silanol renders the system sensitive to humidity.

Depending on the size of the alkoxy group, the reactivity of the alkoxysilane can be tailored. Larger alkyl groups decrease the reactivity. The reactions are catalysed by acids, bases and organometallic systems, such as dibutyltin diacetate. Each type of catalysis leads to a different polymer microstructure which impacts on the properties of the coating systems. Acid catalysis (typically hydrochloric acid) leads to rapid hydrolysis and slower condensation reactions. Base catalysts tend to promote faster condensation reaction than hydrolysis, leading to more compact structures as the slowly formed monomer reacts with existing silicate clusters.

This difference in kinetics is further complicated by the so-called substitution effect where the reactivity of alkoxy silane groups depends on whether neighbouring groups have reacted or not. A good example is octaethoxy trisiloxane (**8**), where under acid conditions the terminal Si–OEt groups are more reactive than the internal ones:

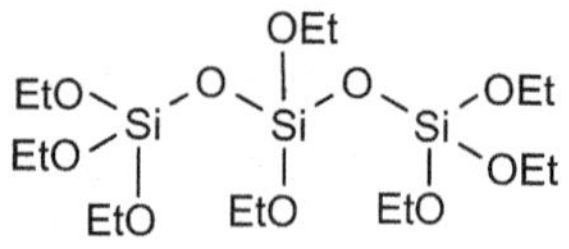

(8)

9.2.2.2 Network Formation in Siloxane Chemistry.

One of the big contrasts between the crosslinking polymerisation of silicic acid ($Si(OH)_4$) and network formation in organic systems is the heterogeneity of the products. As the silicate structures are formed, they rapidly lose solubility and precipitate. The precise nature of the product is more dependent on the reaction conditions from nanometre dimensions to large colloidal particles. If the particles are small enough, transparency can still be maintained. This is in contrast to the much more homogeneous organic systems.

A further difference arises from the non-random nature of the crosslinking of alkoxy silanes. Flory–Stockmayer theory applied to the self-condensation reaction of silicic acid suggests that the system should gel at 33% reaction of the silanol groups. Under acid catalysis, the conversion at gelation has been measured at greater than 83%! This is an enormous discrepancy and demonstrates graphically the difference between conventional organic polymer crosslinking and these highly functional very compact inorganic crosslinking systems. The substitution effect (where the reaction of a silanol group on a molecule adversely affects the reactivity of the remaining groups) can only explain part of the discrepancy, raising the gel conversion to a maximum of 50%. Random formation of inelastic loops during polymerisation likewise is insufficient to explain much of the difference.

The key difference between these inorganic systems and organic systems is the tendency of the former to undergo *non-random* cyclisation reactions to yield cage-like intermediates which then undergo further random polymerisation to yield the gel and ultimately the crosslinked network. The formation of cage-like intermediates such as (9) or (10) can explain the

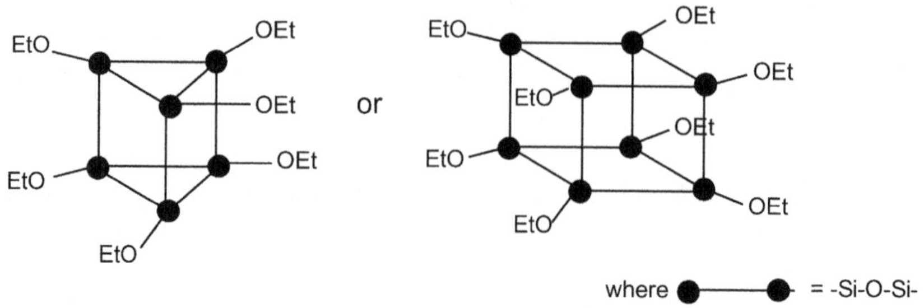

where ●——● = -Si-O-Si-

(9)   (10)

observed gel point. In both the triangular prism and the cube, 75% of all the available groups have reacted to yield a six functional species and an eight functional species, respectively. In order to gel, only 20% of the remaining six groups in the triangular prism are required to cause gelation by random reaction – this would give an overall conversion to gelation of 80%. When the substitution effect is considered and the possibility of random cycle formation, the observed gel point is not difficult to explain. This behaviour presents a major challenge to those who would wish to develop models of network formation and use them to predict the properties of new monomer combinations.

Where mixtures of alkoxy silanes are used, and where substitution effects operate or the esters are based on different alkyl groups with different reactivity, it can be difficult to ensure that the reaction conditions selected will lead to indiscriminate co-condensation of the materials. It may be that the kinetics of the condensation reactions is sufficiently different that a simple blend of two separate siloxanes is formed. Thus, any successful modelling approach must include the influence of the kinetics on the structure formation process.

Iler was therefore not exaggerating when he stated '...there is no analogy between silicic acid polymerised in an aqueous system and condensation-type organic polymers...'. Luckily, the lack of a robust set of 'rules' for the design of inorganic polymer systems for coatings industry has not prevented the more empirical development of inorganic–organic systems.

Where the reactive functionality is less concentrated on a single silicon atom, but is distributed much more evenly along a longer backbone, a behaviour much more akin to the random crosslinking reactions of functional organic polymers is to be expected, but this has not been explicitly proven yet. DT resins rich in the D component might perhaps exemplify this more conventional crosslinking.

Like Q materials, T species also have a great tendency to undergo cyclisation rather than polymerisation in the early stages of the reaction. T materials that possess organic functionalisation can yield reactive silsesquioxanes, exemplified by (**11**):

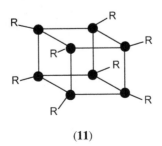

(**11**)

These organofunctional T-based cage structures represent another example of non-random reaction leading to the delay of crosslinking. *All of the alkoxy silanes have reacted to form a soluble low molecular weight cubic cage with pendent organic functionality.* The system is predicted to gel by Flory–Stockmayer theory at only 50% reaction!

A range of functional silsesquioxanes is available. These relatively costly materials are known as polyhedral oligomeric silsesquioxanes (POSS), and are promoted as additives to plastics and coatings to dramatically upgrade the thermal and physical properties, including hardness, impact strength, scratch and abrasion resistance. The functional groups on POSS cover the whole range of organic chemistry. POSS are receiving much attention as they represent the smallest silica nanoparticles available and have good transparency and solubility. They are claimed to increase the T_g of powder coatings without any viscosity penalty at up to 10% addition.

Intermediate in size scale between the mineral silicates and the monomeric siloxanes or silsesquioxanes are the silica sols, small particles bearing surface silanol groups, often present as salts of alkali metals. Such systems also have importance in coatings technology.

9.2.3   Other Silicon-Based Polymers

Polysilanes are polymers whose backbones are entirely comprised of silicon atoms. The Si–Si bonds allow unusual 'sigma' electron delocalisation and the materials find use as photoconductors, free radical initiators and in non-linear optical devices. They may be obtained as homopolymers (**12**) or copolymers (**13**) but as yet find little application in coatings:

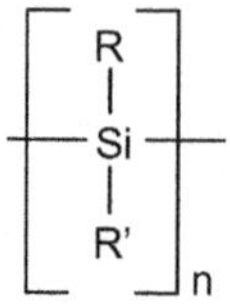

(**12**) 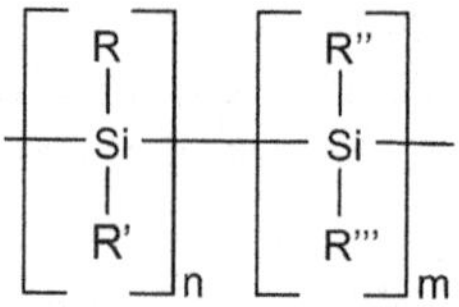   (**13**)

Polysilazanes, on the other hand, are the nitrogen analogues of poly(organosiloxanes) and have an Si–N repeat unit. They are made by the reaction of ammonia with an alkyldihalosilane (Scheme 9.9).

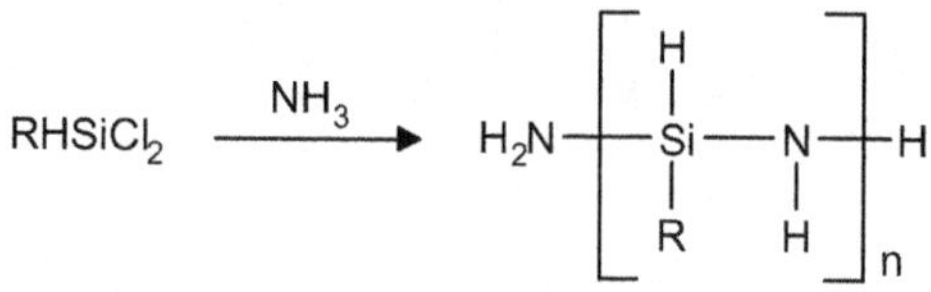

Scheme 9.9

By suitable processing, cyclic systems or ladder polymers may also be obtained. The secondary amino groups are reactive and give the possibility of curing reactions or organic modification. There is much interest in these materials at present for their potential as heat resistant coatings and corrosion protection coatings. Modified systems such as (**14**) or (**15**):

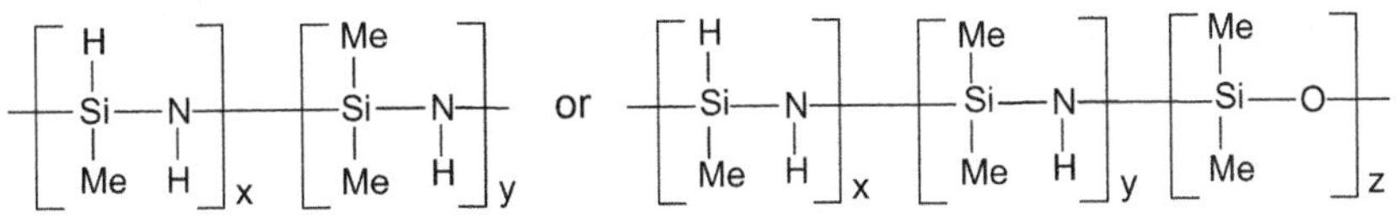

(**14**)   (**15**)

are available as low viscosity liquids.

9.2.4 Sulfur-Based Polymers

Polymeric sulfur itself is a purely inorganic polymer and is readily made by ring opening free radical polymerisation of rhombic sulfur, which consists of eight-membered rings. Depolymerisation of the chains will occur if the temperature exceeds around 175 °C. Sulfur chains have received much interest as they are the simplest possible structure – all chain atoms are identical and bear absolutely no substituents. End capping the sulfur chains leads to 'sulfanes', RS_xR, that have been used as coating additives, but are not common. Selenium and tellurium will also form chains similar to sulfur.

Poly(sulfur nitride) is an important polymer (though not in coatings), again prepared from the cyclic tetramer $(SN)_4$ – a species prone to detonation! The ceiling temperature is around 150 °C. The polymer conducts electricity and at 0.26 K is a superconductor.

Polysulfides, on the other hand, are very important. They are somewhat ill-defined from the point of view that they can have a variable number of sulfur atom sequences in the repeat units, separated by various types of organic segments. The repeat units can be based on hydrocarbons, ethers or formals. The most important polymers are made by the reaction of sodium sulfide with dichloroethyl formal (Scheme 9.10).

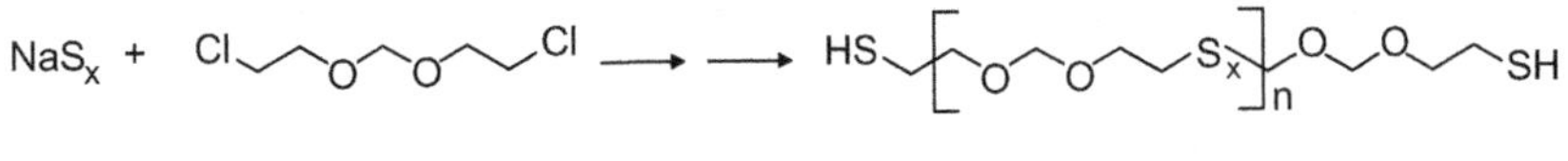

Scheme 9.10

Branching can be introduced by including a fraction of 1,2,3-trichloro-propane depending on the level of branching required. These high molecular weight polymers are often subject to controlled depolymerisation using mercaptans to yield thiol-terminated high sulfur content liquid resins. These systems form the basis of all high performance fuel tank sealants in the aero-space industry. Their flexibility at low temperatures and integrity at working temperatures in excess of 100 °C, coupled with their fuel resistance, render these somewhat malodorous polymers ideal aerospace elastomers. Their crosslinking is commonly accomplished by means of oxidative cure, usually with manganese dioxide, or else through base-catalysed reaction with epoxy.

9.2.5 Boron-Based Polymers

Many boron compounds, if the reader recalls his or her chemistry teaching, have an unfortunate habit of igniting in contact with air. Society would rapidly lose its respect for coatings technologists if such materials were routinely used in paint formulations. A good example is the linear, elastomeric poly(boron fluoride). Carboranes, similarly, can be made into interesting polymer structures, but have so far not yet penetrated the coatings world.

The boroxines are analogues to alkoxy functional cyclosiloxanes. These materials have been used to cationically polymerise epoxy resins to typically give coatings with improved thermal stability and flame retardancy. Trimethoxy boroxine (**16**) is common.

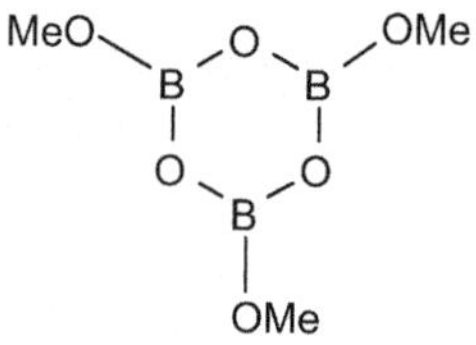

(16)

The reaction with epoxy is self-propagating in the bulk as it is strongly exothermic, but when performed isothermally in thin films at low tem-perature, very poor curing performance is observed. As with alkoxysilanes, the boroxines are sensitive to moisture.

The nitrogen-based analogues to boroxines are known as borazines and polymers are known (**17**), but the formation of cyclic by-products (**18**) is a common problem.

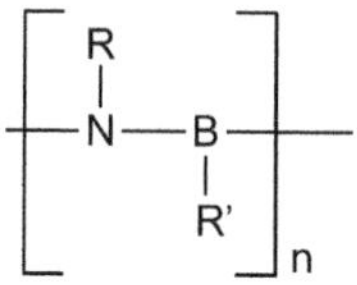

(17) 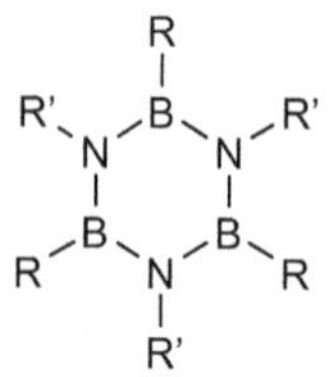   (18)

The resins are sensitive to water but have very good thermal stability. They are often used as precursors to boron nitride ceramics. The polymorph of boron nitride $(BN)_x$ with the diamond-like structure is one of the hardest substances known to man, and is highly resistant to water and oxygen. Unfortunately, like silica, it requires very high temperatures for manufacture – calcination in excess of 1000 °C is often necessary. Boron nitride finds use in crucibles and other refractory materials, but not much application in coatings.

9.2.6   Phosphorus-based Polymers

Many polymers that can react with acids can be easily phosphated to provide for a variety of new properties, including wetting ability, adsorption, adhesion and flame retardance. There are numerous examples of conventional organic polymers modified in this way to provide useful coatings binders.

However, some of the most rapid advances in inorganic polymer research lie in the field of phosphazene chemistry. Like silazanes, phosphazenes comprise an alternating sequence of phosphorus and nitrogen atoms, with two substituents attached to the P atoms (**19**):

(**19**)   (**20**)

The polyphosphazenes are very varied in their properties, ranging from very low T_g elastomers, biomaterials, hydrogels, liquid crystalline materials, semiconductors to flame retardant polymers. This range in properties derives from the great flexibility in choice of substituent groups on the phosphorus atoms. The synthesis methods involve ring opening polymerisation of cyclic phosphazene monomers, usually chlorine substituted such as hexachloro-cyclotriphosphazene (**20**), followed by nucleophilic substitution reactions on the polymers, commonly using alkoxides or amines. Substitution is necessary to stabilise the polymer as it slowly hydrolyses to yield ammonia, hydrochloric acid and phosphoric acid, not the most pleasant combination.

The high chain mobility and flexibility of the phosphazene polymer is very striking. It is perhaps this property which makes the substitution reactions so facile. Normally substitution reactions on polymers are more difficult than their small molecule counterparts, as a result of steric shielding and polymer coiling effects; not so with poly(dichlorophosphazene). The chlorine atoms can even be substituted with relatively non-nucleophilic groups such as those derived from trifluoroethanol. The resulting polymer product (**21**)

$$
\left[\begin{array}{c} OCH_2CF_3 \\ | \\ -N{=}P- \\ | \\ OCH_2CF_3 \end{array}\right]_n
$$

(**21**)

has a T_g of $-66\ °C$ and retains its flexibility at temperatures in excess of $200\ °C$. The fluorinated side groups have been claimed to render the resin more water repellent than teflon or silicone resins.

The literature on the use of such resins in coatings is quite sparse, but is increasing. Their synthetic flexibility suggests it will not be long before useful applications become commercial reality.

9.3 SPECIFIC EXAMPLES OF INORGANIC COATINGS BY END USE

The previous section discussed both mainstream inorganic systems widely used in coatings through systems whose potential is still largely untapped, to systems which are purely of academic interest (at present). There are a large number of inorganic-based coatings available, so it is not possible to be exhaustive. This section gives the reader, it is hoped, a flavour for the types of coating available and some details on specific systems of importance.

In Chapter 12 zinc and iron phosphate-based conversion coatings as pre-treatments have been mentioned. The zinc system is an integral part of many automotive paint schemes while the iron-based systems are commonly used on domestic appliances. In some respects these systems could be regarded as inorganic coatings but they will not be discussed further in this chapter.

9.3.1 Corrosion Protection Coatings

Protection against corrosion is probably the single most important performance attribute of functional coatings. The technology has seen many technical milestones, many of which are no longer used in modern products for one reason or another. The declining use of red lead in alkyds and coal tar pitch in epoxies are good examples of the impact of greater health, safety and environmental concerns on the product offerings of paint companies. It may surprise readers to learn that according to biblical sources, pitch was applied to Noah's Ark, although not of course as an anti-corrosive. Clearly, the survival of humankind was more important than the risk to health due to the presence of polycyclic aromatic compounds. In terms of inorganic coatings, there are numerous examples of which only two will be dealt with in detail.

Purely inorganic coatings find widespread use in corrosion protection and two examples spring immediately to mind. These are:

- metal spray coatings and
- zinc silicate coatings

Metallisation arc and flame spray of zinc and aluminium have been used since 1922 to protect structural steelwork. These coatings are mainly used for steel exposed to windborne salt spray, or in immersion conditions experienced in coastal regions and in the marine environment. They are particularly useful for protection of splash zones on offshore oil rigs and production platforms, where the continual cycle of wetting and drying is especially aggressive. These costly high performance coatings can permit the use of cheaper substrates, providing long-term corrosion protection and wear resistance. Nowadays, the application of a wide range of metals is possible, including nickel, copper, molybdenum, zirconium and many others, as well as alloys. For those anodic protecting metals such as nickel, it is vital to ensure a defect-free coating; otherwise, base metal corrosion will be strongly accelerated.

As indicated, there are two main methods of application – weld surfacing and thermal spray. Weld surfacing can be carried out using all the standard welding techniques, including manual metal arc, metal inert gas, tungsten inert gas and submerged arc welding. High application rates can be obtained, producing an exceptional bond strength to the substrate. Weld surfacing has the disadvantage that it often leads to metal distortion because of the high temperatures involved, and thickness of less than 1 mm is hard to achieve.

Thermal spray generally leads to coatings of inferior quality due to their higher level of porosity and lower bond strengths. Methods of application include flame spray, arc spray, plasma spray and high velocity oxygen fuel spray. Many of these techniques can only be used in a factory setting. Often the coating process is extremely noisy – often in excess of 100 dB.

Metal spray is also used to protect vulnerable substrates such as chemical tanks from chemical attack, providing an impermeable barrier to many chemicals and avoids the problem with organic coatings which often absorb the contents of the tank and contaminate the subsequent material stored in the system. In order for this to be successful, the spray process must not lead to porous films. Metal spray can also provide friction and wear control.

However, such systems cannot really be regarded as coatings in the traditional sense of a pigment dispersed in a polymeric binder with solvent, water or air as carrier. Much more conventional are the inorganic zinc silicates.

Zinc silicate coatings have been used for many years in the heavy-duty coatings markets. The materials exist in both solvent and water-borne forms and provide outstanding corrosion protection to metal substrates (such as mild steel) by means of the galvanic action of the zinc. In the marine industry zinc silicates are primarily used as blast or shop primers and are applied to freshly blasted steel in an automated process, to provide corrosion protection during storage and construction of pre-assembly blocks and prior to application of an epoxy anti-corrosive primer coat. Inorganic-based materials are crucial for this purpose as organic polymers can lead to weld porosity when the steel plates are joined together during block construction.

Zinc silicate coatings are also used in many other applications where corrosion protection is vital, including the oil and chemical industries, where the coating is often used as the primer coat in a multi-coat scheme. The zinc silicate coating is generally applied as a relatively thin film – excessive film thickness can lead to extensive cracking as the material cures over time and the diffusion of volatile components out of the film causes shrinkage stresses which the film cannot tolerate.

The solvent-borne variants comprise solutions of pre-hydrolysed tetra-ethoxysilane (TEOS) polymer, commonly in alcohol, condensed to a certain degree prior to formulation with the zinc pigment and other ingredients. These low organic content binders do not cure particularly well under low humidity conditions.

The coatings are normally formulated to a pigment volume concentration (PVC) above the critical PVC. The dried coatings are therefore porous. On a relatively superficial level, their corrosion protecting properties arise from the combination of the electrochemical potential of zinc relative to the steel substrate and the presence of a conducting pathway through the coating *via* the particles of zinc. The zinc therefore provides a sacrificial anode to prevent corrosion of the steel, at least in the short term. In addition, it is believed that the zinc corrosion products ('white rust') block the pores in the coating and contribute to the overall corrosion performance *via* a barrier mechanism. The zinc apparently also participates in the network formation process, possibly through the presence of Zn–O bonds (**22**):

$$-\overset{\displaystyle |}{\underset{\displaystyle |}{Si}}-O-Zn-O-\overset{\displaystyle |}{\underset{\displaystyle |}{Si}}-$$

(**22**)

The system is therefore more complex than simply a percolating network of zinc particles embedded in a silica matrix. The presence of the difunctional zinc in the network presumably yields lower crosslink density systems. The porous nature of the films renders them very friable, so mechanical damage tends to be highly localised and does not propagate under the film. With zinc in contact with the metal substrate, such damaged areas tend not to be critical as the zinc will provide cathodic protection.

The coatings are very high in VOC, and usually stabilised against premature condensation in the can by the presence of alcohol. On application of the thin protective layer, the volatilisation of the solvent and the ingress of water promote further silicate hydrolysis and condensation reactions. Taken to extremes however, excessive crosslinking can lead to very brittle coatings whose fracture properties are very poor. The resultant general cracking resembles the appearance of a dried-up river bed during a severe drought (mud-cracking). This can also occur on over-application of the coating.

The water-based variants have commonly been based on the soluble alkali silicates. These systems suffer from the major disadvantage that if the silicate is overcoated without washing out the alkali metal salts and then immersed in water, catastrophic osmotic blistering and failure of the coating can occur. A sealer or tie coat may be necessary prior to application of the topcoat if blistering is to be avoided. This tie coat often contains a carboxylic acid functional polymer to neutralise the alkali metal salts still present in the shop primer coat.

Due to the porous nature of the zinc silicate basecoat, which can often lead to blistering in its own right, some products require the application of a very thin 'mist coat' when overcoating. This avoids too much of the topcoat soaking into the porous basecoat which can lead to pinholing and blistering. The tendency to blister is especially severe under high humidity conditions. However, zero VOC water-based systems with excellent resistance to blistering without washing or the need for a tie coat have been produced. This is a major technological advance and amply demonstrates the role that even limited understanding of design principles and the inherent chemistry can play in the product development process.

Coating systems based on purely silicate binders should not be used in situations where exposure to alkali is likely. There are a number of common applications where this is the case. Seawater, for instance, has a pH in excess of 7. The outer hulls of ships and ballast tanks are commonly additionally protected against corrosion by means of sacrificial zinc anodes, which leads to formation of hydroxide ion at the cathode as the zinc corrodes. Chemical cargo tanks often have to carry hot sodium hydroxide.

In non-immersion situations, there is enormous interest and effort in the development of hybrid coating technologies, comprising mixtures of organic polymers and inorganic systems, usually alkoxy functional silanes. Many of these materials are known as 'nanocomposites', implying the existence of separate phases on a very small size scale. There has been an explosion of publications on so-called nanocomposites in recent times and it is increasingly hard to distinguish the really inventive good ideas from simple variations on the theme. The morphology of the composite material is key to the development of properties.

Metal alkoxides, often as mixtures of more than one metal, are also commonly used as wash primers for aluminium in the aerospace industry.

9.3.2 Thermal Resistance Coatings

Another very demanding end use for inorganic-based coatings is in the long-term protection of pipe work and other high temperature surfaces in industrial, oil or chemical installations. Often the temperatures can reach 400 °C and vary widely between temperature extremes on relatively short time scales. The result is an exceptional challenge to the coatings formulator. Atmospheric corrosion problems are often the most severe in areas where high temperature surfaces are subject to cyclic conditions. Often the cyclic conditions cause changes to the coating film which causes them to crack, flake or even delaminate. Where insulation is involved, this can lead to an even more aggressive environment. Mishandling often damages the cladding used to prevent ingress of moisture. When this occurs, water can penetrate the system and ideal corrosion conditions can pertain. The situation can be even worse in offshore or coastal installations where seawater, containing salts, can be used for hosing down or for fire control (which is often regularly tested).

For applications involving temperatures of less than 120 °C or so, organic coatings may serve the purpose well, but for higher temperatures there are few options beyond inorganic systems. Many of these inorganic systems are themselves applied over inorganic zinc silicates of the type described above but this has led to such severe problems (the so-called 'polarity reversal' effect) under insulation which becomes wet that zinc silicate coatings are not recommended by some professional bodies. Others prefer to seal the zinc silicate coating with an aluminium pigmented siloxane system so as to obtain the full benefits of corrosion protection by the zinc silicate during construction. These aluminium coatings often have to experience a high temperature in order to develop their properties fully, and are limited in the film build which can be achieved before the water generated during cure causes blistering and/or adhesion loss.

Inorganic moisture cured siloxane coating systems have been developed which allow application of higher film builds and give good performance without requiring any heat treatment. Multiple applications can be beneficial to protection. However, performance is usually much better when applied over the zinc silicate pre-construction primer. This gives concerns under conditions of high temperature and under wet insulation. One preferred type of siloxane resins is the phenyl-substituted system. These give better thermal stability than the purely methyl-substituted siloxane resins. The pigmentation is often of the barrier type – particularly plate-like high aspect ratio pigments, like aluminium, which can act to curtail the effects of shrinkage stresses during cure. In order to reduce this effect, largely TEOS-based systems are not common – resins of lower functionality density are more widespread. Most of the major suppliers of siloxane resins can provide a range of options, depending on end-use requirements, and where solvent emission is an issue.

Somewhat closer to home, fluoropolymer and silicon-based powder coating technologies for 'non-stick' applications such as cookware are increasing in importance.

9.3.3   Water Repellent Coatings

These materials are almost exclusively based on siloxane chemistry and find widespread use in protecting masonry and in general building protection. Whilst silicone rubbers are used to seal masonry, the water repellent coatings do not block pores or capillaries but instead reduce the water absorbency of the mineral substrates by binding to the substrate and presenting a hydrophobic organic surface to any incoming moisture. The coatings consist of reactive silicone materials – usually alkoxy silane functional – sometimes formulated with non-functional silicone fluids. In general, for use on impervious surfaces such as concrete, the coatings are rich in organosilanes (essentially monomeric T units) whereas for highly absorbent substrates, the coatings are richer in oligomeric siloxanes (D units). Synthetic resins are often added to improve oil repellence as well. It is emphasised that the pores and capillaries remain open to the ingress of moisture. The alkoxy silane groups bind to the substrate while the organic functionality acts as a water repellent. Thus, the ability of the substrate to 'breathe' is not impaired. The coatings are available in solvent-borne form and as emulsions in water.

9.3.4   Chemical Resistant Coatings

Coatings technologists rationalise chemical resistance properties as due to two factors:

- crosslink density
- thermodynamic interactions between coating and chemical

In general, crosslinked coatings will not dissolve in a liquid, but rather they swell, the crosslinks providing a restraining influence on the desire of the chains to be separated by the solvent. The greater the density of crosslinks, the greater the restraining force and the lower the tendency to swell.

The second factor defining the degree of swelling is the degree of thermodynamic interaction between the liquid and the polymer segments of the crosslinked coating. The more favourable the interaction, the greater the swelling. Entropic considerations will always favour swelling but if there is a strong specific enthalpic interaction between the coating and the liquid, swelling will be enhanced. If a tank is designed to carry a variety of different liquids, then depending on the particular details of the system there may be a greater or lesser tendency to absorb the liquid into the coating. After the tank is emptied and refilled with another liquid, there will be ingress of the new liquid into the coating and egress of the previous liquid into the contents of the tank. Contamination therefore occurs and this is a very serious problem. In some cases strict cargo sequencing is necessary to minimise the contamination issue.

There has been much interest in developing coatings with both higher crosslink densities and with much lower specific interactions. The use of hybrid inorganic–organic polymers in this regard has been quite successful. In one particular commercial product, an epoxy functional material bound to silica particles was described and the whole system crosslinked by means of conventional amine-curing agents. The end result was a hybrid system – although by no means a nanocomposite – with much reduced absorption of certain organic liquids compared to the conventional epoxy–amine coatings.

9.3.5 Abrasion Resistant Coatings

Hybrid coatings based on siloxane polymers and epoxy resins (and other organic polymers) hold much promise for a new generation of highly abrasion resistant coatings. Such coatings will find use in the cargo holds of ships, the outer shell of ice-breaking ships and many other applications where abrasion resistance is important.

Various types of hybrid structures can be envisaged, none of which may be represented by a truly homogeneous copolymer network. Instead, a phase-separated structure is the norm. Impregnated hybrids may be prepared by polymerising organic species in a preformed porous inorganic network. Entrapped hybrids are prepared simply by dissolving or dispersing an organic polymer in a sol–gel solution. One of the most important methods in coatings technology is the formation of chemically bonded hybrids involving *in situ* polymerisation of both inorganic and organic components. Coupling agents can also optionally be used particularly if the functionality on the inorganic matrix and the organic polymer modifier are not chemically compatible. It is well known that the properties of a composite material comprising a matrix and a dispersed phase are very often enhanced by ensuring covalent bonding between the two phases. Covalent bond format-ion between the organic chains and inorganic clusters can help prevent macroscopic phase separation and result in the formation of 'homogeneous' and transparent nanocomposites.

Key to the successful development of the required properties is the morphology of the hybrid system. In general, improved properties are obtained if the phase separation of the silica takes place through a spontaneous thermodynamic process rather than an activated nucleation and growth process. The former leads to a much more intimate contact and reinforcement of the organic phase compared to the latter where particles are invariably formed. Often gelation or vitrification arrests the phase sepa-ration process. However, if the 'co-continuous' morphology (as it is known) is formed, tremendous improvements in coating hardness and scratch or abrasion resistance can be made.

9.3.6   Durable Finishes

A range of coatings technologies has been used over the years to provide outdoor durable systems. Alkyds are essentially the 'workhorse' of the industry while aliphatic isocyanate or alkylated alkoxy melamine cross-linked acrylic polyols have traditionally provided high gloss coatings with superior resistance to weathering. The failure modes of exterior durable coatings are described in Chapter 5. Where service lives of 20 years or more have been required, compromises have been necessary – for example, the use of acrylic modified poly(vinylidene difluoride) gives very long colour and gloss stability although the absolute values of the gloss are lower and cost can sometimes be prohibitive.

Inorganic modification has been used to improve performance. One modern innovation, although perhaps not strictly involving inorganic polymers, is to modify the coating with nanosized particles of titanium dioxide or zinc oxide. These materials are transparent and absorb much of the harmful UV radiation, and have the advantage of high temperature stability, negligible leaching and do not degrade with time like other UV stabiliser technologies. The formation of radicals at the surface of the pigment (the so-called photo-catalytic effect), which can lead to performance limitations in the more durable resin systems, is minimised.

One of the best known examples is silicon-modified alkyds. Early variants simply involved the addition of a silicone resin to the alkyd towards the end of the alkyd synthesis, with little, if any, chemical linkage between the alkyd and the silicone. The extent of the improvement in durability depends on the extent of modification, with levels of around 30% being very common. The silicone resins used tend to have phenyl groups, methyl and higher alkyl substitution in order to enhance compatibility. Those silicone resins with greater phenyl contents tend to give more rapid drying. A good example is given by (**23**), a product of Dow Corning, whose nominal structure is:

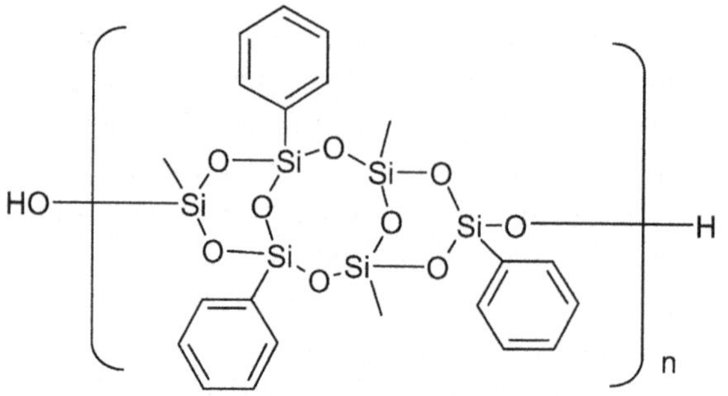

(**23**)

Further performance gains can be made by ensuring the silicone resin and the alkyd actually react together through the unreacted hydroxyl groups on the alkyd resin. Modification of polyester and acrylic resins is also known. There are many Si–OH containing silicone resins available for modification. One issue is the tendency of the silanol functional material to self-condense compared to co-condensation with the organic resin. Titanium alkoxide catalysts have been found to be excellent at promoting co-condensation. If water is present, then zinc or bismuth octanoates are preferred as the titanium alkoxide is rapidly hydrolysed to titania.

Often it is preferred that the modifier is a silanol ester and the co-condensation reaction is a 'transesterification' reaction between an alkoxysilane and an alcohol (Scheme 9.11).

Scheme 9.11

This is the basis for the 'ESCA' technology (epoxy silanol curing acrylic) of Kansai Paint Co. of Japan. In the ambient curing variant, the silanol group was found to be necessary, but in the elevated temperature cure, the alkoxy silane was appropriate. According to Kansai, silanols may also react with epoxy groups although their technology was limited to cycloaliphatic epoxies. Excellent durability performance was demonstrated but the system

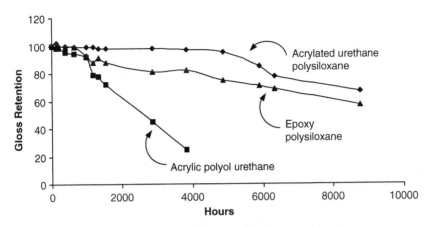

Figure 9.1   *Gloss retention of polysiloxane coatings after QUV-A exposure*

has not really caught on, for a variety of reasons. Perhaps the hydrolytic stability of the Si–O–C bond is the limiting factor.

Recently, so-called polysiloxane coatings have been developed for the ambient cure protective coatings market. These are not pure inorganic systems as suggested by the name but are inorganic–organic hybrids. One-pack variants and two-pack variants are known. The one-pack systems are commonly based on acrylic copolymers containing trimethoxysilylpropyl methacrylate and an alkoxy silyl functional resin. Problems with premature crosslinking due to adventitious water in the formulation (*e.g.* from pigments) and catalytic effects of the pigments result in limited shelf stability. The two-pack systems are more robust, and variants based on durable epoxy functional resins and acrylated urethane resins formulated with aminopropyl tri(m)ethoxy silane and other alkoxysilane functional resins form the current state of the art. Figure 9.1 gives a comparison of the gloss retention data for the acrylic polyol urethane systems with the epoxy and the acrylated urethane polysiloxanes.

It is common practice to compare the time to 50% gloss loss in QUV-A as representative of the relative resistance to photo-oxidation of various paints. A discussion on the merits of this practice is beyond the scope of this book. It is clear, however, that inorganic modification has raised the performance level of exterior durable finishes to a new level. Many more innovations and improvements are expected in the years to come.

9.3.7   Anti-fouling Coatings

The use of methyl/phenyl silicone resins and elastomers for use as non-biocidal anti-fouling coatings has really come into its own recently following worldwide legislation outlawing the use of tributyltin esters as the most common biocidal anti-fouling system despite the existence of commercial formulations for several decades. This is simply indicative of the cost-benefit analysis of the situation. In order to deliver the required performance, multicoat schemes are required. This is because the silicone elastomers show poor substrate adhesion to materials other than themselves. The coating scheme first involves one or two coats of an anti-corrosive primer as the hydrophobic silicone elastomer coating is not impermeable. If the substrate is a ship previously coated with a TBT-based anti-fouling, a primer-sealer coat is applied which seals the underlying tin-based anti-fouling. This is a simpler and more environmentally friendly alternative to blasting the hull of the ship to remove the spent toxic tin coating. This is then followed by a tie coat which is designed to stick to the primer and present a surface to which the silicone anti-fouling topcoat can adhere. The pigmented finish coat is then applied to produce a very smooth, durable hydrophobic coating. It is believed that the coating resists fouling simply by dint of its very low surface energy. Acrylic resin variants based on trimethylsilylpropyl methacrylate are more expensive than the silicone systems. At present, the non-biocidal systems are not wholly resistant to the settlement of fouling organisms, but where fouling occurs, it is very easily removed from the non-stick silicone surface. Thus, the coating system is more effective with ocean going vessels, which remain at sea for long periods and travel at speeds where it is difficult for the organisms to settle. For slower ships or yachts which remain moored for long periods, fouling will occur, but can usually be cleaned off with low pressure water spray.

Fouling control is also important in offshore structures, mainly between sea level and depths of around 30 m. Fouling can reduce the stability of the platform and increase the wave loading. Seawater intake systems are a major problem, particularly in power plant cooling water systems where fouling effectively reduces the diameter of the pipe, reducing cooling efficiency. Environmental legislation also pertains to these structures. Fouling can also be responsible for corrosion problems in these areas.

Current silicone elastomer coatings generally have long lifetimes as they do not rely on leaching of a biocide for performance. The main issue with silicone elastomer coatings is their relatively weak mechanical strength and their poor abrasion resistance. Indeed the severity of the abrasion environment is the limiting factor in determining the lifetime of the fouling control coating. Developments are underway to improve both these aspects of the technology as well as provide for more effective fouling control.

Fluorinated polymers and fluorosilicone hybrid systems have also been investigated but no specific advantages have yet been realised, other than improved abrasion resistance. Fouling performance is generally substantially inferior to that of the silicone elastomers.

9.4   SUMMARY

This chapter has attempted to describe the types of inorganic polymers available, their incorporation into organic coatings and specific applications of these systems. An indication of the different challenge inorganic-based coatings present to the coatings technologist has been given. It is hoped that the reader will be persuaded that there are significant advantages to be gained by delving into inorganic chemistry. It is certain that this important technological area will become more and more widely exploited and (hopefully) understood by its practitioners.

9.5   BIBLIOGRAPHY

1. *Hybrid Organic–Inorganic Composites*, ACS Symp. Ser. 585, 1995.
2. R.K. Iler, *The Chemistry of Silica*, Wiley, New York, 1979.
3. I. Manners, *Synthetic Metal-Containing Polymers*, Wiley, New York, 2003.
4. J.E. Mark, *Acc. Chem. Res.*, 1985, **18**, 202.
5. J.E. Mark, H.R. Allcock and R. West, *Inorganic Polymers*, Prentice Hall, Englewood Cliffs, NJ, 1992.
6. P.C. Painter and M.M. Coleman, *Fundamentals of Polymer Science*, Technomic Publishing, Lancaster, PA, 1997, Chapter 9, p. 307.
7. See, for example, M.H. Swindler, *Ancient Painting*, Yale University Press, New Haven, CT, 1929.
8. C. Schoff, *Coating Film Defects*, FSCT Monograph.

CHAPTER 10

Coatings Components Beyond Binders

ALAN GUY

10.1 INTRODUCTION

Earlier chapters have described the central function of polymers in the chemistry of film formation. However, a successful coating usually requires properties that cannot be provided by any one component alone. The role of the paint formulator is to bring together the required constituents in a stable, cost-effective composition that can be conveniently applied to the substrate. No small challenge!

Paint consists of a dispersion of a pigment or a mixture of pigments, extenders, *etc.*, in a binder or polymer. Other materials may be present to achieve specific properties. They may be organic solvents or water to give the required viscosity, suspending agents to keep the paint in good condition during storage, driers and accelerators which provide for rapid cure of the polymer, flow aids, and so on. These materials will be described in this chapter.

10.2 PIGMENTS

Pigments were in use before the birth of civilisation. Coloured minerals, and materials such as charcoal, were used to colour the bodies and living spaces of primitive man. As his skill grew, the use of minerals expanded to include the colouring of pottery, ceramics, and eventually glass, and the preparation of mixtures with media such as natural oils, and other resinous materials, to make paints.

From these early beginnings of the use of paint for decorative and artistic purposes has come the highly technical and commercially important paint industry of today.

A pigment is a coloured or non-coloured, black or white, particulate compound which can be dispersed in a medium, resin, or polymer, without being dissolved or appreciably affected chemically or physically. When paint is applied as a thin film over a substrate, the dispersed pigment will absorb and scatter light.

Dyes or dyestuffs, on the other hand, are usually soluble in paint media and give transparent or translucent films. This property obviously limits the utility of dyes in the coatings industry to products such as inks and stains.

10.2.1   Properties Imparted by Pigments

10.2.1.1   Colour.   The primary use of coloured, black or white pigments is to produce films which when applied to substrates such as metal, wood or concrete give the substrate a uniform distinct colour. The colour is selected for aesthetic reasons and must be durable, without fading or darkening, and the film must last for some considerable time when exposed to a wide range of different environments.

10.2.1.2   Opacity.   To give a uniform colour in a minimum thickness, the pigment must scatter, reflect or absorb light to prevent it reaching the substrate. White pigments opacify media by scattering and reflecting light, whilst coloured pigments also rely on the absorption of light to give colour and opacity. The degree of scattering, hence opacity, is controlled by difference between the refractive index of the pigment and that of the medium.

10.2.1.3   Mechanical Properties.   By careful selection of a pigment and the amount dispersed in the medium, tough, hard, yet flexible films that resist abrasion, impact and other mechanical insults can be formulated. This is particularly important for coatings, for example, on the internal and external areas of ships, oil rigs and platforms, storage tanks, the under-parts of vehicles and the exteriors of aircraft.

10.2.1.4   Durability.   Pigments play an important role in the protection of the paint media when films are exposed to atmospheric weathering. They can absorb or reflect UV radiation that would otherwise cause the breakdown of polymer systems. This can be shown by the comparison of the durability of clear films with that of pigmented films of polymers, such as alkyds. The clear films lose gloss and degrade much more rapidly.

10.2.1.5 Barrier Properties. Pigments also improve the water resistance of paint films, by increasing the distance a water molecule has to travel to penetrate to a given depth. Depending upon the particle shape and the amount of pigment in the coating, the passage of water molecules and ions through the film can be restricted to very low levels. Lamellar (plate-like) pigments such as aluminium flake, mica, micaceous iron oxide (MIO) and others are particularly useful in coatings to protect substrates immersed in water for long periods or subject to extreme weather conditions, such as ships, bridges and other structures.

10.2.1.6 Anti-corrosive Properties. Almost all substrates will deteriorate if left exposed to the natural environment for any length of time. This applies to ferrous and non-ferrous metals, concrete and wood. Only in exceptional circumstances and locations can such substrates be left unprotected. It is desirable, if not essential, that all industrial substrates be adequately protected so that the structures maintain their strength and performance.

Inhibitive pigments in binders can protect metals such as iron, steel, zinc and aluminium. Without such protection the many technical and industrial uses of these metals would not have been possible. Ships, vehicles, aeroplanes, bridges and industrial plant would not work for long if they were not given the protection afforded by anti-corrosive paints. Metallic pigments can act sacrificially as the anode in the corrosion mechanism and thus prevent electro-chemical dissolution.

Reinforced concrete will corrode and the structure will degrade if not protected by special pigmented coatings. Spalling of the concrete overlayer caused by the expansion of the corroding steel reinforcement is prevented by coatings that delay the ingress of water, carbon dioxide and ions into the concrete. This can be achieved by understanding the relationship of the selected pigments with the media used for these coatings. Likewise, the coating of the steel reinforcement with anti-corrosive systems preserves concrete structures.

10.2.1.7 Biocidal Properties. Special pigments with biocidal properties are used in anti-fouling and fungicidal coatings. These are present in the paint film to produce an environment hostile to the growth of disfiguring algae, animal species and fungi. This is particularly important for paints used on the immersed hulls of ships, the exterior of buildings in tropical areas and industrial plant such as food factories and breweries, where conditions are conducive to the growth of algae and fungi.

10.2.1.8 Chemical Resistance. The chemical resistance of paint films can be improved by the judicious use of pigments. This applies to coatings that have to resist the degrading properties of solvents, oils, fats, acids, alkalis and other chemicals.

10.2.1.9 Fire Resistance. Fire resistance can be provided by the use of pigments such as ammonium phosphate that cause the film to intumesce. This expansion and the char produced when the protective layers are exposed to very high temperatures can insulate the structure for sufficient time to allow the evacuation of buildings, and also to give time for the destructive fire to be fought, and the construction to be saved. Fire resistance is described in more detail later in this chapter.

10.2.1.10 Rheological Modification. The rheological properties of liquid paints play an important role in their storage and application properties. This can prevent the settlement of paint ingredients over long periods, and also allow the application of thick films of paint without the unsightly sagging, running and excessive flow that can happen with badly formulated coatings. Any dispersed particulate material can affect the viscosity of a composition, but certain pigments have a profound effect on rheology (*cf.* Chapter 3).

10.2.2 Types of Pigment

Historically, the pigments first used in paint were essentially minerals produced from natural deposits. These were mined, crushed and ground, sometimes washed, and then chemically modified as knowledge and inventions improved their properties.

Coloured pigments often took names of the area where they were first discovered and processed, and indeed to the present time names such as Sienna, Venetian Red, Umber and Spanish Ochre are used to describe a colour or property of a particular pigment.

Today, many compounds are still obtained from natural sources and treated to give pigments for paint and other uses. The vast majority of such pigments, however, are nowadays the so-called extender pigments. These extenders, or fillers, as they are sometimes known, play a very important part in the formulation of paint, and although they are less costly than coloured pigments or other special pigments, their total contribution to the properties of paint must not be ignored.

The 'prime' pigments, including coloured pigments, anti-corrosive pigments, and other manufactured pigments, are produced by the chemical industry as inorganic metallic salts, metallic and non-metallic organic compounds. Coloured organic compounds, in general, have come from the dye industry that has developed and is continuing to develop dyes for the textile and wool industries. It is a straightforward matter to design insoluble pigment molecules and the manufacture of pigments is now a worldwide business of huge economic importance.

In order to understand the role of pigments in coatings formulations, it is necessary to classify the pigments into groups and sub-groups. However, before we can do this we must consider the use of each particular layer of paint or coating in a protective coating system.

10.2.2.1 The Structure of a Paint Film. In simple terms, a typical coating system might be comprised of a primer, a build coat and a finishing coat. There are variations where primer and build coat are formulated as one, or where the build coat and finishing coat are formulated as one coat, but for simplicity we will consider the three-coat system.

The first coat, or primer, must offer protection to the substrate, give a uniform opacifying colour, and above all have good adhesion to the substrate and offer adhesion to the subsequent coats of paint. If they are needed, anti-corrosive pigments are included in the primer. As they control the corrosive processes at the substrate, they must be as near as possible to the surface of the metal they are protecting. It would be a pointless exercise to formulate build coats or finishes with the anti-corrosive pigment, as this would be a waste of expensive pigment. Primers may also be formulated using metallic pigments that act as sacrificial anodes.

The primer should also act as a barrier coat if possible. The barrier effect slows down the ingress of water molecules and ions. This is achieved by the use of extender pigments chosen because of their hydrophobic and lamellar properties. The colour of primers is not important, except that it must be uniform and distinct from the substrate. Less costly pigments, such as iron oxide, are widely used as colouring pigments for primers.

The build coat gives thickness to the protective coating system. It helps to eliminate imperfections in the surface being coated so that the minimum film thickness necessary to give long-term protection is obtained. The optimum thickness of a build coat largely depends upon the environmental conditions and the structure being coated, but the thickness should be at least that which gives a uniform film, free of imperfections, free of pinholes and which can fully coalesce.

The build coat also provides additional barrier properties giving maximum impermeability. Barrier properties of the film are dictated by the type and volume of extender. The build coat also contributes to a uniform colour and good opacity.

The finishing coat of a system gives the decorative effect, and as it is the coat that is constantly visible, the effect achieved must be satisfactory in every way. The colour and gloss must be correct and uniform, and should not change over a long period of time.

Coloured pigments used in the finishing coats are those with proven opacity and durability. They are used at the lowest effective concentrations, bearing in mind that most coloured pigments are very expensive. Extender pigments are also used in finishing coats to control flow, improve the toughness of the film and lower the gloss of the finish if this is required.

The whole system applied as three coats must retain its integrity without loss of adhesion, without blistering, with good colour retention, and above all it must protect the substrate from degradation. In the protective paint industry pigments are classified into the following groups.

10.2.2.2 Anti-corrosive Pigments.   As the name implies, these materials are used to prevent the corrosion of the substrate. Corrosion occurs because metals are not homogeneous materials. Heterogeneities are introduced at grain boundaries, by stress and surface contamination and by differences in composition. When in contact with an electrolyte, areas of higher potential behave as anodes and those of lower potential as cathodes, thereby creating a corrosion cell. Metal ions are formed at the anodic areas and dissolve into the electrolyte. The electrons produced pass through the metal to the cathodic areas for subsequent reaction and so the process continues. Interrupting or slowing down any aspect of this process reduces the rate of corrosion.

There are two ways in which this can be done for ferrous metals. Either the pigment can act as a sacrificial anode and protect the substrate, which becomes the cathode in a corrosion cell, or the corrosion reactions can be inhibited.

Pigments which act as sacrificial anodes are essentially small particles of zinc metal or alloys of zinc of 4–7 μm in size. Zinc-rich primers are formulated with media such as inorganic and organic silicates, epoxy, chlor-rubber and other inert polymers. The essential point about these formulations is that the zinc content of the film must be high so that life of the primer is long. In other words, there is sufficient zinc available to be 'sacrificed' to protect the steel substrate. As the process is electrochemical, the particles must be in electrical contact so that electrons can pass easily and this is also achieved by high zinc concentrations.

Many workers have suggested that the zinc content of primers should be in excess of 90% by weight in the dry film to give good long-term protection. The actual concentration of zinc metal used in zinc primers is dependent on the particle size of the zinc and the media used in the primer formulation.

Passivating oxide layers are extremely effective at inhibiting corrosion. These layers form naturally on some metal surfaces but for various reasons they do not provide adequate protection. In the case of aluminium, a passivating oxide layer is readily formed at the surface and this prevents corrosion over a pH range of 4.5–8.5. It will, however, corrode rapidly outside of this range. An oxide layer also forms on iron and steel but this is porous and not well adhered leading to very poor protection. Effective corrosion inhibition can be achieved in ferrous and non-ferrous metals by modifying the metal surface using inhibitive compounds. Inhibitive pigments are partially soluble in water and they work by releasing oxidative and other species into the film close to the metal surface. These react with the metal surface to form a tightly adhering impermeable film that prevents dissolution of the metal at the anode and slows down the rate of corrosion.

Over the years a considerable number of inhibitive pigments have been developed, marketed and used by the paint industry. Inhibitive pigments for ferrous metals include oxides, chromates, phosphates, molybdates, borates and complexes formed from these families of compounds. Most are salts of lead and zinc with some strontium, calcium and barium compounds.

For non-ferrous metals the most useful inhibitive pigments are chromates, mainly of zinc, but barium and strontium chromate have been used.

Before abrasive blasting techniques became established as the normal treatment of steel used for the building of structures, tanks, ships, bridges, *etc.*, the method of removing millscale, the oxide layer on new steel, was to allow the steel to corrode. The millscale became detached, leaving a rusty surface that was then wire brushed or flame cleaned, leaving steel with an adhering layer of oxide on the surface. It was to this surface that priming paints were applied. For these conditions the most widely used inhibitive pigment was lead sub-oxide, Pb_3O_4 known as red lead. It was found that this pigment when dispersed in a medium, essentially linseed oil, gave excellent protection to steel cleaned by the processes described above.

The virtue of a red lead-in-oil primer is that the lead oxide reacts with fatty acids present in linseed oil to produce soaps and these help in giving water resistance. The paint, based on drying oils, has excellent wetting properties, which ensures that the rusted surface is saturated with the priming paint. It is also thought that ions such as sulfates, which take part in the corrosion reaction cycle, are removed by lead oxide to form lead sulfate, and thus the corrosion process is stifled. Adding all of the possibilities together, red lead-in-oil primers are very effective in preventing the corrosion of steel subjected to normal weathering.

However, as with most things, it has disadvantages. Red lead has an extremely high relative density, which leads to severe settlement problems on storage of paint. It is also difficult to apply on a large scale. Linseed oil is very slow in drying, which leads to overcoating problems and, due to the reaction to form soaps, thickening and gelation can occur on storage. The most important disadvantage is that red lead, like all lead compounds, is a toxic material and its commercial use is increasingly being discouraged, sometimes by legislation, in many countries.

Red lead was a very important pigment, it was widely used, and this prompted the development of many other pigments to overcome its deficiencies and disadvantages.

Lead silicochromate was widely used, especially in the USA. It acted in a similar manner to red lead and had a lower relative density. It consisted of tribasic lead silicate and monobasic lead chromate on a silica core to control their rate of dissolution. It has, nevertheless, fallen into disfavour since it contains lead and hexavalent chromium.

Calcium plumbate is a pigment developed particularly for use in primers for galvanised steel but has suffered the fate of all lead-based anti-corrosive pigments.

Zinc molybdate, calcium zinc molybdate and zinc molybdenum phosphate are anodic inhibitors and have been promoted and used as anti-corrosive pigments. These, together with compounds such as calcium borosilicate and barium metaborate, are free of the toxic hazards associated with lead, and are therefore attractive as red lead replacements.

Chromates of zinc have also been used as anti-corrosive pigments for steel but they are more effective as inhibitors for aluminium and other non-ferrous alloys. Besides which, hexavalent chromium presents health risks at least as great as those associated with lead compounds.

The most useful and manageable pigment developed as an anti-corrosive pigment for steel is zinc phosphate, $Zn_3(PO_4)_2 \cdot 2H_2O$. It has been shown that zinc phosphate in a drying oil medium offers long-term protection to steel and because of the many advantages over red lead it has become widely used in the paint industry, especially in Europe.

Zinc phosphate is easily prepared from relatively cheap raw materials, waste zinc metal and phosphoric acid. It has low toxicity, it is white, and thus has the advantage of being able to produce light coloured primers, and it has a low relative density. The crystalline shape is lamellar and it acts as a barrier pigment. It can be formulated with many types of polymer as priming paints, and is quite stable in media such as alkyd, epoxy, chlor-rubber, vinyl, urethane and water-based systems.

The anti-corrosive properties of zinc phosphates are somewhat difficult to prove by classical studies but because of its barrier properties, colour, stability and relative low cost, zinc phosphate has achieved predominance in the market. Many attempts have been made to improve the anti-corrosive properties of zinc phosphate by mixing it with materials such as zinc molybdate, or co-precipitating it with this and other compounds. The higher activity of zinc phosphate molybdate has been attributed to the inhibitive effect of the molybdate ions.

Aluminium zinc phosphate, also produced by co-precipitation techniques, is claimed to have improved inhibiting properties due to its higher phosphate content and higher solubility. Organic modification of zinc phosphate is also possible, and compounds have been introduced as improved anti-corrosive pigments. Nevertheless, few alternatives have gained a leading position in the market place.

Anti-corrosive pigments specifically designed to function by an ion-exchange mechanism are now available. These materials are high surface area inorganic oxides in which the surface hydroxyl groups have been exchanged with ionic corrosion inhibitors. These remove ions such as chloride and sulfate, which are necessary for the corrosion process to occur. The acidic or basic properties of the oxide determine whether a material is used as an anion or cation exchanger. For example, aluminium oxide is used as a support for anions and silica for cations.

For the protection of aluminium and its alloys, basic zinc potassium chromate, $3ZnCrO_4Zn(OH)_2K_2CrO_4$; zinc tetroxychromate, $ZnCrO_4 \cdot 4Zn(OH)_2$; and strontium chromate, $SrCrO_4$ are used with various media. Strontium chromate in a cured epoxy binder has particular use in the aircraft industry.

Zinc tetroxychromate is used almost exclusively in the formulation of etch primers. These are based upon polyvinylbutyral (PVB) and phosphoric acid, the latter being added just prior to the application of the primer. Films of etch primers are applied at only a few microns in thickness, and these adhere extremely well to ferrous and non-ferrous metals. They form the base for other primers that would not necessarily adhere well to the metal surface. The components of the etch primer complex with the metallic substrate producing a surface that is more amenable to organic coatings subsequently applied. Considerable work is currently underway to formulate chromate-free etch primers.

10.2.2.3 Barrier and Extender Pigments. The binders or polymers used in protective coatings dry or cure at ambient temperatures either by solvent evaporation, air oxidation, or by chemical reaction of components mixed together before application. The morphology of these polymers is complex, and although they form continuous films, they are invariably porous, to a greater or lesser degree, to oxygen, water, chlorides, sulfates, carbon dioxide, *etc.* In other words, as clear films, they are not entirely suitable for long-term protection of substrates against corrosion. They have to be reinforced by pigments. Pigments particularly suitable for reducing permeability are those with plate-like, or lamellar structures. They can be metals, oxides, silicates, *etc.*, and are known for their barrier protection.

Aluminium flake is manufactured from high purity metal by milling aluminium powder in a solvent such as white spirit, with stearic acid as a lubricant. The stearic acid coats the aluminium flake and causes the aluminium to 'leaf', that is to float to the surface of a paint to form a layer where the overlapping flakes are orientated parallel to the surface to give a bright film with high lustre. If an alternative lubricant such as oleic acid is used in the manufacture, non-leafing aluminium flakes are produced. These, when distributed in a polymer, remain dispersed in the body of the film and do not migrate to the surface.

Leafing aluminium flakes are used in finishing paints, whilst non-leafing aluminium is used in primers and build coats. The particle size of the flakes ranges from about 1 to 100 μm in the longer dimensions and 0.05 to 2 μm in thickness.

Aluminium flakes reduce the permeability of polymers by causing ions or molecules to follow a long, tortuous path to the substrate, as compared with a direct path through an unfilled polymer.

In finishing paints, leafing aluminium flakes can be used with other colouring pigments but the predominant pigment is the aluminium, which also has the advantage of excellent resistance to ultraviolet radiation. So-called metallic finishes are widely used in the automotive industry where their characteristic ability to show a different colour when viewed from different angles, the so-called 'flop', has a strong bearing on body design. Less pronounced relief can be introduced into body panels with the same aesthetic effect.

In primers and build coats, non-leafing aluminium flake is mixed with other pigments such as MIO, silicates, silica, and filler pigments such as barium sulfate. The optimum amount of aluminium pigment used in the film is usually determined from permeability studies.

Thin flakes of chemical-resistant glass can also be used in a similar manner to aluminium flakes. They are also manufactured by a milling technique, the flakes tending to be larger than aluminium flakes. Glass flakes are formulated into coatings that are applied by specialised spray techniques to give very thick films having excellent resistance to abrasion. They are particularly useful for the repair of the insides of large storage tanks, the legs and undersides of oil rigs, and the external submerged surfaces of barges and ships which have suffered steel loss due to corrosion and abrasion.

The media used for these coatings are essentially solvent-free materials, with two-pack epoxy and peroxide cured polyester being preferred.

Glass flakes with low density are not easily handled during manufacture of these compositions, and care has to be taken to ensure that the flakes are well distributed in the media and thoroughly 'wetted out'.

One of the most important pigments used in coatings for the protection of large structures is MIO. This material is specular haematite (approximately Fe_2O_3 90%), a natural mineral originally found in England and Austria. Other sources have been found, but the Austrian mine still provides the mineral with the best lamellar structure. The term 'micaceous' refers to the mica-like structure of the iron oxide with flakes up to 100 μm in size and approximately 5 μm thick. MIO has excellent heat resistance, and good chemical and water resistance. Also, it does not degrade under the influence of UV, and thus protects the media of paint films. It has a low oil absorption and this is important in allowing high loadings in paint. MIO can be formulated with a wide range of binders such as oleo-resinous materials, alkyds, chlor-rubber, epoxies; vinyls and urethanes to produce primers, build coats and finishes. The only disadvantage it has in finishes is its dark colour but it can be mixed with coloured pigments and aluminium to produce attractive films that sparkle in bright sunlight.

Many structures, some going back to Victorian times and including the Eiffel Tower, have been maintained with MIO-based coatings since they were built, with excellent results. German and British railways specified the pigment for many years.

In the formulation of MIO-based coatings, the extensive addition of other, cheaper, extending pigments must be avoided as these detract from the orientation of the iron oxide. High loadings of MIO are essential but there are optimum concentrations for each medium to obtain films with the best impermeability. High film thicknesses must also be applied to give good orientation of the MIO. A synthetic grade of MIO has been developed for use in high performance coatings.

Probably because of the absence of local sources of MIO in the USA, the largest use has been in Europe, the Middle East and Far East. In the USA mica has been used extensively as a lamellar pigment.

Mica, an aluminium potassium hydrated silicate ($Al_2KSi_3 \cdot 12H_2O$), is found in several forms, the most important for the paint industry being muscovite. Mica particles have a high aspect ratio, *i.e.* diameter to thickness ratio, of greater than 25, but the pigment has a high oil absorption and this limits its loadings in paint films to relatively low levels.

Additions of mica to zinc-based primers have been shown to improve the performance of these products. Other lamellar pigments marketed for the paint industry include stainless steel flake.

There are several other minerals that have been used in paints for many years. These were called extenders, and were considered as cheap materials that reduced the cost of the finished paint. However, experience has shown that the so-called extender pigments are of great value in giving improved water resistance, improved durability, easier application and good storage stability to paints, as well as improving commercial viability by lowering costs.

Extender pigments are generally white, naturally occurring minerals with very low solubility in water and preferably inert to the action of acids and alkalis. This does not exclude coloured minerals such as iron oxides that have use in primer formulations as extending pigments.

The minerals are won by mining operations and are crushed, cleaned and segregated as products with particles from sub-micron to a few microns in size. The minerals cover a wide range of chemical compounds such as sulfates, oxides and silicates. Carbonates of calcium and magnesium are used extensively in paints for decorative purposes but because of their solubility, especially in acid solutions, are not formulated into coatings requiring high durability and high water resistance.

Magnesium silicate, sometimes associated with other minerals such as aluminium silicates and quartz, is found in many parts of the world as talc ($3MgO \cdot 4SiO_2 \cdot H_2O$) and is the most widely used extender pigment. It can have a fibrous or needle-like structure, be plate-like, or amorphous, depending upon the source. Talcs give good reinforcing properties and improve the impermeability of films. The fibrous structures confer good storage properties on paint by preventing settlement of pigments, but care has to be taken to exclude asbestiform fibres because of health considerations. (Inhalation of asbestos fibres of a certain size can lead to asbestosis and lung cancer.) Talcs are generally hydrophobic, have good colour and are easily dispersed in paint media.

Barytes, the mineral form of barium sulfate, is an inert compound with a low oil absorption and as such can be used in high loadings in build coats and primers. It has a relatively low cost and finds use in chemical resistance coatings. The precipitated form of barytes is Blanc Fixe which, being purer, is used as an extender for coloured pigments. Barium sulfate has the disadvantage of a high relative density and this can lead to settlement problems.

Synthetic silica, especially fumed silica prepared by flame hydrolysis of silicon tetrachloride, is an important product having an extremely high surface area and small particle size. Its large surface area results in a very high oil absorption and thus makes it suitable as a matting agent for coatings. It also gives structure to media to prevent settlement and to improve sag resistance in films applied at high thickness.

Wollastonite, a form of calcium silicate, is an acicular (*i.e.* needle shaped) pigment and performs as a reinforcer in a paint film. It is available in a range of particle sizes from 10 to 100 μm and has a low oil absorption, which makes it very useful in the formulation of high solids paints. Surface-treated grades are also produced and these materials provide better corrosion resistance than the untreated varieties.

Kaolin or China clay is a hydrous aluminium silicate, $Al_2O_3 \cdot 2SiO_2 \cdot 2H_2O$. The calcined form has found considerable use as an extender in water-based systems. It is a lamellar pigment with a relatively high oil absorption and low relative density.

Nepheline syenite is a nodular form of potassium sodium aluminium silicate. It has a low oil absorption, similar to Wollastonite, and finds use in high solids anti-corrosive primers.

Extender pigments are formulated with anti-corrosive and barrier pigments to give a wide range of products for the protective coating business.

10.2.2.4 The Properties of Coloured Pigments. Colour, the measurement of colour, and an appreciation of the whole subject of the use and application of coloured pigments is a study in its own right, and can only be mentioned in a broad sense in a review such as this.

The colour of a paint film is controlled by several factors, apart from the colour of the pigment itself. The amount of pigment present has an obvious bearing on colour: a high gloss film will have a different colour from a low gloss film; the degree of dispersion of the pigment influences the colour of a film. The primary particle size of coloured pigments is small, generally less than 1 μm, and these particles are present as agglomerates. It is the extent of the breakdown of the agglomerates during the dispersion of the pigment which controls the ultimate colour attained when a pigment is 'ground' into a paint binder.

The colour of a paint binder or polymer also has an effect on the finished colour of paint. Acrylic media are more or less colourless, whilst some alkyds and oleo-resinous media are pale yellow when dry.

The properties of coloured pigments are briefly as follows:

Mass colour	The colour given to a film by a pigment at high opacity.
Tinting strength	The degree to which a pigment gives colour to a white base. A pigment with high tinting strength needs smaller amounts to give a colour compared with a pigment with low tinting strength. For a white pigment the converse applies.

Chroma	The intensity of 'saturation' or purity of a particular colour when compared with a neutral grey.
Hiding power	The degree of opacity of a coating, *i.e.* ability to obliterate the colour of the substrate.
Light fastness	The ability of the colour to remain unaffected when exposed to natural weathering. This is particularly important when coloured pigments are used to tint white paints to off-white colours, but pigments in the mass colour also may change colour on exposure to sunlight.
Durability	Coloured paints may appear to fade on exposure to UV radiation. This, however, may be entirely due to the oxidation of the polymer. Pigments can protect the polymer by absorbing the radiation and thus improve the durability of the coating. On the other hand, some pigments can speed up this degradation.
Bleeding	Coloured pigments can be soluble in the solvents used in paints. This causes problems on storage as the pigment dissolves and then crystallises out of solution and the colour of the paint changes. Also, the pigment in an undercoat can bleed into the topcoat, changing the colour of the topcoat.
Chemical resistance	For example, the resistance to acid rain is also important together with the reactivity of the pigment with media.

10.2.2.5 The Colour Index System.   The colour index system is a coding system developed under the joint sponsorship of the Society of Dyers and Colourists in the United Kingdom and the Association of Textile Chemists and Colorists in the United States. The colour index (CI) identifies each pigment by giving it a unique colour index name (CI name) and a colour index number (CI number). It describes all classes of coloured compounds, irrespective of end use or chemical composition. To illustrate how the system operates, phthalocyanine blue has the CI name Pigment Blue 15 and the CI number 74160.

The number of coloured pigments used in the paint industry as a whole is quite large, but as the industry is split into several businesses, it does not necessarily mean that the same pigments are used by all of the sub-sections of the industry. For example, for the automobile industry many pigments are specially developed and processed for that business. The following is a description of some coloured pigments used in protective coatings. For simplicity, they are divided into colours – white, red, blue, yellow, and so on.

White Pigments. A white pigment is not coloured – it is, in fact, a transparent material. It appears to be white because the small size of the particles of pigment causes light to be scattered and reflected back, and the eye receives the whole spectrum of reflected light. The best light scattering is obtained with the maximum difference in refractive index between pigment and medium. It is customary to think in terms of a highly refractive pigment, but it is also possible to produce a white coating by contrasting a medium with air introduced as pigment-sized glass bubbles.

The white pigment used by the paint industry is titanium dioxide. It is the most important pigment used in the industry and has had an immeasurable effect on all aspects of decoration used by the industrialised nations.

Before the development of titanium dioxide in the 1920s and wide scale use in the 1950s, white pigments used in paint were compounds such as lead carbonate, zinc oxide, antimony oxide, lead silicate, lead sulfate and co-precipitated forms of these compounds.

Apart from zinc oxide in various forms, all other white pigments have become obsolete. In any case, many would now be considered toxic and could not be used in most paints.

Titanium dioxide is manufactured in millions of tonnes per annum by two methods, the sulfate process and the chloride process. It is made from natural ores found as ilmenite, an iron titanium dioxide ($FeTiO_3$), or rutile, a natural form of titanium dioxide.

In the sulfate process the ilmenite ore is reacted with sulfuric acid to produce titanyl sulfate, which is hydrolysed to give hydrated titanium dioxide. This is then calcined to give the pigment titanium dioxide.

In the chloride process, rutile ore is treated with chlorine to produce titanium tetrachloride. This is then vaporised and oxidised to produce titanium dioxide and chlorine, which is reused in the process.

The sulfate process can produce two crystalline forms of titanium dioxide, anatase and rutile. The chloride process only produces the rutile form, which is a more compact tetragonal crystal. This gives it the better opacity due to its higher refractive index.

The rutile form of titanium dioxide (Pigment White 6) is generally preferred in high performance coatings and, in many respects it is almost an ideal pigment. It is considered non-toxic and can even be used in foodstuffs, such as instant mashed potato where it provides the attractive white colour. It also has the highest refractive index of all white pigments (2.76 compared to 2.4 for diamond) and it can be produced as a very small particle, approximately 0.2 μm diameter, and this gives excellent opacity. Its oil absorption is relatively low and its relative density is not too high permitting the manufacture of stable, fluid dispersions. The tinting strength of titanium dioxide is also good.

Its disadvantages are that it can cause degradation of polymers by being photochemically active and that it does absorb some light at the blue end of the spectrum, which makes it a yellow toned white. For most purposes, however, this has little effect on the appearance of white paints.

In terms of photoactivity, titanium dioxide is both a UV-activated oxidation catalyst and a UV absorber. The behaviour of titanium dioxide as an oxidation catalyst is well documented. Free radicals are formed at the surface of titanium dioxide and these then oxidise the binder by photocatalytic degradation. This reduces the gloss and produces a friable layer on the surface of the paint film – a process called 'chalking'. Free radicals are formed according to Scheme 10.1.

$$H_2O + O_2 + h\nu \xrightarrow{\quad TiO_2 \quad} OH^{\cdot} + HO_2^{\cdot}$$

The hydroxy and peroxy radicals oxidise and degrade the binder.

$$3OH^{\cdot} + 3HO_2^{\cdot} + 2(-CH_2-) \longrightarrow 2CO_2 + 5H_2O$$

Scheme 10.1

Coating the particles of titanium dioxide with a layer of one or more inorganic materials significantly reduces the damaging effects of free radicals. In its simplest form, the coating prevents contact between the titanium dioxide surface and the binder. Non-durable grades have a coating of hydrous alumina and durable grades have an additional coating of silica. 'Superdurable' rutile titanium dioxides are encapsulated in silica shells. Some coatings also contain zirconia, which acts as a free radical scavenger and thereby reduces the photocatalytic activity of the pigment. Organic compounds, such as amines or silicones, can also be applied to improve the dispersibility and durability.

As the particle size of titanium dioxide is reduced, it becomes ineffective at scattering visible light and its opacifying power is lost. At the same time, it becomes better at scattering UV radiation. Coated titanium dioxides with particle sizes of less than 0.1 μm are virtually transparent in coatings and can act as highly effective UV stabilisers in certain paints.

Red Pigments. Red pigments can be divided into two groups, one consisting of various grades of iron oxide, the other consisting of organic compounds.

The most obvious differences between these two groups are that the organic compounds have a higher chroma and thus give brighter, purer, colours compared with the dull reds produced from iron oxide.

On the outskirts of the small village of Roussillon in Provence in Southern France, in an area of only a few acres, there can be observed and explored the residual piles and strata of a defunct mineral quarry. In this small area, in the bright Provencal sun, can be seen a glorious range of colours from the palest of yellows through bright yellow to orange, reds, pinks, browns and almost black. Indeed, it is said that as many as 19 different coloured forms of iron oxide were mined from this small area until commercial pressures from other sources caused the mine to close. This is a fine example of the range of coloured pigments obtained from one chemical compound by variations of particle size and minor variations of chemical composition. In reality, these bright colours are relatively dull in that they have low chromas compared with other pigments such as the organic reds.

The iron oxides used as red colouring pigments are mainly 'synthetic' iron oxides, which are produced by a number of processes, one of which is direct precipitation. Calcination of synthetic yellow iron oxide, $Fe_2O_3 \cdot H_2O$, removes the combined water to give red Fe_2O_3. Calcination of black iron oxide, $Fe_2O_3 \cdot FeO$, which is produced synthetically, also gives red forms of Fe_2O_3.

By careful control of the processes a range of red coloured pigments of varying particle size can be produced. These range from yellow toned to blue toned reds. They are also designed to give easy dispersion in media.

The important properties of red iron oxide (Pigment Red 101) are low cost, and excellent light fastness and heat resistance. It absorbs UV radiation to a certain extent and this gives good durability. It also possesses good hiding power, good chemical resistance, and it is insoluble in paint media and solvents.

The disadvantages are a poor tinting strength and a dull shade but this is not so important when the decoration of large structures is considered.

For bright red coloured coatings a range of organic compounds is available, these are:

Toluidine reds (*e.g.* (**1**)) are azo pigments manufactured by coupling, for example, diazotised *m*-nitro-*o*-toluidine with β-naphthol. They have very high chromas, producing bright colours with good light fastness in the full tone but poor light fastness when reduced to tints. They also bleed in aromatic solvents. They find application in decorative paints that have general-purpose use as, for example, air drying alkyd-based coatings containing white spirit solvent.

By changing the amine and introducing chlorine-containing amines, a range of products such as Para red (**2**) and Parachlor reds (**3**) can be produced. These have various shades of red and find some use in the paint industry but are very important in the ink industry.

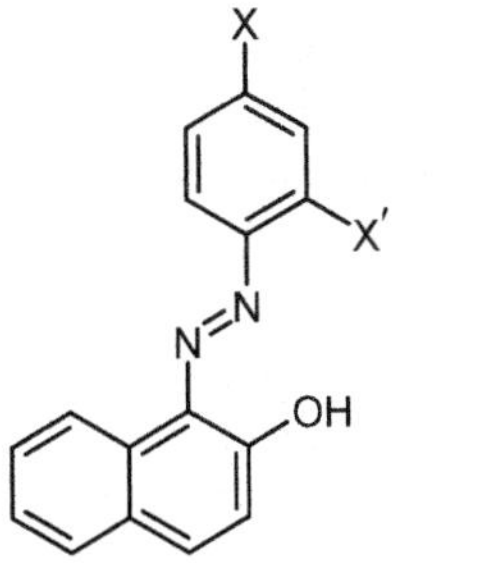

	X	X'
(1)	CH$_3$	NO$_2$
(2)	NO$_2$	H
(3)	Cl	NO$_2$

Other red pigments used are Naphthol reds (**4**), where the β-naphthol is replaced by hydroxynaphthoic acid anilide in the coupling reaction with diazonium compounds. These again are used in decorative paints, especially in full shades.

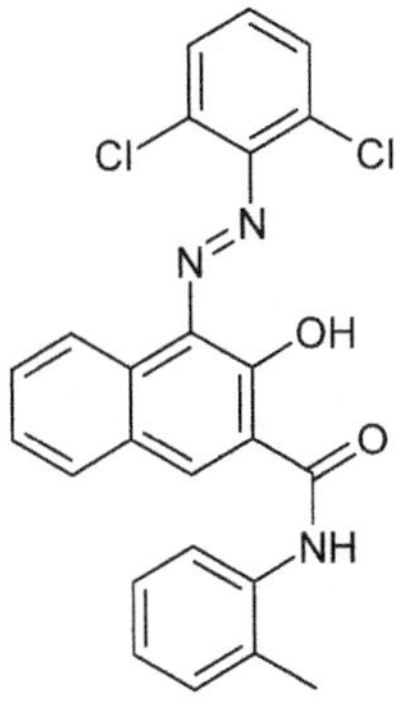

(4)

High performance quinacridone reds and violets (*e.g.* (**5**)) have excellent brightness, good durability, good solvent resistance and good tint stability. Unfortunately, they are very expensive.

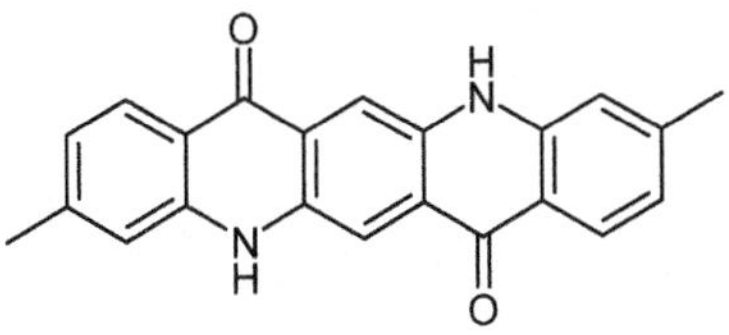

(5)

Perylene reds (**6**) have found use in the automotive market, especially in metallic finishes. They provide pure transparent shades with good resistance to bleeding through. In addition, they have good thermal stability, excellent light fastness and generally excellent chemical resistance.

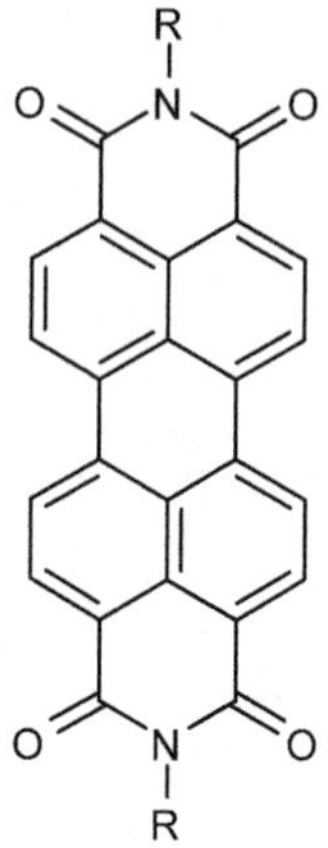

R = H, CH3

Perylene Reds

(**6**)

Benzimidazolone-based reds (**7**) are azo compounds that are mainly used in the plastics industry but they are now finding use in coatings. They have excellent light fastness, good weatherability and outstanding heat resistance. They find use in industrial finishes such as powder coatings, coil coatings and automotive refinish.

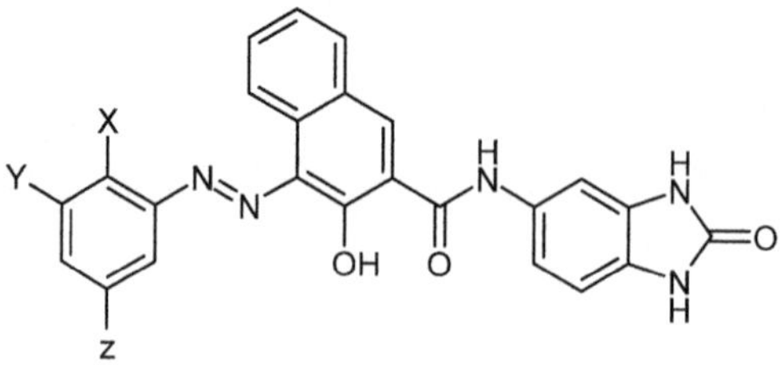

Benzimidazolone Reds

(**7**)

More recent introductions are based on 1,4-diketo pyrrolo(3,4-c) pyrrole (**8**) and on nickel complexes with pyrazoloquinazolone derivatives (**9**).

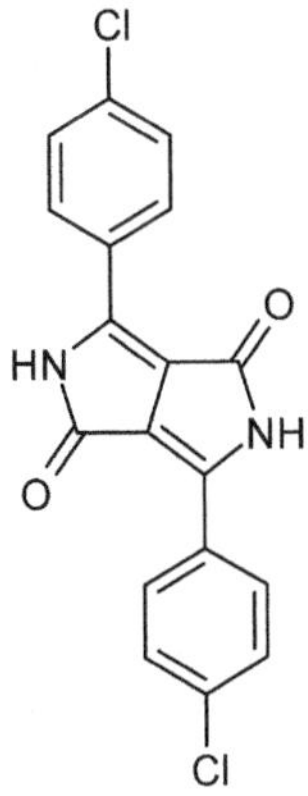

Pyrrolo-pyrrole (PR 254)

(8)

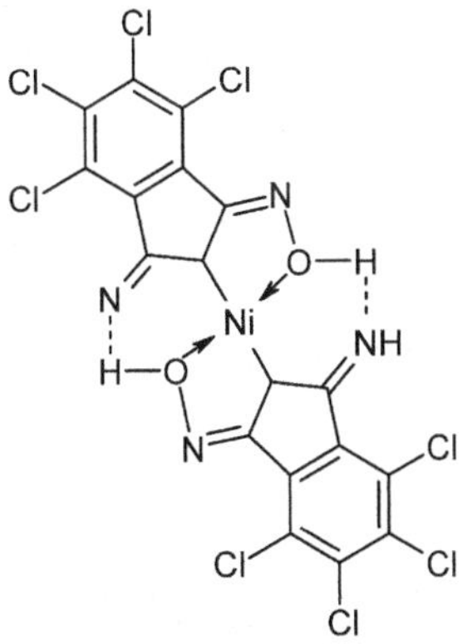

PR 257

(9)

Blue Pigments. The blue pigment which has universal application in the paint industry and which ranks with titanium dioxide as the most successful pigment in its class is Phthalocyanine blue.

Phthalocyanine ('phthalo') blue, $C_{32}H_{16}N_8Cu$, **(10)**, is reported to have been discovered accidentally in a chemical plant where phthalic anhydride was being reacted with ammonia in a copper vessel. This discovery led to an industry that has proved to be very important for the paint, printing, plastic, rubber and fibre manufacturing industries.

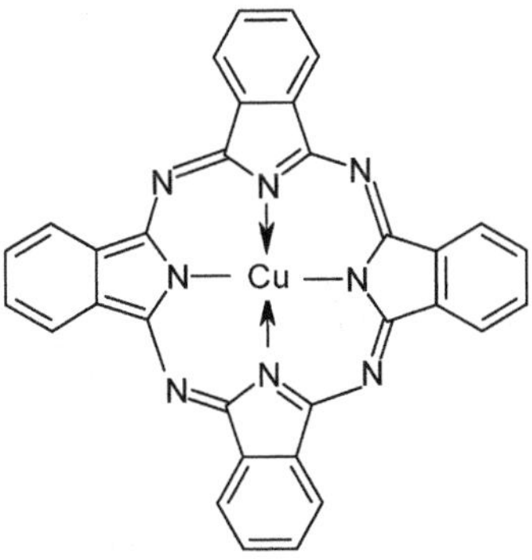

(10)

Phthalo blue is now made by reacting phthalic anhydride with urea, a copper compound and a catalyst in a high boiling solvent. The resulting crude pigment is then processed by one of two methods, acid pasting or salt grinding, to give a wide range of pigments of different particle size and crystal form. Copper-free phthalocyanines and chlorine modification of the molecule are also possible to produce other grades of the pigment. The pigment can exist in a range of shades from red blues to green blues, in flocculating and non-flocculating forms.

The essential properties of phthalocyanine blue are its light fastness in full and reduced tone, its excellent chemical and solvent resistance, resistance to crystallisation and flocculation in paints.

Phthalo blues are rarely used in the mass tone as they are rather dark but give superb colours when diluted with titanium dioxide and yellow pigments. They have replaced all other blue pigments such as Ultramarine blue (a sulfide of an aluminium silicate complex) and Prussian blue (potassium ferro-ferric cyanides) in virtually all coatings.

Green Pigments. The discovery of blue phthalocyanines gave impetus to the search for pigments with similar chemical structures but having other colours. This has resulted in chlorinated copper phthalocyanines and chlorobromo copper phthalocyanine pigments that are green in colour. These are commercially available and have similar properties to the blue phthalocyanines.

Green pigments are now dominated by phthalocyanine compounds either as green phthalocyanines or as mixtures of blue phthalocyanines with safe yellow pigments.

Cheaper greens formulated from blends of lead chromate and Prussian blue have more or less disappeared from the market due to the toxicity of lead compounds.

The most inert green pigment is chromium oxide, Cr_2O_3, Pigment Green 17. This dull grey/green coloured pigment has excellent chemical resistance, good heat and light stability, good opacity, and is of low toxicity. It reflects infrared radiation and finds application in deck paints and in camouflage coatings for military purposes.

Yellow Pigments. The yellow pigments are an important class of colours as they can also be mixed with other primary colours, reds and blues to produce orange, browns and greens. As with the red pigments, they can be divided into inorganic compounds and organic compounds. The inorganic section contains yellow iron oxides, lead chromates, cadmium yellow (though this pigment is now considered too toxic for general use), and pigments such as zinc chromate, but this is mainly used for its anti-corrosive properties.

Yellow iron oxide is produced synthetically from scrap steel in a range of dull yellow shades. $Fe_2O_3 \cdot H_2O$ has an acicular or needle-shaped crystal and it is the size of the crystal that determines the colour. Yellow iron oxide has all of the attributes mentioned for red iron oxide, apart from heat resistance as it loses the water molecule on heating to approximately 180 °C to give red iron oxide. It is a relatively cheap pigment and is used extensively to produce cream tints with titanium dioxide, browns and dull shades used in industry.

Lead chromates are manufactured by precipitating the pigment as lead chromate or co-precipitating lead chromate and lead sulfate to give primrose chromes, lemon chromes and middle chromes with a red/ yellow tone.

Chromes are bright pigments with excellent opacity and are relatively cheap. They are fast to heat and are insoluble in solvents. They unfortunately contain lead and hexavalent chromium and are not recommended for certain purposes in decorative coatings because of their toxic properties.

Lead chromates, if left untreated, darken on natural weathering. To improve this property, lead chromes are surface treated with compounds such as silica and alumina. They are also affected by alkalis and can fade when exposed to chemical fumes. As a group these pigments have been important but their use is declining.

Associated with the lead chromes are mixed crystals of lead chromate and lead molybdate. These are the so-called scarlet chromes or molybdate orange pigments and they are available in colour range from orange to scarlet. The scarlet chromes are useful in blending with organic pigments such as organic reds and maroons. Blending scarlet chromes with quinacridone reds produces a range of colours that can be used in full shade as cheaper pigments with good quality.

For many years, the toxicity of several paint raw materials, including pigments, has been a cause for concern. In particular hexavalent chromium and lead compounds have been extensively studied. Current data show hexavalent chromium to be carcinogenic by inhalation and probably also by ingestion. Lead is classified by the World Health Organisation as a cumulative general poison, and exposure to high levels may produce a number of health problems. It is also implicated in adversely affecting the intellectual development of children.

Despite the toxicological information, the use in paints of lead and hexavalent chromium is currently not prohibited. Additionally, the labelling requirements for paint containing such materials vary considerably by country. Generally though, paint manufacturers have voluntarily decided to develop, evaluate and adopt less toxic alternatives.

Monoazo yellow pigments can be prepared in a similar way to the toluidine reds. These were introduced under the trade name of Hansa yellows, the most popular being a compound manufactured by coupling 2-nitro-*p*-toluidine with acetoacetanilide.

Hansa yellows (now known as arylamide yellows (**11**)) have been largely superseded in other industries, but they still find favour in the paint industry for decorative coating use. Like the toluidine red pigments the yellow arylamides have good durability in mass tone but have poor durability as tints. They tend to bleed in aromatic solvents and are only used in paints containing weak solvents. They are lead-free and replace lead chromes in full shades where lead-based pigments are restricted by legislation or for other reasons.

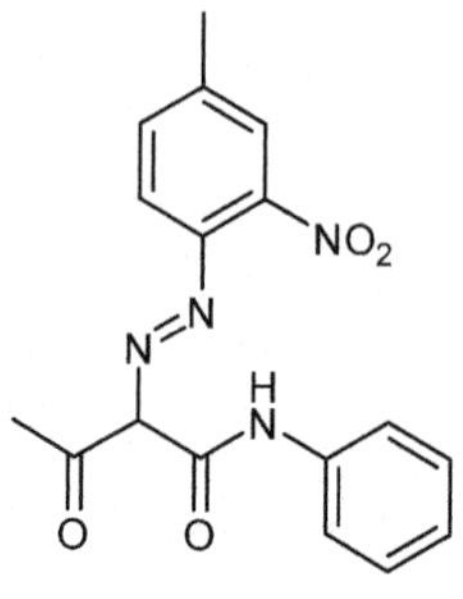

(11)

Diazo yellow pigments, which were known as benzidine yellows and are now known as diarylide yellows (12), have improved tint strength, better solvent resistance, and give good opacity compared with monoazo yellows. Unfortunately, their light fastness is not so good as that of monoazo pigments. They are used extensively in the USA, especially in printing inks. The diamine used in the diazotisation is dichlorobenzidine with modified acetoacetanilides acting as the coupling reagents.

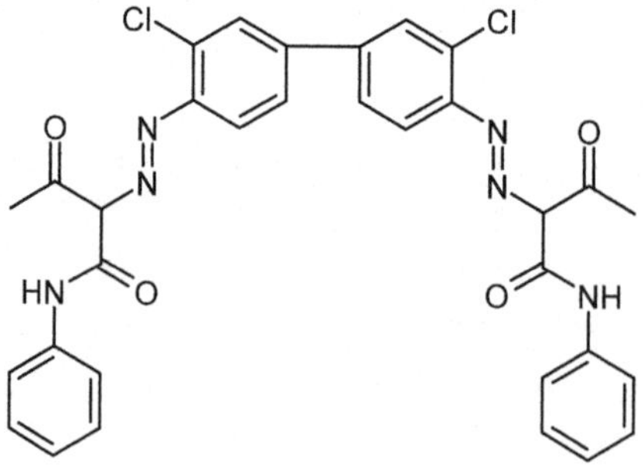

(12)

There is also a range of organic pigments known as Vat yellows (13), which have good light fastness in mass tone and as tints. They are based on polycyclic aromatic compounds such as flavanthrone and anthrapyrimidine but because of the high cost of production find use in a limited range of paint products, especially for the automotive industry.

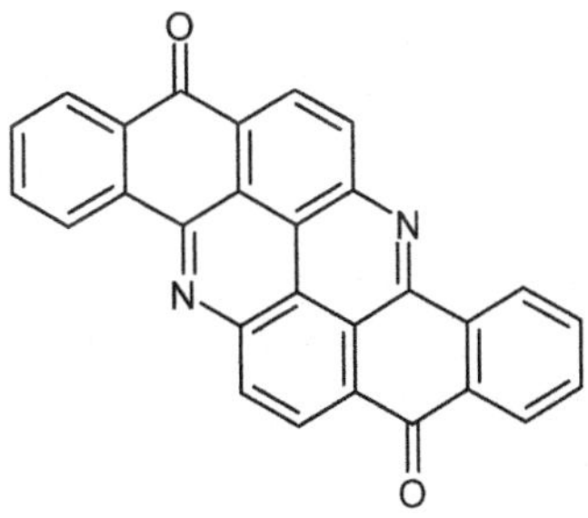

(13)

Benzimidazolone yellows (14) are monoazo compounds similar to their red counterparts. They share the same excellent weatherability, heat stability and light fastness and find use in the same industrial finish applications.

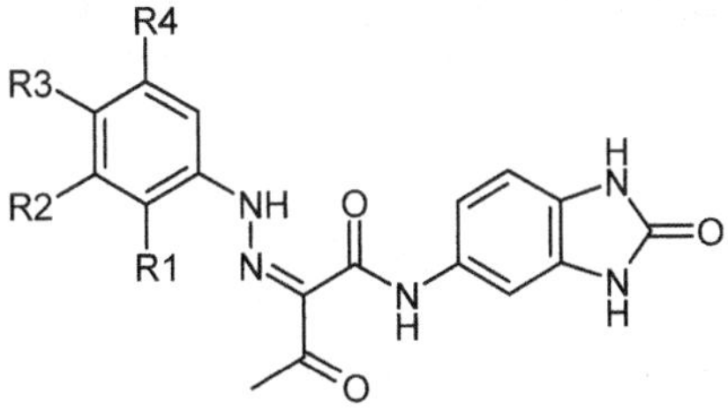

(14)

Black Pigments. The black pigments used in the protective coatings industry are of two types, black iron oxide and carbon black. Black iron oxide (magnetite), $Fe_2O_3 \cdot FeO$, is available as a mineral but is also made synthetically. It is relatively cheap, inert, with good light fastness and chemical resistance and it has lower oil absorption than the carbon blacks.

Black iron oxide is rarely used to give good colour in finishes, except perhaps to tint white pigments to grey colours. It finds use in primers and build coats.

Carbon black is made by heating oil, gas or acetylene at high temperatures with limited air to give the carbon pigment. It is produced in large quantities for the rubber industry as a reinforcing agent but it is used in the paint industry as a pigment to give a superior black colour.

Carbon black is a pigment with a very small particle size (0.01–0.1 μm). In general, the smaller the particle size, the better the colour obtained. However, a consequence of the small particle size is a very large surface area, and this can cause severe problems with dispersion of the pigment.

In turn, this can lead to poor gloss and poor colour. It is also difficult to make stable paints with carbon black in air-drying media, and when used with other pigments such as titanium dioxide, it can flocculate changing the colour on storage.

Vegetable blacks, which were made by burning vegetable oils in a limited supply of air, have a larger particle size than carbon blacks but have poorer colour in that they appear dull and less brilliant than a good carbon black when formulated in gloss paints. They have limited use in tinting with other coloured pigments.

Miscellaneous Pigments. Pearlescent pigments are thin flakes coated with one or more layers of a metal oxide such as titanium dioxide or iron oxide. The layers are thin, approximately 100–150 nm depending on the coating, and may comprise one or more oxides. Light diffraction and interference occur in these layers producing a pearlescent effect, in much the same way that light striking a film of oil on water is diffracted. Mica is used as the substrate due to its plate-like structure, and its transparency adds to the optical effects. Alternative substrates, such as silica and aluminium oxide, are also being investigated. The colours produced are dependent on the type and thickness of the applied oxide layer. Thin layers of titanium dioxide produce a silver colour and increasing the thickness moves the colour from violet, to blue, green yellow, red and finally to white. Addition of iron (III) oxide or chrome (III) oxide to titanium dioxide will make golden and green colours, respectively.

Other pigments that give refracting and interference effects are based on an aluminium core coated with reflected layers of magnesium fluoride and chromium.

Opalescent effects can be produced using sub-pigmentary titanium dioxide dispersed in binder. Opalescent pigments cause the perceived colour to change with the angle of observation and this effect is employed in automotive paints. The particle size of sub-pigmentary titanium dioxide is in the range 0.02–0.05 μm compared to 0.2 μm for the conventional pigment.

Fluorescent pigments are materials that fluoresce in daylight and are used mainly in safety applications. They are typically made by polymerisation of a fluorescent dye into a suitable polymer matrix.

10.3 SOLVENTS

Most resins used as media for paints are solids or highly viscous materials, which have to be diluted with a liquid to lower their viscosity so that they can be manufactured into paint and the paints then applied to substrates.

The liquid can act as a true solvent for the resin, or as in the case of water can act as a diluent in water-based latex paints.

All solvents for paint resins are low molecular weight organic compounds, most of which can be classified as aliphatic or aromatic hydrocarbons and oxygenated organic compounds.

The most important property of a solvent is to dissolve the resin completely to give a clear solution that can be used to disperse the other ingredients of the paint formulation. The pigment dispersion is corrected for viscosity by further addition of solvent so that the paint can be applied. After application of the paint the solvent evaporates from the film to give a surface that is smooth, uniform, of the correct gloss, and free from imperfections (*cf.* Chapter 4).

The role of solvents is very significant in that they control many factors associated with the properties of paint films. The physical and chemical characteristics of solvents that control paint properties are as follows.

10.3.1 Solvency

A true solvent should dissolve a resin completely to give a clear solution over a wide range of concentrations, and, if the solution is applied as a film, the solvent should evaporate to give a transparent film free from imperfections.

The molecular forces affecting solvency are dispersive forces (London forces), polar forces (Keesom and Debye forces) and hydrogen bonding. These forces also control solvent properties such as boiling point, surface tension and latent heat. The solubility parameter, which is a function of the cohesive energy density of liquid, is also an important factor in the appreciation of the solvent power of solvents. For solvents and polymers to be miscible they should have very similar solubility parameters. The effect of polarity of the solvent and polymer is very important in the relationship of their solubility parameters.

In practical paints it is usual that more than one solvent is used in a formulation. In many cases, mixtures of a true solvent and a diluent are used as the 'solvent', a diluent being a liquid that is not a true solvent. The reasons why mixtures of solvents are used can include the control of evaporation, control of flash point, cost, and also to reduce health and safety risks. Nevertheless, whether a paint solvent is a single liquid or a mixture of liquids, it must have high solvent power over the whole time it is present in the paint film. Premature evaporation of the 'stronger' components in a solvent mix would lead to precipitation of polymer.

Addition of too much diluent may produce a θ-solvent whose mixtures with polymer show increased viscosity (*i.e.* polymer–polymer interactions predominate over polymer–solvent interactions, and the chains are pictured as coiling up like the Greek letter). At still higher levels, the polymer will be precipitated.

The phenomenon of latent solvency is of considerable importance in the formulation of solvent-based paints. For example, mixtures of ethanol and toluene or butanol and xylene, none of which is a satisfactory solvent in its own right, show excellent solvency for certain polymers.

10.3.2 Viscosity

Organic solvents are low viscosity Newtonian liquids. Solvency, inter-actions and concentrations being equal, a lower viscosity solvent will produce a lower viscosity polymer solution. However, the typical situation is much more complex, *cf.* Chapter 11.

10.3.3 Flash Point

As solvents are low molecular weight organic compounds, there is a fire and explosion risk when they are used in the manufacture, storage, transport and application of paint. The flash point, which is the lowest temperature at which the vapour of the liquid can be ignited in air under defined conditions, has to be taken into account in the formulation of paints. There are strict regulations governing the transport and storage of paint, and the flash points of paints have to be determined by specific methods to comply with the legislation. With a mixture of solvents azeotropic phenomena come into play and it cannot be assumed that the flash point will be that of the solvent with the lowest flash point. Mixtures of xylene and butanol, for example, have flash points which are lower than those of either component.

10.3.4 Evaporation Rate

A solvent should evaporate from a paint film at a rate suitable for the method of application. For example, the evaporation rate in a brushing paint should be such that the applied film remains fluid for a time so that the 'wet edge' remains open, and areas of wet film can be easily joined without showing signs of poor flow, *i.e.* brush marking.

For spraying paints it is important that one solvent evaporates very quickly and the wet film increases in viscosity to avoid excessive flow and runs (*cf.* Chapter 3). Remaining solvent in the film evaporates more slowly and thus allows the wet film to flow and form a fully coalesced, pinhole free film.

The rate of evaporation of solvent from a paint film can vary widely unless the atmospheric conditions of application and drying/curing are controlled. For factory-applied paint, this is generally not a problem. However, for protective coatings that have to be applied during winter and summer, in dry and humid conditions, the choice of solvents and solvent mixtures can be very important if problems of badly applied and poorly cured films are to be avoided.

The retention of solvents in films is also a factor to be considered. It can lead to blistering of immersed films due to osmotic effects of, for example, retained oxygenated solvents that are soluble in water. It is largely governed by the vitrification of the polymers present (*cf.* Chapter 4), and can be a major problem in systems designed to cure rapidly.

Two classes of coatings constituents that may be loosely regarded as solvents as they significantly reduce paint viscosity are reactive diluents and plasticisers.

Reactive diluents are mobile, reactive low molecular weight materials used to reduce the viscosity of compositions. They are subsequently combined with the cured material rather than being lost to the atmosphere (*cf.* Chapter 7). In reality, molecules small enough to be mobile will have appreciable volatility and their satisfactory design and use depends on maximising reactivity and minimising evaporation rate.

Plasticisers are inert materials of very low vapour pressure. They are intended to remain in a coating film throughout its life to modify its mechanical properties. A plasticiser effectively interferes with the secondary valency forces that hold the polymer chains together. The chains are separated and can move more freely thus reducing T_g, lowering tensile strength whilst increasing the elongation and flexibility. Typical materials are phthalate and phosphate esters and certain polymers. Low molecular weight liquid hydrocarbon resins based on, for example, α-methyl styrene are also successfully used in many applications. Adipates and sebacates are also utilised.

In practice, the lower molecular weight types migrate or slowly evaporate with shrinkage and embrittlement, and it is generally better to modify the film-forming resin to provide the required properties ('internal plasticisation').

10.3.5 Toxicity

Most solvents must be treated as toxic materials, and legislation is in force in all industrialised countries to control their use. They can enter the body by inhalation, skin contact and ingestion. Measures have to be taken to reduce risks to humans to a minimum. These can be the use of simple protective clothing, such as gloves for application by hand, or complete air-fed suits with visors for spray application in enclosed areas that are also ventilated to prevent build-up of vapour.

Toxicity studies provide data to assist in the control of toxic hazards. Official industrial hygienists in various countries publish lists of substances with atmospheric concentrations to which workers can be exposed safely. The widest known values are the threshold limit values (TLVs) produced in the United States. They are weighted averages that take into account the time in which a person can work in an atmosphere containing the solvent. The TLV is expressed as parts per million in air and should not be exceeded. In the UK, a similar system uses occupational exposure limits (OELs). TLVs or OELs can be taken as a working guide to the toxicity of paint solvents but it is important to be aware of changes. For example, the popular solvent 2-ethoxyethanol, which was once considered a relatively benign material, showed severe effects in animal tests. Its TLV was dramatically lowered, and in many industries, 2-ethoxyethanol has been replaced by the less toxic but more expensive alternative, 2-methoxypropanol.

10.3.6 Environmental Properties

The association of solvents with industrial smog in places such as California, the depletion of the ozone layer and the introduction of legislative controls were discussed in Chapter 2. Some products, for example, most chlorinated solvents, will have to be phased out. It is quite apparent that all organic solvents are being considered an environmental problem that has to be greatly reduced. This is forcing the development of water-based and solvent free coatings to be a major priority in the coatings industry.

Legislation in Europe, the Solvent Emissions Directive, is aimed at a reduction in release of all solvents into the atmosphere, irrespective of chemical type. By contrast, the VOC regulations in the USA are currently only aimed at reducing emissions of solvents that are photochemically active in the atmosphere, *i.e.* those that can produce ground level ozone, the primary component of smog. Individual solvents are not all equal in their potential to form ozone, and certain compounds such as acetone and *p*-chlorobenzotrifluoride have been determined as having no appreciable contribution to ozone formation. These materials have been given VOC-exempt status and are excluded from the regulatory definition of VOC. They can therefore be used (in the USA) in paints without contributing to the overall VOC.

Supercritical carbon dioxide has been used to replace part of the solvent in spray-applied paints. The supercritical carbon dioxide and paint are mixed in a specially designed chamber before entering the spray gun. The carbon dioxide reduces the paint viscosity sufficiently to allow good atomisation and evaporates before the paint reaches the substrate. The remaining solvent ensures that proper film formation occurs. This system, called the supercritical fluid spray process, has been shown to reduce VOC emissions by up to 80%, depending on the paint formulation. It has been patented and commercialised under the name Unicarb®. For this system to work satisfactorily, it is important to understand the phase behaviour of ternary mixtures of solvent, polymers and carbon dioxide. One of the limitations of this approach is that it cannot be used with epoxy amine paints. Unwanted reactions between carbon dioxide, the amine curing agents and moisture produce carbamates that result in poor curing and reduced film performance.

10.3.7   Odour

The odour of paint solvents has always been a contentious subject, and efforts are made to reduce the nuisance to a minimum. Several attempts have been made by the oil industry to produce solvents, such as aliphatic hydrocarbons, with very low odour to overcome the problem.

10.3.8   Examples of Solvents

Solvents are manufactured by a variety of processes, but virtually all are derived from petroleum nowadays. Hydrocarbon solvents are fractions obtained at the same time as fuels. Other solvents are manufactured substances that may also have uses as raw materials for the synthesis of organic chemicals, plastics, polymers and pharmaceuticals. Water, when used in coatings, nearly always acts as a diluent with organic solvents or a continuous phase in disperse systems.

Typical aliphatic hydrocarbon solvents are hexane, cyclohexane, 'white spirit' of various distillation ranges and special fractions manufactured for particular uses.

Aromatic solvents are xylene mixtures ('xylole'), toluene and various trimethyl benzenes. Benzene itself is not allowed as a paint solvent because of health hazards. Tetrahydronaphthalene ('tetralin') finds limited use.

The oxygenated solvents are:

Alcohols	ethanol, ethanol and isomers of propanol, butanol and higher alcohols.
Ketones	acetone, methylethyl ketone, methylisobutyl ketone, diisobutyl ketone and cyclohexanone are typical of the class.
Esters	ethyl acetate, butyl acetate and mixed acetates.
Glycol ethers and derivatives	2-ethoxyethanol, 2-butoxyethanol, 2-methoxy-propanol and their acetates.

The volume of solvents used by the paint industry is relatively small compared with the total volume of organic liquids produced mainly as fuels. Paint solvents have only two uses: to allow paint to be manufactured and to allow paint to be applied. Most conventional paint solvents are lost to the atmosphere without change or are burnt as waste products. This causes concern as it is wasteful and an environmental hazard.

The need to minimise the use of paint solvents has been the driving force in the development of alternative water-based coatings and solvent-free coatings, and this will become still more urgent as legislation is tightened.

10.4 ADDITIVES

The major components of paint by mass are generally pigment, resin and solvent. It would be expected that these materials would effectively dominate the manufacture, and subsequent performance of the paint film. In most cases, this is true. However, situations commonly arise where the use of relatively low levels of specialised additives is essential in order to make paint or to achieve the required performance. The effects of these materials can be dramatic, making all the difference between a viable paint and a total failure. Careful selection of additives and use levels, together with an understanding of the potential interactions are extremely important. It should be stressed that although interactions may be well understood in some instances, the system dependency is such that they are not predictable and great care must be taken at all times.

The most common paint additives will be described in this section. Each of the topics is a science or a technology in its own right, so the treatment here can only be brief.

10.4.1 Wetting and Dispersing Agents

Wetting and dispersing agents aid the incorporation of pigments and fillers into paints. The incorporation or dispersion process is one of the most important steps in paint manufacture. The purpose of the dispersion process is to break down the pigment agglomerates to the optimum particle size. If this is not achieved, a variety of defects can occur such as flocculation, poor development of colour and colour shift, flooding and floating, settling and loss of gloss.

Pigments are supplied as agglomerates or groups of particles where the gaps between them contain air and moisture. The individual particles are in contact along their edges and corners and the interactive forces holding them together can be overcome during the grinding or dispersion process. In the first stage, the air and moisture are replaced by the resin solution, which penetrates the space between the agglomerates. Wetting agents reduce interfacial tension between the pigment surface and the resin solution and speed up this process of preferential adsorption.

As the dispersion proceeds, the agglomerates are broken down into their individual particles revealing higher surface areas and at some point the pigment particles will be at their optimum size and evenly distributed throughout the liquid medium. If this dispersion has not been stabilised, the particles will come into contact and may be held together by van der Waals forces producing undesirable flocculates. Stabilisation is achieved by the use of dispersing aids and these operate by charge stabilisation or steric stabilisation, reminiscent of those described in Chapter 8.

In charge stabilisation, the dispersing aid forms an electrical double layer around the pigment particles that extends into the medium. Since each pigment particle has the same charge, they repel and the dispersion is stabilised. Charge stabilisation is an especially useful mechanism in water-based systems.

Stabilisation can also be achieved *via* steric hindrance arising from the adsorbed dispersing aid. The long, resin compatible chains of the dispersing aid become solvated and protrude into the binder producing a steric barrier and preventing particle contact. Effective steric stabilisation needs an adequate stabilising layer thickness (0.01–0.1 μm) and sufficiently dense coverage of the particle surface by the dispersant. Steric stabilisation is an important mechanism in solvent-based paints but it also plays a part in water-reducible systems.

The structures of wetting and dispersing agents differ. Both must wet and anchor to the surface of the pigment particle, whether it is organic or inorganic, but the polymer chain on the dispersing agent must be long enough ($>C_{18}H_{37}$ in solvent-based systems) to provide an effective steric barrier once it has been solvated.

Products are many and varied and cater for both water-based and solvent-based systems. Some allow for a controlled degree of flocculation, and some combine both wetting and dispersing properties in one molecule. They also require different modes and stages of incorporation in paint manufacture, and have different addition levels.

10.4.2 Foam Control Agents

Foam is a gas, usually air, which has been stabilised in a liquid medium. The air is usually incorporated during manufacture and if it is not released, it can reduce grinding efficiency and will cause problems in filling off the finished paint into cans. Foam can occur in both solvented and water-based paints. It can also be generated during application or through the build-up of gas from chemical reactions with binders.

When air is introduced into a liquid paint, bubbles are formed and these rise through the liquid towards the surface. As they rise the bubbles assume a spherical shape. Once at the surface the bubble wall, the lamella, begins to drain and becomes thinner. In a pure liquid, the bubble wall bursts at a thickness of about 100 Å, the air is released and no stable foam is formed. If the liquid is not pure and surface-active materials (surfactants) are present, a stable foam can be formed. The surfactants concentrate in the lamella at the gas–liquid interface. As the lamella drains and the inner and outer surfaces come together, electrostatic repulsion from the ionic groups of the surfactant prevents collapse and foam stabilisation occurs. The lamella is also elastic in nature due to the Marangoni–Gibbs effect. As the lamella is stretched or deformed, the concentration of adsorbed surfactant changes producing a higher surface tension that pulls the lamella back together. This additional foam stabilisation effect is also called Gibbs elasticity.

The terms 'defoamer' and 'anti-foam' are sometimes used interchangeably but strictly speaking, they are not the same. A defoamer destroys existing, stabilised foam at the air interface whereas an anti-foam prevents the formation of stable foam. However, because of the difficulty in distinguishing between the mechanisms, we will continue to use the term 'defoamer'.

An effective defoamer must overcome the foam stabilisation mechanism. Generally, they are low surface tension liquids that have a controlled insolubility or incompatibility with the liquid medium. Once in the system, they enter the foam lamella and spread over the interface, interacting with or displacing the surfactant. The resulting lamella has both reduced surface tension and lower cohesive strength and is consequently destabilised and weakened. This in turn leads to rupture and collapse of the foam.

Chemically, defoamers are either mineral oil, or based on silicone or fluorinated materials. As they are not soluble in the paint, they must be dispersed to the correct droplet size in the paint to be effective. Both aromatic and aliphatic hydrocarbons have been successfully used but care must be taken in their use as they can affect the gloss in some formulations. Organic alcohols such as hexadecanol and esters (tributyl phosphate) also function as defoamers. Pure silicone oils are very effective but again, overdosing can lead to poor cosmetic appearance of the subsequent films. Polysiloxane–polyether (**15**) copolymers have been found to give effective defoaming with negligible effect on final gloss of films. In selecting a silicone defoamer, attention must be paid to the chemical structure as silicones with a relatively short chain can actually demonstrate foam stabilisation behaviour rather than destabilisation.

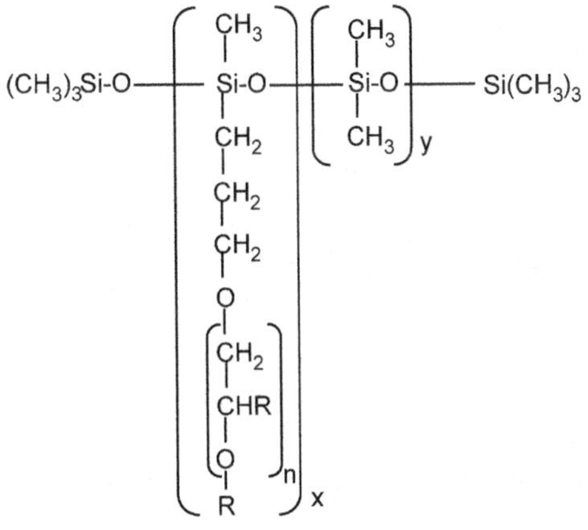

$$(CH_3)_3Si\text{-}O\text{---}\left[\underset{\underset{\underset{\underset{\underset{\underset{\underset{\underset{R}{O}}{\underset{CHR}{\left[\underset{CH_2}{}\right.}}}{O}}{CH_2}}{CH_2}}{CH_2}}{\underset{CH_3}{Si\text{-}O}}\right]_x\left[\underset{CH_3}{\overset{CH_3}{Si\text{-}O}}\right]_y\text{---}Si(CH_3)_3$$

Structure (15):

$(CH_3)_3Si\text{-}O$ — $\left[\begin{array}{c}CH_3\\|\\Si\text{-}O\\|\\CH_2\\|\\CH_2\\|\\CH_2\\|\\O\\|\\\left[\begin{array}{c}CH_2\\|\\CHR\\|\\O\end{array}\right]_n\\|\\R\end{array}\right]_x$ $\left[\begin{array}{c}CH_3\\|\\Si\text{-}O\\|\\CH_3\end{array}\right]_y$ — $Si(CH_3)_3$

(15)

In water-based systems, the use of surface-treated hydrophobic silica particles in combination with a silicone polymer produces a very effective defoamer. It is thought that the silica particles are transported to the foam lamella where they provide a point source for rupture of the bubbles. Particles of hydrophobic polyurea particles are also successfully used as a component in defoamers.

Air can also be trapped in the paint film as a consequence of the application process. Air introduced in this way must leave before the paint film has dried. If it remains, pinholes may be formed and the performance and appearance of the film compromised. Removal of these bubbles is called deaeration and agents that promote this effect are 'deaerators'. Deaerators operate at the interface of the bubbles spread throughout the liquid film and this mode of action is different from that of defoamers. They modify the environment surrounding the bubbles, possibly by reducing the viscosity at the air/liquid interface, and increase the velocity of the bubbles as they rise to the surface. They may also serve to encourage the coalescence of bubbles by destabilising the bubble interface. The bigger bubbles rise to the surface more quickly where they burst. The rate at which bubbles rise in a liquid is described by the following equation derived from the Stokes law:

$$v \propto \frac{r^2}{\eta} \qquad (10.1)$$

where v is the rising velocity, η is the viscosity and r is the bubble radius.

Deaerators need to be of limited compatibility with the paint system and selection of the appropriate agent is system dependent. Many compounds function as deaerators including organic polymers such as polyethers and polyacrylates, polysiloxanes and silicone–polyether copolymers.

10.4.3 Rheology Modifiers

Rheological additives are used primarily to modify storage stability and application properties. They prevent settlement in the can and also excessive sagging during application. This is particularly important in the DIY market where paint may be stored and used over a long period of time and where a one-coat high build application is desirable rather than a two-coat scheme.

Additives used to control viscosity in water-borne systems have been described in Chapter 3, so this section will concentrate on those materials used for solvent-based and solvent-free paints.

Rheological modifiers must impart high viscosity at low shear rates to counteract settlement under gravitational forces, but low viscosity under high shear rates such as those undergone during application. Once the paint has reached the substrate, a high viscosity must be very quickly achieved in the film in order to prevent sagging. An additive providing the appropriate thixotropic rheology will usually perform both functions satisfactorily although it is common to use combinations to get the best results. The most widely used modifiers are those based on clays, silicas and waxes.

Modified organoclays are widely used to produce thixotropic behaviour in paints. These are bentonite ($Al_2O_3 \cdot 4SiO_2 \cdot 2H_2O$) or hectorite (($Mg,Li)_3Si_4O_{10}(OH)_2$) clays modified with a quaternary ammonium salt. Selection of the clay type, the quaternary ammonium chemistry and the processing conditions allow organoclays to be produced for specific applications. The clays themselves occur naturally as sheets or platelets stacked together. The production process breaks up the platelets and covers the faces with a long chain quaternary salt and this treatment tends to keep the individual platelets apart. The edges of these platelets possess hydroxyl groups and these are not affected or displaced by the quaternary ammonium salt. Organoclays are supplied as agglomerated stacks of platelets, and under the influence of a polar activator such as alcohols or water and mechanical dispersion, the platelets become fully dispersed. A gel structure is developed *via* hydrogen bonding through the hydroxyl groups at the edges of the platelets. In some grades, the activator is not required.

Finely divided surface-treated fumed silicas are used to prevent settling. They possess surface hydroxyl groups capable of producing a hydrogen-bonded structure in a paint. Once disrupted, the hydrogen-bonded silica network can take up to 30–60 s to reform and this slow recovery makes silicas unsuitable for use as an anti-sagging agent. The hydrogen-bonded structure can be affected by other components in the paint, especially monofunctional hydrogen bonding species. Alcohols, carbonyl compounds, ethers and amines will all hydrogen bond to the surface hydroxyls and prevent the build-up of the required structure.

Polyamide waxes and hydrogenated castor derivatives can prevent both in-can settlement and confer anti-sagging properties on a liquid paint. They are supplied as powders or as pre-activated dispersions. They are activated by a combination of shear forces and heat and this results in deagglomeration, softening and solvent swelling of the particles. In hydrogenated castor-based additives, a hydrogen-bonded structure is formed on cooling. In the case of polyamide additives, micelles are formed coupled with chain entanglement. This provides associative interactions with pigments and extender surfaces preventing settlement and imparting good anti-sag properties to the paint.

Waxes based on straight chain polyethylene chemistry are also used as anti-settling additives. They are commercially available as fine powders or as pastes in solvent. Their structure as supplied is a coiled crystalline matrix. In common with the polyamide waxes they must be activated by a combination of heat and shear over a period of time. During this process, the initial helical form becomes extended into a straight chain configuration and it is this structure that produces chain entanglement and prevents or reduces pigment settlement.

Calcium sulfonate-based thixotropes have found application in some systems. They contain acicular crystals of calcium carbonate associated with the long chain sulfonate. The hydrocarbon chains (typically C_{12}–C_{30}) associate through weak van der Waals forces producing a stable structure suitable for prevention of pigment settling and imparting reasonable anti-sag properties.

Liquid thixotropes are also available based on modified urea functionality in solution in *N*-methylpyrrolidone. The active material is insoluble in common paint solvents. When added to paint, a controlled precipitation takes place producing very fine needle-like crystals. The crystals form a three-dimensional lattice structure that confers thixotropic behaviour on the paint.

10.4.4 Flow Aids

Flow aids are materials incorporated into a paint formulation to enhance film properties by promoting levelling or flow, and eliminating surface defects such as brush marks, orange peel and craters. Defects are caused by too much or too little flow and the two most important influencing factors are viscosity and surface tension. Surface tension tends to minimise the area of the film to create a level, even surface. The viscosity of the paint opposes this effect, and acts to retard flow. Low viscosity is needed for elimination of brush marks and high viscosity is needed to avoid sagging. A well-formulated paint therefore represents a compromise between these two extremes.

Almost all flow aids function by modifying the surface tension of paint. They can differ considerably in their chemistry with the most common types being based on silicones, fluorinated compounds and polyacrylates. Their efficacy is very system dependent and it cannot be assumed that one level will be appropriate for all resin systems and all chemical types.

Silicone and fluorinated modifiers have significant effects on surface tension, and produce considerable reductions even when present at levels of parts per million. They must therefore be used with care as overdosing can easily result in film defects. It can also give problems on recoating as excess modifier will accumulate at the paint/air interface making it difficult for the next coat to wet out the film fully.

Polyacrylate modifiers (**16**) have a much less pronounced effect on reducing surface tension than the silicone and fluorinated types. Poly-acrylates work by migrating to the surface, forming a monomolecular layer and producing an even surface tension.

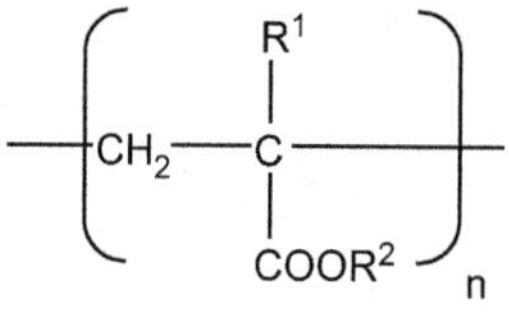

$$\left(CH_2-\overset{\displaystyle R^1}{\underset{\displaystyle COOR^2}{C}}\right)_n$$

R^1 = H, CH_3
R^2 = Alkyl (linear, branched), polyester, polyether, amine salt

(**16**)

10.4.5 Slip and Mar Aids

Slip is a term usually applied to describe the smoothness or feel of a coating. Mar resistance is the ability of the coating to withstand mechanical abuse such as abrasion. In many circumstances the terms are used interchangeably and in practice, an additive that confers better slip to a coating is also likely to give better mar resistance. Waxes and silicone-based oils are by far the most common materials used to improve slip and mar resistance.

The waxes used are incompatible with the resin systems and migrate to the surface of the film where they improve the lubricity and 'feel' of the coating. They are available with a range of hardnesses and it is found that softer waxes give the best slip properties. Polyethylene (PE), modified PE and hydrocarbon waxes are all used at levels of up to 3% by weight of paint.

Pure or modified silicone-based oils are widely used in both solvent-borne and water-borne systems. These migrate to the surface of the paint and reduce surface irregularities giving a much smoother film. In addition, the presence of a thin film of the silicone reduces sliding friction. This results in better feel and improved mar resistance. Care must be taken, however, as overdosing of these materials in a formulation will result in many film defects and possible recoat problems.

10.4.6 Biocides

The organic nature of paints means that they are susceptible to attack, both in the can and also as a dry film, by a number of agents such as bacteria, fungi, yeasts and algae. Addition of a carefully chosen biocide package can effectively eliminate problems caused by microbiological attack.

By and large, in-can preservation is a much bigger problem for water-based paints than for solvented products. Attack by microorganisms often manifests itself as gassing, pH drift and discoloration of the paint. Where cellulosic thickeners are used, the paint viscosity can be lowered and hold-up on application vastly reduced. Anaerobic bacteria can reduce sulfur compounds resulting in offensive odours on opening the can.

Sterile manufacturing conditions and high temperature treatments could prevent much of the problem but this is clearly not practical in a typical paint manufacturing environment. Because of this, the use of biocides to control in-can stability is now commonplace. A wide choice is available, depending on the specific challenge. The required concentration is best determined by experiment and this will allow the selection of the most cost-effective biocide that will confer adequate protection on the formulation.

Effective in-can biocides include isothiazolinone derivatives such as benzoisothiazolinone (BIT). BIT (**17**) is frequently used in alkaline systems at a pH of 9 and above but it can also be successfully used in acidic systems. Mixtures of methylisothiazolinone and 5-chloroisothiazolinone (MIT/CIT) are also employed. These are cost-effective broad spectrum biocides and are effective against bacteria, fungi and yeasts. Formaldehyde adducts (the reaction products of formaldehyde and various nucleophiles) are also effective against a wide variety of organisms either used alone or in combination with other biocides as they release formaldehyde slowly. They range from o-hydroxymethyl compounds (o-formals) to a wide variety of condensation products such as 1,3-oxazolidines. Phenolic compounds such as the sodium

(**17**)

salt of *o*-phenylphenol and chloracetamide derivatives are particularly effective for preserving natural materials such as casein or gelatine.

Use levels for these in-can biocides vary according to the specific challenge but typically range from 0.02 to 0.3% by weight of paint.

Dry films are also subject to microbiological attack by fungi and algae. The result of this can be discoloration of the film and in some situations the film integrity and subsequent performance may be seriously affected. Not all paints are equally susceptible to attack and epoxies, for example, are relatively unaffected compared to oil-based paints. A dry film preservative should ideally have a relatively low water solubility to prevent leaching from the film to maintain effectiveness for the maximum time. It should also be unaffected by the in-service conditions and by the substrate to which the paint is applied.

Film preservatives include carbendazim, octylisothiazolinone, zinc pyrithione, chlorothalonil and substituted cyclopropylamino-*s*-triazine derivatives. As with in-can biocides, the optimum use levels are determined by experiment.

Certain biocides are used in combination in anti-fouling paints to prevent fouling of ships' hulls and to reduce fuel consumption and also to reduce fouling build-up on the legs of oil rigs. Fouling consists of slimes, weed and various animal species, such as barnacles. In anti-fouling paints, the biocides are used at much higher levels than those for in-can and film preservation and are arguably not 'additives'. They are contained within a film and gradually leach out over time producing a toxic surface that discourages growth of marine life on the paint. Amongst the biocides used in this application are zinc and copper pyrithione and copper oxides.

10.4.7 HALS and UV Absorbers

On external exposure, a paint film meets several challenges, not least of which is the damaging effect of UV radiation on the polymer backbone. UV radiation has sufficient energy to break covalent bonds and is readily absorbed by many structures and functional groups in a polymer. Aromatic structures such as those in bisphenol A-based epoxies are known to absorb UV strongly and are very susceptible to degradation. The wavelengths of the damaging radiation are 400–315 nm, UV-A, and 315–280 nm, UV-B. The relative intensity of these bands varies with season and location but, over time, both will adversely affect polymers.

Photochemical degradation takes place *via* radical generation and formation and decomposition of peroxides. The precise mechanism is clearly dependent on the polymer and the other components of the paint. In combination with free radicals, water and atmospheric oxygen produce a particularly damaging environment for paint films. The result of this attack is loss of gloss, colour drift, chalking, cracking, flaking, and ultimately detachment of the paint film from the substrate.

Strategies to prevent degradation involve the use of materials (UV absorbers) to selectively absorb the damaging wavelengths, in combination with other compounds (hindered amine light stabilisers, HALS) that can scavenge those free radicals that do get produced.

UV absorbers are generally substituted aromatic species, for example, hydroxy-substituted benzophenones, hydroxyphenylbenzotriazoles and substituted acrylonitriles. These materials are often tautomeric and readily absorb UV radiation. In effect, they compete for the incident radiation with the polymer system and thereby reduce the adverse effects on the film. When this radiation is absorbed, the enol form of the molecule is rearranged to the keto form (Scheme 10.2). The keto form is converted back into the enol form by dissipation of the absorbed energy as heat. This conversion process does not degrade the molecule and can continue essentially indefinitely. UV absorbers are expensive and use levels are typically up to 3% on total formulation weight, although this is formulation dependent.

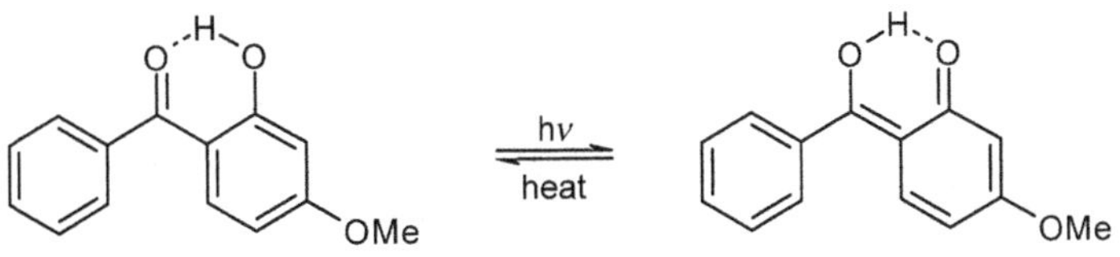

Scheme 10.2

HALS are often used in combination with UV absorbers to give maximum protection to a paint film. UV absorbers cannot be totally effective in capturing all incident UV radiation. The small amount that is not absorbed will generate damaging free radicals and begin the degradation process. HALS scavenge these free radicals and inhibit the degradation. In this process, the tertiary amine is oxidised into a nitroxy radical that can react with free radicals to produce an amino ether. The amino ether can then react with peroxy radicals thereby regenerating nitroxy radicals and so the cyclic process continues, as shown in Scheme 10.3. Newer HALS are amino ethers and therefore do not require the initiation step to enable them to scavenge free radicals. HALS are used at low levels in paints, typically up to 3% by weight, but as usual, this is system dependent especially in pigmented systems.

Used together, UV absorbers and HALS act synergistically, and combinations are often used in demanding applications such as high performance finishes.

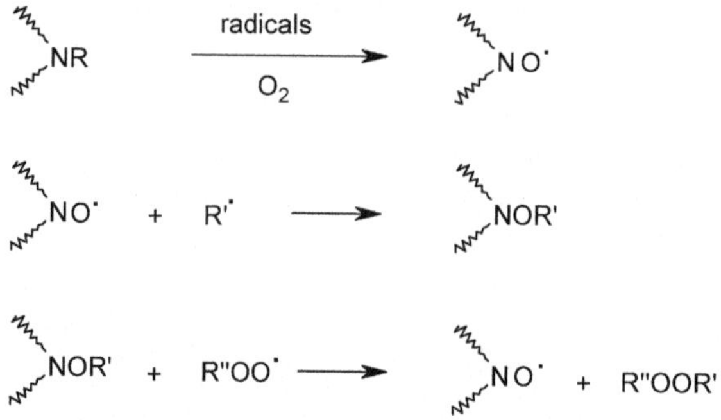

Scheme 10.3

10.4.8 Emulsifiers

An emulsion is a stable mixture of two or more immiscible liquids, such as oil and water, held in suspension. All emulsions are comprised of a continuous phase and a disperse phase and those that contain water can be described as either water-in-oil or oil-in-water emulsions. The mixture is stabilised by surface-active materials called emulsifiers and these function by reducing the surface tension at the interface of the suspended particles. Emulsifiers are ubiquitous in water-based paints and their selection is critical to their successful manufacture as described in Chapter 8. Soaps – salts of long chain fatty acids – are examples of emulsifiers.

10.4.9 Flash Rust Inhibitors

Flash rusting is an effect seen in water-based systems applied directly to steel. In this situation all of the components necessary for corrosion of steel – water, oxygen and ionic species – are present. It manifests itself as small brown rust spots, sometimes accompanied by blistering, that appear very soon after application of paint to the steel substrate. The problem is made worse when the steel surface is 'active' such as when it has been freshly blast cleaned. A pH of <7 will encourage flash rusting and control of pH during film formation is a critical factor in preventing the problem. Fortunately, most of the common water-based paints for these applications are alkaline and this reduces but does not completely eliminate corrosion. Flash rusting is relatively easy to control on flat steel plate but it is much more difficult over welds where the complex metallurgy encourages corrosion.

Flash rust inhibitors are typically based on ammonium and sodium nitrates and nitrites, benzoates, metaborates and phosphates occasionally in combination with materials such as morpholine and amino methyl propanol. More complex organic salts are also used, such as (2-benzothiazolythio)-succinic acid amine salt.

Use levels are low, typically 0.1–1.0% on the paint and will vary according to the system and substrate.

10.4.10   Agents to Control pH

The pH of water-based paints must be carefully controlled, both to maintain stability of the polymer and also to help control flash rusting when the paint is to be applied directly to steel. An alkaline pH is preferred and amines are most commonly used to maintain pH within acceptable limits, often >9. These include ammonia, triethanolamine, diethanolamine and AMP, 2-amino-2-methyl-1-propanol. The latter compound is widely used as it offers pH control without the excessive odour associated with the use of ammonia.

10.4.11   Coalescing Solvents

Film formation in a latex paint involves the fusion of small particles of a high molecular weight thermoplastic polymer. For this to take place, the molecules within the polymer particles must be able to move sufficiently to facilitate coalescence and the formation of a continuous, integral film. This can be achieved if the polymer has a sufficiently low T_g, but in the absence of a crosslinking mechanism, the resulting film will be soft and tacky. Alternatively, the T_g can be reduced by the incorporation of high boiling solvents. Once coalescence has occurred, the solvents evaporate slowly leaving a hard film.

Coalescing solvents may be partially or completely miscible with the continuous phase. They may also migrate into and swell the polymer particle and in this case, care must be taken not to add too much as the polymer particles may coagulate.

Glycol ethers and their acetates and alcohols such as hexylene glycol have been widely used as coalescing solvents. Glycol ethers of the alkoxyethanol type (except butoxyethanol) have been largely superseded in view of their toxicology.

10.4.12   Freeze–Thaw Stabilisers

When cooled below 0 °C, some water-based systems will freeze and the paint will be destabilised. To help alleviate this problem, materials are added to depress the freezing point of the paints. Alcohols are particularly good in this respect and ethylene and propylene glycols have been widely used in the past. Glycol ethers have also been successfully used. Coalescing solvents described above also behave as freeze–thaw stabilisers.

10.4.13 Adhesion Promoters

Adhesion, particularly wet adhesion, is one of the key properties of any protective coating. Adhesion promoters are materials designed to improve the adhesion between a polymer film and a substrate. In most applications, this substrate is a metal or a plastic but in some situations the substrate could be an aged coating. Polymer to substrate adhesion can be improved by mechanically roughening the surface, for example, by sanding or by grit blasting. Other methods involve solvent washing, chemical etching, flame and plasma treatment, and UV irradiation. Clearly, there are many situations where these methods are not possible and improvements in adhesion must be achieved through careful selection of components in the paint formulation.

In general, all adhesion promoters work through a common mechanism related to their structure. Basically, they can be considered as one molecule containing two different functional groups, one of which can react with the substrate and the other with the polymer. The result is a chemical bridge between the two that improves the overall adhesion and subsequent performance of the coating.

Commercially available adhesion promoters are typically organofunctional silanes (**18**), titanates, zirconates, zircoaluminates, alkyl phosphate esters and metal organic complexes. Different classes of material have their own individual advantages and disadvantages and these would have to be taken into account during the selection process. Silanes, titanates and zirconates are available with a wide range of organic modifications and have excellent solubility in organic solvents. However, they suffer from varying degrees of water sensitivity. In particular, silanes are very water sensitive and will even hydrolyse to silanols in moist air. Zircoaluminates, alkyl phosphate esters and metal organic complexes are less water sensitive but are not as compatible with organic solvents.

$$Y\text{-}R\text{-}Si\text{-}X_3$$

Y = polymer reactive group
X = hydrolysable group, -OMe, -OEt
R = hydrocarbon

(**18**)

Clearly, the choice of material will depend on the nature of both the polymer and the substrate to be coated. Characterisation of the substrate is obviously key to the selection and in the case of metals, relatively straightforward. This is a much more difficult prospect when overcoating an aged coating as the surface is complex and can change with the age of the coating. Use levels vary but are typically up to 3% by weight on resin solids in a clear system. The situation is more complicated where pigments are present, as the promoter may also migrate to the polymer–pigment interface thus reducing its concentration and effectiveness at the substrate–polymer interface. It is sensible therefore to use an increased loading to offset this competing effect.

The mode of incorporation is also different from product to product. This can vary from simple stir-in under low shear, or addition to the grind charge in the case of titanates, zirconates, zircoaluminates and the alkyl phosphates.

10.4.14   Driers

Driers are used to promote the oxidative crosslinking of oil-based or alkyd resins either at ambient or elevated temperatures. They are essentially solutions of metal soaps usually in an aliphatic hydrocarbon solvent. Oil-based media or alkyd resins are manufactured from triglyceride oils or C18 fatty acids and contain unsaturation – typically from oleic, linoleic and linolenic acids. Drying takes place by oxidative crosslinking through the double bonds present in these unsaturated fatty acids. Oxygen is absorbed initially, followed by the formation and subsequent decomposition of hydroperoxides or peroxides. Intermolecular combination involving the hydroperoxides and peroxides takes place on the pendent fatty acid side chains, leading to a crosslinked structure as described in Chapter 6. This process is catalysed by the metal component of the driers.

Driers can be made from a wide variety of metals and acids. It is the metal that effects the drying and the acid that confers the solubility and compatibility in solvents and paint media. The metals themselves usually, but not always, have variable valency and the most commonly used in ambient-cured paints are cobalt, lead, calcium, manganese, zirconium, lithium and zinc. Driers based on other metals, such as cerium and iron, are also available for heat-cured systems.

The earliest, traditional driers were based on naphthenic acid, a material derived from crude oil. These had excellent compatibility with a range of media and solvents. As naphthenic acid became increasingly scarce, alternative acids were sought. Most driers today are based on branched chain synthetic acids such as 2-ethyl hexanoic acid and neodecanoic acid.

Driers are supplied commercially as solutions, often in white spirit. The solution may also contain various stabilising additives such as alcohols that are included at low levels to help prevent precipitation. Driers themselves are generally described by their metal content, *e.g.* cobalt octoate 10%, means that 100 g of the drier solution contains 10 g of cobalt metal by weight. Knowing the metal content is important as it is used to calculate the required amount of drier solution to be added to a paint or resin. Levels of addition are very low, typically 0.01–0.6% by weight of metal on resin solids, depending on the drier metal.

Driers are usually classified as being either 'primary' or 'auxiliary'. Primary driers are oxidation catalysts and they encourage surface drying in an alkyd film producing tack-free films very quickly. They promote the uptake of oxygen and the decomposition of hydroperoxides and crosslinking of the film. They are also known as surface or top driers. Drier metals in this group are cobalt, manganese and vanadium. Cobalt is the most widely used giving the shortest drying times with the least discoloration of films. Manganese, often in combination with 1,10-phenanthroline, is almost as effective as cobalt but it leads to staining problems. Vanadium behaves much the same as manganese in both drying performance and tendency to produce staining.

Auxiliary driers promote the polymerisation reaction that leads to gradual hardening of the film. They are also known as through-driers. The most widely used are those based on zirconium, barium, calcium, lithium and zinc. Lead driers are very effective in combination with cobalt and calcium but toxicity concerns have all but eliminated it as a drier metal.

Aluminium organic compounds can also perform as driers. They are based on ethyl acetoacetate and may also contain alkoxide or hydroxide groups attached to the aluminium atom. These materials function by interaction of the aluminium organic with the carboxyl and hydroxyl groups on the alkyd producing coordination linkages. In effect, they act as through or auxiliary driers and are also called coordination driers. Aluminium complexes are used together with a primary drier such as cobalt to give optimum drying performance. In order for the aluminium complex to be an effective drier, the alkyd resin must contain sufficient acid and hydroxyl groups. As the crosslinking mechanism is independent of oxygen, coordination driers have been reported to reduce in-can stability and shorten the shelf life of the paint.

Primary and auxiliary driers are normally used in combination to achieve the desired drying performance. The exact levels of each metal will depend on the paint formulation and the end use requirements and would normally be determined by experiment. Optimised combinations are sold as 'mixed driers'.

Table 10.1 shows typical and maximum loading for the most common driers.

Table 10.1   *Metal loading of driers in alkyd compositions*

Drier metal	Typical concentration*	Maximum concentration*
Cobalt	0.06	0.2
Manganese	0.03	0.1
Zirconium	0.2	0.6
Calcium	0.2	0.4
Lithium	0.03	0.04
Zinc	0.2	0.3
Barium	0.2	0.4
Aluminium	0.5	1.0

*Expressed as % by weight on resin solids.

10.4.15   Anti-skinning Agents

Anti-skinning agents are used in air-drying alkyd systems to prevent the formation of a skin on storage in the can. As alkyds dry by reaction with atmospheric oxygen, the same reaction will take place with the oxygen in the headspace in a can. This is particularly noticeable in part filled cans.

To prevent this, various oximes are added to the paint formulation. Oximes are thought to function by complexing with the driers thereby interfering with the drying process. Oximes with similar volatility to that of the paint solvents are preferred. Once the paint has been applied to the substrate, the volatile oxime can evaporate from the film allowing the expected drying process to take place.

Methylethyl ketoxime is probably the most widely used anti-skin. Butyraldoxime and cyclohexanone oxime both find use, the latter when low odour is required. Typical use levels range from 0.05 to 0.3% based on the weight of the paint.

10.4.16   Matting Agents

There are applications where high gloss is not the preferred option and satin or low gloss finishes are desired. Matting can be achieved by using a high PVC or by introducing incompatibility into the film. However, by far the most common method is the use of additives and this allows better control of the effect. These additives include silicas, polymethyl ureas and waxes.

Silicas are the most widely used matting agents in the coatings industry and are available mainly as precipitated and fumed grades. Diatomaceous silicas are also used. Physically, they are of low particle size, less than 10 μm, and they can be chemically surface treated to improve compatibility with all major resin systems from solvent-borne to water-borne to UV and electron beam cured systems. They function by introducing micro-roughness into the surface of the film and this scatters rather than reflects the incident light. Silicas can be difficult to incorporate and their effectiveness can be reduced by overdispersion. In addition, they tend to increase the viscosity of the system. The refractive index of silica is very close to that of many resin systems, and consequently, they can be used in clear varnishes to produce matt surfaces without haziness. They are very effective and are used at levels of up to 3% by weight on paint.

Micronised polymethylurea resins, PE and PTFE modified PE waxes are all used as matting agents. They work in the same way as silicas by migrating to the surface and producing micro-roughness.

10.4.17 Fire Retardant and Fire Resistance Additives

The majority of polymeric binders will readily combust if heated in the presence of air. The exceptions to this are highly chlorinated binders such as alkyds made from chlorendic anhydride or chlorinated rubbers. As the coating burns, flames and potentially toxic gases are produced. If the coating is protecting steelwork, heat will be transmitted to the steel and this can result in structural failure. To be of most value, a coating intended to protect steelwork should therefore ideally not burn and it should insulate the steelwork from heat.

Fire retardant coatings do not support combustion and this can be achieved by the use of chlorinated binder and additives that interfere with the radical nature of the combustion process. Halogenated materials such as chlorinated paraffins and brominated diphenyl are often used in synergistic mixtures with antimony trioxide and triaryl phosphate esters. In a fire, aluminium trihydrate decomposes liberating large quantities of water of hydration thereby inhibiting combustion by absorption of heat and exclusion of air.

To provide maximum protection to the steel from the damaging effects of heat, intumescent coatings are used. In addition to being flame retardant, intumescent coatings expand when heated to form a thick insulating char around the steelwork protecting it from the heat and maintaining the structural integrity for longer. This foam can be up to 2–3 cm thick when fully formed.

Intumescence is achieved by incorporating three basic materials into the binder: a material to produce carbon, generally a polyol such as dipentaerythritol, an acid releaser such as ammonium phosphate and a blowing agent such as melamine. Under the action of heat, the decomposing phosphate produces phosphoric acid. This reacts with the polyol and the resulting ester subsequently decomposes forming a char and regenerating the phosphoric acid. At the same time, the blowing agent decomposes producing large volumes of nitrogen and the char expands, producing thick insulating foam around the steelwork.

Pentaerythritol and starch have been used as the polyol, melamine phosphate as the acid producer and urea and dicyandiamide as the blowing agents.

10.4.18 Moisture Scavengers

Scavengers are important in moisture sensitive systems such as those containing isocyanate and also zinc- and aluminium-rich paints. Moisture can get into the paint by a variety of routes. It is found dissolved in solvents and also absorbed onto the surface of pigments. To avoid potential problems in moisture-sensitive systems, carefully dried materials must be selected and for added security, it is sensible to use scavengers to remove any residual traces of water.

In a urethane system, the reaction between isocyanate and water will produce carbon dioxide and amines. The presence of even trace quantities of moisture will cause gassing in the can and this could result in the lid blowing off. In addition, the contents will increase in viscosity and may ultimately gel. To overcome this problem, aluminium silicate-based molecular sieves and finely divided silica gel are used to consume the water. Monomeric isocyanates, triethyl orthoformate and oxazolidines have also been used to good effect.

Some paints contain finely divided zinc and aluminium metal and in certain circumstances, these metals will react with water to produce hydrogen gas. This potential hazard is controlled either by inhibiting the metal or by removing the water by use of a moisture scavenger. Again, molecular sieves and silica gels have both been successfully used to prevent the problem.

10.4.19 Conductivity Modifiers

Particles or droplets of paint that are applied by electrostatic spray are given an electric charge at the gun. The charged material is then attracted to the earthed surface of the object to be painted. If the resistance of the material is too high, the result will be poor application. The resistance of a liquid paint formulated from non-polar solvents will be too high for good electrostatic spray application. It has been found that a resistivity of between 10^4 and

$10^7 \, \Omega \, m^{-1}$ is necessary for good transfer efficiency. This can be achieved by the incorporation of small quantities of quaternary ammonium salts, occasionally in conjunction with a polar solvent for an improved effect. Care must be taken when determining the correct level, as too high a conductivity will result in charge leakage at the gun, reduced electrostatic field and lowered transfer efficiency.

10.4.20 Anti-skid Additives

Areas subjected to heavy traffic, such as decks of ships, oil rigs and floors are often coated with hard abrasion-resistant epoxy-based paints. The unmodified surfaces of these paints can be extremely slippery, especially if they become wet. In order to increase the coefficient of friction and reduce the risk of accident, large particle size fillers or aggregates are employed in the coating. They can be added during manufacture or sprinkled on the wet coating. Many materials from polymer beads to sand and bauxite can be used and they range in particle size up to approximately 2.5 mm depending on the application.

10.4.21 Deodorants

Deodorants, also known as reodorants, are occasionally added to paints to mask the smell of solvents or of by-products from the drying process. These include natural products such as pine oil and limonene.

10.4.22 Hammer Finish Additives

The hammer finish effect seen in paints is actually a controlled surface defect. This defect is deliberately introduced by incorporating very high molecular weight polydimethylsiloxanes into the paint formulation. These materials are incompatible with the paint binder and migrate to the surface. Lower molecular weight silicones have positive effects on levelling and flotation, and as the molecular weight increases, the silicones become useful defoamers. At molecular weights of around 300,000, the polymer produces the desired hammer finish effect at the surface.

10.5 BIBLIOGRAPHY

1. T.H. Durrans, *Solvents*, 8th edn, Chapman & Hall, London, 1971.
2. R. Lambourne, *Paint and Surface Coatings, Theory and Practice*, Ellis Horwood, Chichester, 1987.

3. P.A. Lewis (ed), *Pigment Handbook, Vol. 1, Properties and Economics*, 2nd edn, Wiley, Chichester, 1988.
4. H.F. Payne, *Organic Coating Technology*, Vol. 2, Wiley, Chichester, 1967.
5. D. Satas and A.A. Tracton (eds), *Coatings Technology Handbook*, 2nd edn, Marcel Dekker, New York, 2001.

CHAPTER 11

The Science and Art of Paint Formulation

ALAN GUY

11.1 INTRODUCTION

In general, formulation involves a large matrix of issues, obstacles and compromises. Though firmly rooted in scientific understanding, its practice often seems more like an art form and the skills required cannot be easily or quickly transferred. However, the development of computerised formulation systems is generally making it easier for the inexperienced formulator to control the many variables.

11.2 BASIC PAINT FORMULATION

The components of paint are illustrated in Figure 11.1. It should be appreciated that not every paint will contain all of the types of ingredients shown. Also, the same ingredients combined at different levels can produce both paints and films with quite different properties.

The formulation process begins with establishing the criteria that will determine the choice of the raw materials and the way in which they are put together. Amongst these are:

Performance	Paint has to provide the necessary degree of protection or decoration for the desired time period at the recommended film thickness.
Application methods	The way in which paint is applied must also be taken into account – for example, whether the paint is applied by hand brushing or by automated spray equipment.

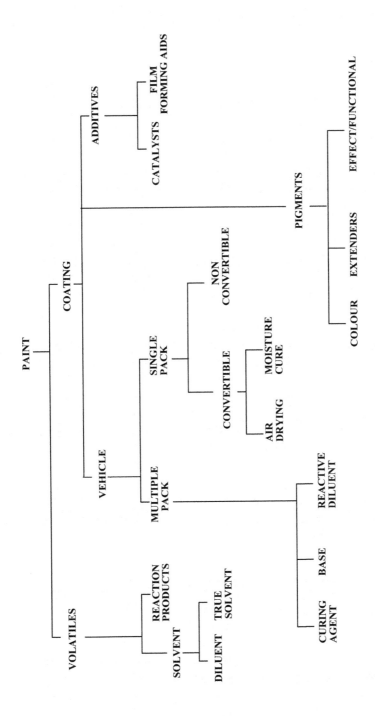

Figure 11.1 *Components of paint*

| Legislative issues | The selection of raw materials is dictated by existing and future legislation, particularly that relating to solvent emissions. The use of toxic materials is often not prohibited by legislation, but is voluntarily regulated by paint companies. |
| Cost | The situation rarely arises where cost is not important and a viable paint must be cost competitive. |

In practice, all of these requirements are interlinked. Paint is never formulated simply on the basis of low cost and neither would it be formulated for application without taking into account the performance expectation. The result is that, inevitably, all paint formulations to some extent are compromises.

Paint formulation is carried out within a set of guidelines defined by simple mathematical equations and the principles are the same no matter whether the paint is a powder coating or household emulsion paint. It is possible to formulate for all aspects of performance, although it must be said that the trick of realising all the desired properties at the same time usually eludes the formulator.

The formulation process itself is initially concerned with the production of a compliant, cost-effective liquid or solid paint suitable for application, and secondly, and perhaps more importantly, the design of a suitable dry film with the required properties.

The main calculations for liquid paint relate to cost, regulatory compliance (VOC), spreading rate, relative density, mass solids, volume solids and batch size. The calculations are mostly handled with the aid of computer programs that range from simple spreadsheets with manual entry of data, to more complex systems linked to factory raw material databases with automatic updating.

Stoichiometry is of critical importance in two component systems such as an epoxy/amine paint. In terms of the dry film, the pigment volume concentration (PVC) is fundamental.

11.2.1 Pigment Volume Concentration

The thickness and composition of the dried film are the main factors in determining performance. The film usually has a different physical state and can also be chemically different from the paint due to crosslinking and/or loss of solvent. It comprises mainly pigment and polymer and the volume relationship between these two is important.

The relationship of pigment to binder in a film is described by the PVC, which is normally expressed as a percentage:

$$PVC = \frac{\text{Volume of Pigment}}{\text{Volume of Pigment} + \text{Volume of Binder}} \times 100 \qquad (11.1)$$

At a PVC of 100%, a 'film' would consist entirely of pigment, whilst at 0% PVC, the film is devoid of pigment and consists entirely of polymer. In between these two extremes a range of films can be produced with different properties depending upon the PVC. It can be seen that as the volume of polymer in the film decreases, a composition will be reached that contains just enough polymer to coat the particles and fill the voids between them. Any further reduction will result in incomplete coverage. This point is termed the critical pigment volume concentration or CPVC, and below it the paint film properties change significantly (Figure 11.2).

It is important to be able to estimate the CPVC, and this can be achieved by measuring the oil absorption of the pigment. Linseed oil is added dropwise to a known weight of pigment and the two are mixed with a spatula on a glass slab. The end point of the titration is reached when the pigment particles are held together to form a coherent mass rather than a crumbly solid. The amount of oil required to achieve this state is measured and the oil absorption expressed as the weight of oil absorbed by 100 g of pigment.

The oil absorption can be thought of as approximating to the CPVC of the pigment in linseed oil as it marks the point where the pigment is just 'wetted out'. However, strictly speaking, the oil absorption value determined using linseed oil should not be translated to other polymers, as they will have different dispersion properties. In addition, most paints use mixtures of different sized and shaped pigments and extenders, and the way these disperse and pack in real films can be quite different from their individual

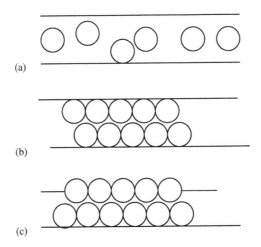

(a)

(b)

(c)

Figure 11.2 *Paint film: (a) below the CPVC, (b) at the CPVC and (c) above the CPVC*

behaviour in oil absorption measurements. Nevertheless, oil absorption figures can provide chemists with a useful parameter with which to work when formulating coatings. Approximate values of the CPVC can be generated from oil absorption figures, but the true CPVC of paint is best determined by ladder experiments where the effect of composition on physical properties is studied. In general, pigments or blends with high oil absorption values give low values for the CPVC, whilst those with low oil absorption values give high values for CPVC.

The CPVC can be estimated from the oil absorption by the use of Equation (11.2):

$$\text{CPVC} = \frac{1}{\left(1 + \dfrac{\text{OA}\,\rho}{93.5}\right)} \times 100 \tag{11.2}$$

where OA is the oil absorption and ρ is the specific gravity of the pigment.

Paint films formulated at the CPVC have optimum values for some properties mainly those related to durability of build coats or primers. At the CPVC, there is a minimum of free binder in the film, which gives maximum resistance to the transmission of ions and molecules through the film. Hence, at the CPVC, it would be expected that there would be maximum resistance to corrosion and to blistering of the film.

Although the concept of the CPVC is important in obtaining the desired properties of impermeability, blister resistance, adhesion and film strength of anti-corrosive primers, other considerations have to be taken into account. They are the correct volume of the anti-corrosive pigment to give good protection and long life, the relationship of volume of the anti-corrosive pigment to other pigments and the amount of colouring pigment necessary to give colour and opacity. The type of barrier effect pigment is important and the total cost of the primer is crucial as it must be competitive in the marketplace. Formulations are best determined by using a statistical approach and tests such as salt spray resistance, accelerated weathering studies, immersion tests in solvents and solutions, adhesion and mechanical strength measurements, together with overcoating examinations. Ideally, tests should be recognised industry or international standards to ensure that they are reproducible and the information is accepted by industry, though their correspondence with real service conditions is often questioned.

11.2.2 Opacity of Paint Films

Whilst the CPVC is important for anti-corrosive and barrier coatings, for finishing paints (a.k.a. cosmetic finishes, topcoats), particularly those with a high gloss, pigment properties that provide colour and opacity are of greater importance. In the case of a typical pigment, such as titanium dioxide, these properties are determined by the refractive index and the particle size of the pigment.

The effect of refractive index can be shown in an equation derived by Fresnel:

$$\text{Reflectivity of a coating} = \frac{(\eta_p - \eta_r)^2}{(\eta_p + \eta_r)^2} \qquad (11.3)$$

where η_p is the refractive index of the pigment and η_r is the refractive index of the medium (Figure 11.3).

The higher the difference is between the refractive index of the pigment and that of the medium, the greater the reflectivity of the film. Most organic materials including the polymers used in the paint industry have a refractive index of about 1.5. Titanium dioxide at 2.76 has the highest refractive index of any white pigment. Zinc oxide has a value of 2.0, whilst that of white extenders, such as barium sulfate, is around 1.6. Using Equation (11.3) and taking the relative opacity of titanium dioxide as 100, zinc oxide will give a relative opacity of ~25, and barium sulfate ~1. In practical terms this means that films containing titanium dioxide will be opaque, whilst those containing barium sulfate will be translucent.

The situation is different above the CPVC, when air is present in the film. As air has a refractive index much lower than that of the binder, and remembering that it is the difference in refractive indices that produces opacity, extenders can give opaque matt films in certain situations.

The particle size of a pigment controls light scattering, and this in turn affects its capacity to produce opaque films. Good dispersion at the correct PVC is therefore important in order to achieve good opacity.

Cosmetic finishes are formulated with PVCs much lower than the CPVC to produce a high gloss. A clear, unpigmented film will give maximum gloss and it would be expected that adding any pigmentation would result in

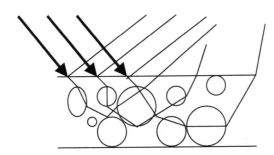

Figure 11.3   *Scattering of light from a pigmented paint film*

lower gloss. However, for most practical purposes the gloss of pigmented films remains constant with increasing PVC, until the volume of pigment reaches the point at which the particles of pigment disturb the surface of the film producing micro-roughness. This slow change in gloss at low PVC is due to the presence of a clear layer of polymer at the surface of the film.

Maximum opacity is obtained at high PVC, whereas low PVC favours high gloss. A compromise therefore has to be made with high gloss films in the relationship between opacity and gloss. Other influencing factors are the degree of dispersion of the pigment, stabilisation of this dispersion against flocculation and the thickness of the applied film. The PVCs of gloss paints are typically 10–25, whilst those of barrier coats and anti-corrosive primers are much higher, nearer 40. Some paints such as anti-foulings have even higher PVCs, above the CPVC, in order to control the dissolution of pigments and biocides from the film.

11.2.3 The Effect of Pigments on Paint Viscosity

Paints containing pigments are two-phase systems with the pigment as the dispersed or internal phase and the resin or resin solution as the continuous phase. The viscosity of these systems is dependent on many things including the volume of the dispersed phase and the relationship can be described using the Mooney equation (Equation (11.4)):

$$\log \eta = \log \eta_e + \frac{k_E V_i}{2.303(1 - (V_i/\theta))} \tag{11.4}$$

where η is the viscosity of the total system,
η_e is the viscosity of the resin solution (external phase),
V_i is the volume fraction of the pigments (internal phase),
θ is the packing factor
K_E is a constant that depends on particle shape.

This equation is valid in cases where there are no particle–particle interactions. The internal phase volume V_i includes the volume of both the pigment and the absorbed polymer layers on the pigment surface.

An increase in the volume of the internal phase results in an increase in the system viscosity. In high solids systems, the main factors affecting viscosity are the extent of pigment flocculation and the thickness of the adsorbed stabilising layers on the pigment particles. An increase in the degree of flocculation results in an increase in viscosity. This is attributed to the immobilisation of the continuous phase around the irregularly shaped aggregates and to the packing or crowding of aggregates within each other.

The absorbed layer provides steric stabilisation to the pigment particles and thereby prevents particle flocculation. An increase in the thickness of this layer results in an increase in the system viscosity. It is important to keep the absorbed layer thickness at a level that maintains stability whilst minimising the effect on the paint viscosity.

11.3 FORMULATION FOR PERFORMANCE

'Performance' is not one single property, it encompasses a wide range of attributes such as mechanical properties, anti-corrosive performance, aesthetics, *etc.* In paint formulation, it is usual to attempt to maximise one aspect of performance whilst minimising the loss of other properties.

11.3.1 Mechanical Performance

Coatings used to protect substrates are subjected to various types of mechanical abuse ranging from a grazing impact with a shoe, to severe, high rate, high energy impact with lumps of iron ore during the loading of cargo into the holds of ships. Where in-service conditions are harsh, protection is afforded either by hard epoxy coatings or elastomeric polyurethane or polyurea coatings.

Epoxy-based coatings are typically pigmented with hard fillers such as calcined bauxite and provide excellent protection over many years through a combination of abrasion resistance and anti-corrosive performance. However, epoxy systems that offer maximum abrasion resistance have low flexibility and limited impact resistance.

In environments such as mineral slurry processing, pneumatic conveying and mineral transport applications such as rail cars and hoppers, coatings based on polyurethanes and polyureas are used, but at substantially higher thicknesses than an epoxy coating – typically 2 mm and above. In these applications, the films must be able to withstand the stresses and physical movements of the substrate. They also have to resist the gouging and tearing forces caused by the mechanical abuse. Thick films are necessary to produce the elastomeric properties needed to protect the substrate and thin films do not display the same behaviour. Once damaged though, these coatings are more difficult to repair than an epoxy coating.

The mechanical properties of clearcoats used in the car refinish market are improved by the incorporation in the resin system of materials such as silica with particle sizes of between 0.1 and 15 μm. This gives improved mar and scratch resistance to the coatings.

11.3.2 Aesthetics

11.3.2.1 Weathering. The appearance of paint is very important when used as a finish coat. In particular, the initial colour and gloss and the retention of both properties are critical. To maintain good appearance or aesthetics, the film must be able to withstand the challenges of the in-service conditions encountered. Some of these are UV and IR radiation, water, ozone, pollution and salt spray. Coatings can be exposed to temperatures from −60 °C for an aircraft flying at high altitude, to around 80 °C for a dark blue paint on a yacht in a sunny tropical location. Thermal cycling also induces potentially damaging stresses within the film. The combination of thermal oxidation, photo-initiated oxidation and chemical attack causes coatings to break down.

The polymer systems preferred for use in topcoats or finishes are transparent to UV radiation and as such, they do not interact, and should not degrade. The most commonly used polymers are acrylics, vinyls and two-pack polyurethane acrylics. Coatings based on fluorocarbons are also used but they are very expensive.

Even so, in practice, some interaction always occurs between components in the film (often the pigmentation) and the incident UV radiation. To overcome the potentially damaging effects, paint formulations will usually contain a UV absorber in combination with a HALS as described in Chapter 10. High-performance cosmetic finishes that contain the colouring pigment at low PVC will also include flow and levelling aids to produce a smooth surface, free of defects and with maximum gloss.

UV absorber and HALS packages are particularly important in wood varnishes where the underlying wooden substrate can be degraded on exposure, even though the coating is untouched, resulting in detachment of the film.

Wood stains protect the substrate by a different mechanism. They contain fine particle size transparent iron oxide pigments made by co-precipitation techniques under carefully controlled conditions. These pigments are transparent to visible light but scatter UV radiation and this helps protect the wooden substrate.

Even when adopting the best combination of approaches to minimise breakdown, organic coatings will degrade on exposure *via* the mechanisms outlined above. The result will be a loss of gloss, changes in colour, embrittlement and possible adhesion loss. By contrast, inorganic coatings based on silicon chemistry are much more resistant to these degradation mechanisms. This is primarily due to the strength of the Si–O bond ($452 \, kJ \, mol^{-1}$) compared to that of a C–C bond ($350 \, kJ \, mol^{-1}$) which makes them more resistant to heat and UV degradation. In addition, the Si–O bonds are already oxidised, which clearly limits the scope for further oxidation.

Initial applications of silicon-based coatings were limited by the need to heat the films to ensure crosslinking. Ambient cure is now possible with a crosslinking mechanism based on the hydrolytic polycondensation of alkoxysilyl functional polyorganosiloxanes. Organic–inorganic hybrid coatings have been developed based on polysiloxane resins with various organic modifications (*cf.* Chapter 9). Hydrogenated epoxies and acrylic and polyurethane acrylic modified siloxanes show outstanding resistance to external weathering. Polysiloxanes with hydrogenated epoxy modifications outperform an aliphatic acrylic urethane (based on time to 50% gloss retention using a QUV-A weatherometer) by a factor of 3, and a polyurethane acrylic modified polysiloxane could produce a fivefold increase in lifetime.

11.3.2.2 Self-cleaning Coatings. Coatings for the exteriors of buildings collect dirt and are difficult and costly to clean and considerable effort has been directed towards the development of self-cleaning coatings. Self-cleaning can be achieved by producing a hydrophobic microtextured surface from which dirt and water are easily removed. This has been termed the 'Lotus Effect'$^{®}$. The surface texture is formed by very small particles secured on a carrier at the surface. The particles themselves have a fissured structure with a peak to trough height in the 50–200 nm range. The particles are typically hydrophobic, high surface area silicas (BET surface area from 50–200 $m^2 g^{-1}$) and they can be used alone or in combination with other nano-sized particles to achieve the required balance of surface hydrophobicity. The surface topography minimises the contact area between the paint surface and solid particles of dirt and drops of water and also lowers the adhesive forces. This combination of effects produces the self-cleaning properties. The structured surface produces increased diffusion of light, and as a consequence, high gloss finishes are not possible with this technology.

Self-cleaning effects can also be produced in coatings by the use of hydrophobic effects. One system uses silicone-modified crosslinkable hydroxy functional polyacrylate additives at levels of approximately 4% w/w to modify the surface tension of the coating. In particular, they reduce the polar component of the surface tension by up to 97% producing a highly hydrophobic surface that easily sheds water and dirt.

Fluorinated resins in architectural applications also provide excellent dirt-releasing properties, in addition to superior durability. Coatings are typically based on partially fluorinated resins, such as Fluorobase Z$^{®}$ (**1**), containing perfluoropolyether (PFPE) blocks for durability and dirt shedding, along with a hydrogenated component that offers solubility in conventional solvents. They are crosslinked with isocyanates through terminal hydroxyl groups.

$$HO-CH_2-CF_2-O-(CF_2-CF_2-O)_p-(CF_2-O)_q-CF_2-CH_2-OH$$

$$p{:}q = 0.6{:}1.2$$

Fluorobase Z$^\circledR$

(1)

The surface of the film is highly hydrophobic and of low surface energy. This is thought to arise from some limited mobility of the PFPE segments producing a surface layer enriched in fluorine compared with the bulk.

11.3.2.3 Rust Hiding Coatings. Protective coatings applied to steel can break down, especially when subjected to mechanical abuse. Once exposed, the underlying steel will corrode resulting in unsightly rust staining of the coating. Coatings containing pigments such as calcium editronate are used to disguise rust staining. When steel rusts, ferrous ions are initially produced and later oxidise to brown ferric ions, which appear as the familiar, unsightly rust stain. Calcium editronate (**2**) masks rust by reacting with the ferrous ions before they convert to ferric ions. The resulting ferrous editronate is a white compound and is easily washed off the paint by rain, thus maintaining the appearance of the structure. Anti-staining agents, in addition to being able to form white complexes with iron, must have a solubility in water in the desired range. By adjusting the PVC and plasticiser level in the film, the anti-staining compound can be released from the film at an appropriate rate to provide the required coating lifetime. Resin systems are typically high molecular weight acrylics.

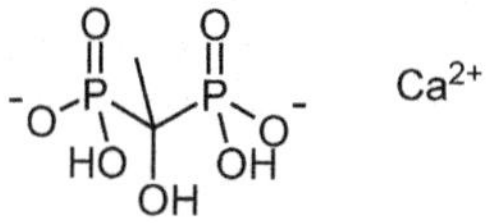

(2)

11.3.3 Anti-corrosive Paints

Corrosion is an electrochemical process driven by a number of factors such as heterogeneities introduced at grain boundaries, stress and surface contamination and by differences in composition. In contact with an electrolyte, metals corrode when areas of higher potential behave as anodes and those of lower potential as cathodes, thereby creating a corrosion cell. Metal ions are formed at the anode and dissolve into the electrolyte.

The electrons travel through the metal to the cathodic areas for subsequent reaction and so the process continues.

The corrosion of steel occurs according to the following scheme:

Anodic reaction:

Iron dissolves at the anode into the electrolyte, releasing electrons:

$$Fe \longrightarrow Fe^{++} + 2e^-$$

These migrate to the cathode through the metal producing a corrosion current, I.

Cathodic reaction:

Electrons react with H_2O and O_2 to make OH^- ions:

$$2H_2O + O_2 + 4e^- \longrightarrow 4OH^-$$

These react with ferrous ions producing ferrous hydroxide, which is subsequently oxidised to hydrated ferric oxide that we know as rust:

$$Fe^{++} + 2OH^- \longrightarrow Fe(OH)_2 \longrightarrow Fe_2O_3 \cdot H_2O$$

A typical corrosion cell is shown in Figure 11.4.

In the corrosion cell, the corrosion current (I) is related to the total resistance of the cell (R) and to the potential differences between the anodic and cathodic sites (V) by Ohm's law:

$$I = V/R \qquad\qquad (11.5)$$

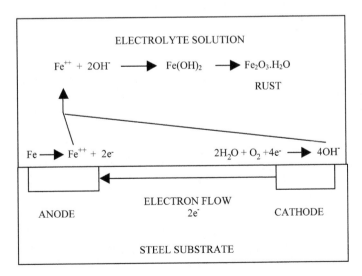

Figure 11.4 *An illustration of a typical corrosion cell*

For a coated metal, this equation becomes:

$$I = V/(R_a + R_{af} + R_c + R_{cf} + R_e) \qquad (11.6)$$

where R_a is the resistance of the anode,
R_c is the resistance of the cathode,
R_{af} is the resistance of the anodic film,
R_{cf} is the resistance of the cathodic film and
R_e is the resistance of the electrolyte.

Corrosion protection is achieved by interfering with the corrosion mechanism, and anti-corrosive paints can be divided into two main types, differentiated by their mode of action – inhibitive primers and barrier anti-corrosives. Zinc-rich primers protect *via* a sacrificial mechanism that can be considered as a form of cathodic protection.

Inhibitive primers contain anti-corrosive pigments specifically designed to have controlled solubility. When a film containing an anti-corrosive pigment is exposed to moisture, ions from the pigment travel to the substrate and either form a protective monomolecular film or reinforce the oxide layer on the metal surface. This effectively passivates the metal surface and retards the corrosion process. The pigments artificially increase the polarisation of the anodic or cathodic areas and act as controlled release agents supplying inhibitive ions dissolved in the diffusing moisture. Inhibitive pigments are available for a range of metal substrates.

As the protection afforded is determined by the rate of release of the ionic species, the optimum level of anti-corrosive pigment in the dry film must be determined by experimentation. Other pigments and fillers are present in the film and the relationship between them and the soluble pigment is important as it affects the rate of release of the active material. Typical PVCs in these paints are of the order of 40% with the anti-corrosive pigment representing approximately 5%.

Zinc-rich primers protect steel by a sacrificial mechanism. In a well-formulated coating, zinc particles and the steel substrate are in full electrical contact and the zinc oxidises in preference to the steel. To function in this way, the coating must contain 85–95% w/w of zinc metal; otherwise, there is insufficient electrical contact with the steel throughout the film. In addition, the zinc salts formed in the oxidation process tend to block the pores in the coating and improve the barrier properties of the film. Barrier anti-corrosives contain no inhibitive pigmentation. Instead, they have high film resistance and function by slowing down the movement of ionic species through the paint film and between the anodic and cathodic areas on the metal substrate. This significantly reduces the electrical conductivity of the film and reduces the corrosion current to very low values. It has also been proposed that a coating reduces the transmission of oxygen through the film, thus starving the corrosion reaction of one essential component.

Factors affecting film resistance include crosslink density and homogeneity in thermoset polymers, and the degree of crystallinity in thermoplastics. Many barrier anti-corrosives are based on epoxy resins as they possess high crosslink density and low inherent conductivity as well as strong adhesion and high resistance to hydrolysis. Other polymer systems can also be used but highly polar systems (such as those containing carboxyl groups) are more conductive, especially on immersion, and this leads to an increased rate of corrosion. The pigmentation is usually laminar in structure, and materials such as mica, talc, aluminium flake, and micaceous iron oxide are used. The function of the pigment is to provide a tortuous path to slow the ingress of water. The pigments are inert and they must have low water-soluble contents to avoid the possibility of osmotic blistering in service. Typical PVCs of barrier anti-corrosives can be up to 35–40% and the appropriate levels are determined by experiment. Higher values can lead to defects in the film and premature breakdown of the coating.

Water-soluble solvents are not normally used in barrier anti-corrosives, as they may not fully leave the film during the drying process. Any solvent remaining in the film has the potential to cause osmotic blistering and failure of the coating.

Some land-based structures such as pipelines and immersed structures, such as ships and oil rigs, use cathodic protection systems to maintain the integrity of steelwork in the event that the coating is damaged. At a break in the coating, the cathodic protection system produces an extremely alkaline environment – pH 14 typically – and the film must be able to resist this attack. The excellent chemical resistance of epoxy resins often makes them the system of choice in these instances.

11.3.4   Anti-fouling Paints

Paints are applied to the underwater hulls of ships in order to prevent the colonisation of the surface by a range of marine organisms such as weed, slimes and animals, such as barnacles. When a ship is fouled, the result is not only unsightly but it significantly increases the fuel consumption of the vessel – in the worst cases by up to 40%. Coatings that prevent fouling are therefore of great value in the marine industry (*cf.* Chapter 1).

Coatings can be divided into two main types, those that release biocidal materials into the water to prevent attachment – anti-fouling coatings, and those where the fouling is only loosely attached and is removed once the vessel is underway – foul release coatings.

Anti-fouling paints can be classified as follows – soluble matrix, insoluble matrix and self-polishing copolymer. They contain a soluble biocide package and it is the way in which this is released into the environment that differentiates the paints. The paints generally contain a resin or matrix, a plasticiser, a biocide package and inert pigmentation. The biocide release rate is governed primarily by the type of resin and the PVC, in addition to the solubility characteristics of the biocides.

With soluble matrix or conventional anti-fouling paints, the biocide is simply mixed into the resin system, which is typically rosin (abietic acid (3)) or a rosin derivative. On immersion in seawater, the resin matrix dissolves releasing the biocide into the water. Continuous dissolution of the surface occurs throughout the lifetime of the coating ensuring constant release of biocide. The paints often contain cuprous oxide for additional biocidal effects. The coating lifetime tends to be short, of the order of 12–18 months.

(3)

Insoluble matrix anti-foulings are based on acrylic, vinyl or chlorinated rubber polymers, and as might be expected, these are not soluble in seawater. As with soluble matrix paints, the biocides are physically mixed into the polymers. When seawater enters the film, the biocide is released by a dissolution and diffusion mechanism from within the insoluble matrix. The initial release of biocide is very high but tails off exponentially as the outer layers of the coating become depleted and the biocide has to travel further through the film. In addition to an organic biocide, insoluble matrix anti-foulings will often contain cuprous oxide as the primary biocide.

Self-polishing anti-fouling paints are based on a binder, typically an acrylic copolymer containing a hydrolysable group, in combination with a biocide package and a plasticising resin. These copolymers are thermoplastic and not crosslinked. In seawater, the leaving group is hydrolysed at a predictable rate leaving a polymer that has a limited solubility in seawater. Water penetrates the film and the biocide leaches out and this deters the settlement of fouling. The remaining polymer matrix at the surface of the film is termed the 'leach layer' and it controls the rate at which water can enter and leave the film. This in turn determines how quickly the biocide is leached out and consequently the useful lifetime of the coating.

As the vessel moves through the water, the leach layer dissolves and smoothes the paint surface, thus reducing drag even further. This is termed 'polishing' and is the origin of the description 'self-polishing copolymers'.

Important formulating variables for anti-fouling paints are polymer composition and molecular weight, plasticiser type and level, pigmentation and soluble pigmentation (type and level of biocide) and the PVC.

Organic biocides are not equally effective against every fouling species. For example, one may be extremely effective against barnacles but relatively ineffective against weed. In this case it must be blended with another biocide for optimum performance. The biocides used, for example, Preventol® A5-S (**4**), decompose in seawater to harmless products and they do so at different rates. They are under increasing scrutiny regarding their presence in the environment and the most persistent are banned from use in some areas of the world. Biocides with short half-lives in seawater are therefore the most commonly used.

Preventol® A5-S

(**4**)

'Non stick' or foul release coatings contain no biocide and they possess a low energy surface that deters the settlement of fouling. These coatings are generally based on polysiloxanes boosted with a silicone oil, or fluoropolymers. Fouling organisms secrete a polar fluid as an adhesive, and they have difficulty in attaching to a non-polar surface. These coatings are most effective on fast-moving vessels where currents or turbulence provides sufficient force to remove fouling once the vessel is underway. They are, however, easily damaged and can be difficult to repair.

11.3.5 Paints with Tailored Thermal Properties

Coatings can be formulated with tailored infrared properties for a variety of applications. Infrared reflecting coatings applied to the exterior of buildings or to roofs keep the interior cool by minimising the transfer of heat.

They are also used in marine applications to cut down the demands on air conditioning systems in superstructures of cruise liners. Infrared reflecting coatings are also termed low solar absorption coatings.

Organic binders do absorb in the near IR but most absorption is due to the pigmentation. Dark colours absorb heat more than light colours and certain pigments such as carbon black absorb intensely in the near IR. Replacing carbon black with black iron oxide significantly reduces the absorbance and results in less heating of a painted structure.

In contrast, coatings have also been designed to absorb infrared radiation and these are used to paint pipes to heat water in solar panels, amongst other applications. Ideally, the coatings must absorb solar energy effectively but radiate thermal energy poorly. This balance is achieved by incorporating metal powders, which have low thermal IR emittance, and pigments such as carbon black, which effectively absorb solar radiation.

11.4   FORMULATION FOR APPLICATION

The application of paint involves the transfer of either a liquid or a solid onto a substrate in a controlled fashion to produce the desired degree of protection or appearance. The choice of application technique is dependent on many factors such as the size and shape of object, the applicator, the nature of the substrate, the speed of application, cost, the type of paint – solid or liquid, solvent or water-borne, aesthetic requirements and environmental considerations. In the case of liquid paint application, control of rheology is the most important factor. In the application of solid powder coatings, control of particle size and melt rheology are important.

The overwhelming majority of paints are Newtonian, shear thinning or thixotropic in nature. Shear thickening or dilatant behaviour is rarely if ever formulated into paint as this is not useful in practice – for example, a dilatant household emulsion paint would be extremely difficult to brush but it would simply run down the wall after application. Thixotropic behaviour is most commonly formulated into paint to provide low viscosity for application followed by rapid recovery to prevent sagging.

As discussed earlier in Chapter 10, thixotropy is introduced into the paint by means of rheological additives such as organoclays or polyamide waxes. Alternatively, it can be achieved by chemically modifying certain polymers to produce hydrogen bonding. At the same time flow aids are often incorporated to adjust surface tension to ensure that the film has the good flow and levelling needed for best appearance.

Roller applied paint can be prone to spattering, air entrapment or bubbling, and poor/uneven finish. Spattering can be reduced and surface texture improved by adjusting the thixotropy of the paint. Air entrapment is eliminated or minimised by the use of a defoamer and/or deaerator.

Liquid paints used in the Heavy Duty market or automotive industry are generally applied by some form of spray equipment. Spray application involves the disintegration of a liquid jet or sheet by either the kinetic energy of the liquid or by exposure to high velocity air/gas stream. As the paint breaks up, droplets are formed. The size of the droplets is important for the resulting film appearance and depends on the application method employed. In general, when formulating for spray application, the viscosity, rheology and surface tension of the paint are the most important factors. Poor atomisation will produce large droplets and film properties will be compromised. Spray paints normally have near-Newtonian or thixotropic rheology and viscosities are usually controlled to within well-defined limits, typically 0.1–0.6 Pa s depending on the application technique.

Paint typically contains a mixture of slow and fast evaporating solvents. A solvent that evaporates quickly from the film will produce paint with a faster drying time. However, if it evaporates too fast, then the resulting rapid increase in paint viscosity will not allow sufficient time for the paint to flow and to remove any defects in the film. Slower evaporating solvents are therefore formulated into both spray and brush applied paints to help flow and levelling. This keeps the viscosity of the paint film lower for longer and allows the development of a smooth integral film. Careful blending of solvents produces paints that apply easily and give coatings with the required drying time and appearance.

There can be significant loss of solvent between the spray tip and the substrate, particularly in hot climates, and this in turn can lead to poor flow and film appearance. To overcome this, paint manufacturers supply ranges of paint thinners containing differing blends of fast and slow solvents. So, when applying a paint in a hot climate, a thinner containing a higher proportion of slow solvent would be used.

Pigments and fillers also affect paint rheology and application properties. As the pigment volume fraction in the paint is increased, spray application properties often improve, but at some point the dry film properties are compromised and performance becomes unacceptable. Particle morphology can also influence application properties and high loadings of acicular or needle-like pigments can cause shear thickening behaviour and this adversely affects spray application properties.

Application of water-based paints presents different problems than those of solvent-based paints. Water-borne coatings can be divided into three main classes: aqueous dispersions or latices, colloidal or water-solubilised dispersions and water-soluble systems.

Water-soluble systems are true solutions of the polymer in water and, to that extent, resemble organic polymer solutions. The viscosity of the solution is therefore related to molecular weight. In solution, the polymer chains tend to be extended and acceptable application viscosity can only be achieved at low solids. Colloidal or solubilised dispersions and latex systems are aggregates of particles with diameters of the order of 0.001–1 μm. The system viscosities of both types are less sensitive to the effect of molecular weight. Application viscosity is more easily achieved by addition of water as the continuous water phase effectively dominates the viscosity of the system.

In the case of water-based two-pack paints, and to a lesser extent with one packs, the climatic conditions during application are important. Depending on the ambient temperature and relative humidity (RH), water can be lost too quickly or too slowly compromising film formation and subsequent performance. These problems can be mitigated by the use of slow evaporating solvents that assist particle coalescence, and aid the formation of defect-free films. The solvent must leave the coalesced film at some point; otherwise, it would remain soft. In practice, the situation is managed by combining the optimisation of the blend of coalescing solvents with the specification of a set of application conditions – temperature, RH and air movement – that will ensure good film formation and performance.

11.5 FORMULATION FOR COMPLIANCE

Many materials have been used in paint formulations and over the years some of the ingredients have come under increasing scrutiny regarding their toxicity to individuals or to the environment. Many of the materials of concern are subject to legislation, but some are not. In both cases, paint companies have removed or are in the process of removing these materials from paint as part of ongoing product stewardship initiatives. In particular, its use of organic solvents has made the coatings industry subject to considerable legislation in recent years.

Solvent emissions are controlled by the VOC regulations in the USA and the Solvent Emissions Directive (SED) in Europe (*cf.* Chapter 2). The SED targets reductions in the total solvent emissions from specific installations over a certain size, irrespective of the type of solvent. In contrast, the VOC regulations target only photochemically active solvents such as xylene and white spirit. Most common solvents are photoactive but acetone and some fluorinated materials are less so, and can be used in the USA in paint without being included in the VOC. This approach of using VOC-exempt solvents is not allowed under SED legislation, in view of the greater population densities near many industrial regions in Europe.

The overall effect of the legislation is to force a reduction of the solvent content of paint, and compliance can be achieved in a number of ways. These may involve the development of high solids/solvent-free paints, moving to a technology such as water-based or polysiloxane chemistry, or adopting radiation cure or powder-coating processes.

11.5.1   The Formulation of High Solids and Solvent-free Paints

The most common approach to compliance is to formulate higher solids or solvent-free paints. In these cases, technology similar to that in conventional coatings is used in a different way. The viscosity and rheology must be matched so that the high solids paint can be applied using the same equipment.

The formulation of high solids paints generally requires the use of lower viscosity resins than those currently employed. Polymer viscosity can be reduced by modification of the polymer architecture, as discussed in Chapter 7, but in practice this is normally achieved primarily by the use of lower molecular weight oligomers of the same polymers. This approach does have drawbacks, however, because as the molecular weight decreases, there is a greater likelihood that health and safety issues might arise, such as increased potential for sensitisation. Alternatively, substitution of all or part of the high viscosity resins by low viscosity reactive or non-reactive diluents can also help in lowering binder viscosity (*cf.* Chapter 7).

Formulating with low molecular weight oligomers will result in an increased concentration of reactive crosslinking groups in high solids or solvent-free two-pack paints. In these systems, the rapid build-up of molecular weight and viscosity within the mixed paint leads to a shorter working life and makes application more difficult. Latent or blocked crosslinking chemistry is therefore very desirable for high solids paints.

Low solids paints contain high molecular weight polymers, many of which have T_gs above the curing temperature, and so can form acceptable films simply by solvent evaporation. By contrast, once the solvent in a high solids paint has evaporated, the remaining low molecular weight oligomers are still essentially liquid and there is little inherent resistance to sagging. Sag resistance must be formulated into the paint by the use of thixotropic additives such as polyamide waxes.

Polyamide waxes must be well dispersed and then heated to specific activation temperatures. The combination of dispersion and heat produces micelles coupled with chain entanglement, and this generates the required thixotropic behaviour. Inadequate dispersion or incorrect temperatures will not produce the desired effect. The activation temperature of any particular polyamide wax is dependent on the polarity of the polymer and solvent system, and in general, the higher the polarity, the lower the activation temperature. In paints with high levels of polar solvent, polyamide wax thixotropes can be activated at temperatures of around 55 °C. As the level of polar solvent is reduced, the polarity of the system decreases and the same polyamide wax needs a higher temperature for activation.

Addition of pigments or fillers to a resin solution increases the system viscosity, as described earlier. In general, for a given weight of pigment, the smaller the particle size, the larger the increase in viscosity and this is directly related to the oil absorption of the pigment. Traditional grades of filler often have a very broad particle size distribution and the small particle size fraction, or fines, has a disproportionate effect on viscosity. Newer materials, classified to produce grades with narrow particle size distributions, are now available. Removal of the 'fines' results in lower paint viscosity for the same filler loading, which makes these grades very useful in formulating high solids paints.

Oil absorption can provide a useful guide when selecting pigments and fillers for use in high solids paints. Generally, the lower oil absorption materials will give lower viscosity at the same loading than high oil absorption pigments. Particle shape is also important and lowest viscosities are obtained when materials are used whose geometries approximate to spheres. The use of wetting and dispersing agents will reduce particle–particle interactions and also ensure minimum viscosity.

Even when using the lowest viscosity resin systems in conjunction with narrow particle size distribution fillers, low PVCs must be employed to maintain a workable paint viscosity. For example, a typical PVC of a solvent-free two-pack amine cured epoxy is around 15%, whereas that of an epoxy paint with a volume solids of 70% may be 35%.

11.5.2   Radiation Cured Coatings

Radiation cured coatings are reactive low viscosity materials applied to a substrate and subsequently cured. The curing mechanism is usually free radical in nature and initiated by ultraviolet (UV) light or electron beam (EB), but newer technology is based on cationic cure.

UV light is supplied by either a medium pressure mercury or gallium lamp, emitting radiation in the 200–400 nm band and the formulation must contain a photoinitiator to provide a source of free radicals. In the case of EB cure, the ionising qualities of the incident radiation preclude the need for an initiator. Both UV and EB systems use the same raw materials and the paints themselves comprise unsaturated polymers and monomers. The most important resins are the polyester acrylates, such as hexanediol diacrylate (**5**) although other ethylenically unsaturated materials are being used (*cf.* Chapter 7).

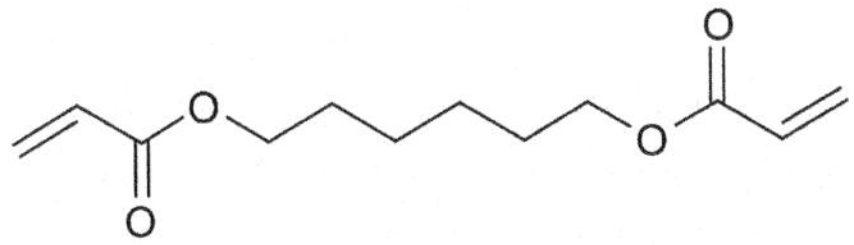

(**5**)

Pigments and fillers can also be included in radical cured systems. The higher energy and greater depth of penetration of EB means that pigmented films can be relatively easily cured compared to UV cured systems. Pigments absorb UV radiation at certain wavelengths and it is important to select a source of UV light that allows maximum penetration of the film. This is achieved by the use of doped lamps where the spectrum is shifted towards the longer wavelengths, typically >380 nm. Photoinitiators for pigmented systems have absorbance properties outside those of the pigments to ensure that they can initiate the curing process and bisacylphosphineoxides (**6**) are typical. UV light is still absorbed by the film and care must be taken to ensure that the film is not too thick to allow full cure to take place.

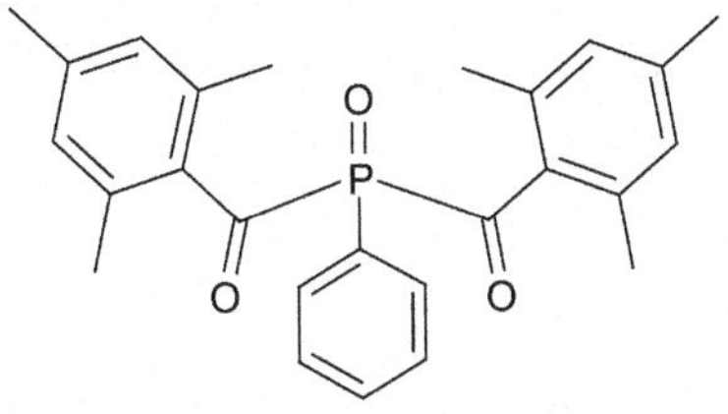

"BAPO"

(**6**)

Advantages of UV and EB cure include zero VOC and extremely fast drying times – of the order of 1 s – which makes the technology suitable for high throughput/volume applications such as coating wood, metal, plastic and paper. Newer technologies are finding increasing use in the car refinish market.

The biggest drawback of UV and EB technology is the difficulty of dealing with three-dimensional substrates as the curing process only takes place in irradiated areas. In addition, the reaction can be inhibited by oxygen, which can combine with free radicals. This inhibition reaction can be overcome by the use of inert gas blankets.

Compared to EB, UV systems are cheaper and easier to use, and cure using a hand-held lamp is possible in some circumstances. UV cure also has an advantage in that little heat is put into the substrate. The coatings produced using EB systems do not contain an initiator that can be degraded by UV light and they consequently have better durability than UV cured coatings.

Radiation cure systems are also based on cationic cure involving the photolysis of onium salts, such as bis-[4-(diphenylsulphonio)phenyl] sulfide-bis-hexafluorophosphate (**7**), and other light-sensitive compounds. Irradiation forms Lewis or Bronsted acids, and these produce the cationic species that initiate the polymerisation of materials such as epoxides, polyols and vinyl ethers. One difference between this and radical cure is that the onium salts generate a long-lived species and cure continues after exposure and also in shadowed areas. In addition, cationic cure systems do not suffer from oxygen inhibition. Dual cure systems are also possible by using cationic sulfonium salt photoinitiators that decompose to produce both cations and free radicals. These will polymerise a mixture of acrylates and epoxy compounds to produce an interpenetrating network.

$$^-PF_6 \quad ^+S(C_6H_5)_2\text{-}C_6H_4\text{-}S\text{-}C_6H_4\text{-}(C_6H_5)_2S^+ \quad ^-PF_6$$

(7)

Many pigments are basic in nature and can neutralise the Bronsted acids formed by the photolysis process. Materials such as amine-treated titanium dioxides, therefore, cannot be used in cationic cure systems.

11.5.3 Powder Coatings

Powder coatings by their nature constitute a compliant technology and little or no volatile materials are given off during their manufacture, application or curing processes. They offer a number of advantages over conventional 'wet' paint. For example, very thick films up to 500 μm can be applied in a single operation. This can be very important where corrosion protection is of prime importance in applications such as pipe coatings. Powder coatings offer extremely good material utilisation of up to 95%, since overspray can be collected and reused, unlike many other paints where this is not possible. The cured coatings are tough, damage resistant and are very chemically resistant due to the very high film T_gs that can be achieved during the curing process. It is also possible to formulate highly durable systems and the process itself is well suited to automated situations.

The technology, however, does have limitations. The specialised equipment needed for manufacture and application means that powder-coating operations must be confined to production lines or fixed installations. Cleaning operations are also difficult and it is not easy to change from one colour to another without lengthy cleandown. Unlike liquid paints where contamination becomes diluted and unnoticeable as the new paint is flushed through, impurities in a powder coating would be immediately visible. Thin films, say 25 μm, are also more difficult to produce with a powder coating than a liquid paint, and thicker films will increase the cost although this can be offset to some extent by the better transfer efficiency. Finally, since powder coatings are cured at elevated temperatures, it is not possible to coat objects that cannot withstand those temperatures.

There are many situations where powder coatings are viable alternatives to solvent-borne paints and research work aimed at developing products with better flow and lower curing temperatures will further expand the use of this technology.

11.5.4   Water-based Coatings

The use of water-based coatings has grown steadily over the past 20–30 years. Originally used in only low-performance applications, the technology has evolved and water-based coatings are now found in much more demanding situations. Almost all types of paint, from single pack alkyds to two-pack epoxies are available in a water-based form and there are many advantages to be gained by choosing this route (*cf.* Chapter 8).

Water-based coatings can be divided into three main classes depending on the type of resin system used. Resins can be aqueous dispersions or latices, colloidal or water-soluble dispersions, or water reducible. They vary considerably in mechanical and physical properties.

Co-solvents are often added to latex and water-soluble systems and they perform a number of functions depending on the formulation and the paint. They can function as coalescing agents, freeze–thaw stabilisers, and flow and levelling agents. Many co-solvents act as coupling solvents. As such, they turn the components of a multiphase system, such as water and an incompatible resin, into a clear homogenous system. Glycol ethers such as triethylene glycol monomethyl ether (**8**) and dipropylene glycol methyl ether (**9**) are commonly used as coupling solvents. It should be appreciated that any co-solvent is a volatile organic compound, and must be taken into account when calculating VOCs of water-borne paints.

$$CH_3(OCH_2CH_2)_3OH \qquad\qquad CH_3O[CH_2CH(CH_3)O]_2H$$

$$(\textbf{8}) \qquad\qquad\qquad\qquad (\textbf{9})$$

Pigments used in water-borne paints must be carefully selected as high water solubility or reactivity can be problematic. For example, anti-corrosive pigments are by design water soluble, albeit to a limited extent, and ionic species from the pigment can destabilise some paints. In addition, the surface of reactive pigments such as aluminium flake must be treated to prevent reaction with water and the potentially dangerous liberation of hydrogen.

Advantages of water-based paints include the following:

- reduced fire risk and associated insurance costs;
- reduced odour and organic vapour emissions;
- existing equipment can be generally used to apply water-based coatings.

Unfortunately, they are not without their disadvantages amongst which are:

- limited shelf life of paints;
- an increased tendency for the resulting films to be water sensitive, due to the presence of surfactants required to achieve dispersion;
- good film formation requires control of humidity during drying. High humidity can cause cracking and very poor film formation, and low humidity can cause film defects such as cratering;
- higher cost than organic solvent-based paints;
- substrates must be clean and free from grease. This is not always possible in a heavy industrial environment;
- flash rusting can also occur when water-based paints are applied to steel substrates, particularly over welds;
- in forced drying applications, there may be increased energy costs associated with driving off the water from the paint;
- they cannot be used in situations where the temperature drops below $0\,°C$.

Despite these limitations, water-based paints are forming an increasingly significant proportion of the paint market in the efforts to produce compliant coatings.

11.5.5 Inorganic–Organic Hybrid Technology

For many years, aliphatic polyurethanes were the dominant technology in the field of cosmetic finishes. The formulation of low VOC ($<250\,g\,L^{-1}$) products using them is extremely difficult due to limitations of the current technology, and can only be achieved by compromising the exterior durability of the coating.

It is now possible to formulate very low VOC cosmetic finishes with outstanding durability using inorganic/organic hybrid technology based on polysiloxanes. Siliconate coatings based on polysiloxane resins blended with various organic modifications were introduced to the Protective Coatings market in the mid-1990s.

The basic polysiloxane resins are of low viscosity and offer the opportunity to produce paints with very low VOCs. As well as low VOC, the technology offers excellent adhesion, good chemical resistance, good health and safety profile, and most importantly, excellent weather resistance for reasons that will be described later in this chapter (also *cf.* Chapter 7).

Polysiloxanes themselves do not form useful films at ambient temperatures and must be modified for use. Hybrids have been formulated with many organic modifications including epoxy, acrylate, vinyl, acetoacetates, fluorinated and phenolic polymers, according to the end use. The organic modifications improve adhesion, flexibility, chemical resistance and barrier properties. The most useful polysiloxane hybrids form polymeric networks with shared chemical bonds. Paints based on polysiloxane technology promise to become increasingly important as the regulatory pressure to remove solvents increases.

11.6   FORMULATION FOR ECONOMY

The paint formulator has a vast range of materials to choose from when formulating paint. On purely economic grounds, it is important to use materials that offer the best value for money, and to use them in the most efficient way possible; otherwise, the paint will not be cost competitive. This is often termed 'value engineering'. However, the performance of the paint must not be compromised since it must match the customer's expectation to be viable.

The two most costly components in paint on a weight basis are generally the polymer system and the coloured pigmentation. Additives are often more expensive per kilogram but they are used at such low levels in a formulation that their impact is limited. On the other hand, using them at higher levels might produce cost savings when it achieves proper dispersion and stabilisation of expensive colouring pigments.

11.6.1   Economic Use of Binders

The performance expectation of binders varies considerably and this, to some extent, dictates the way in which they are treated when it comes to minimising cost. Alkyd resins can produce excellent coatings at very low cost and it is difficult to imagine any material that could be added that would significantly reduce the cost without affecting performance. However, binder systems such as epoxies and polyurethanes are considerably more expensive and they can be blended with low cost extender resins without loss of performance. In these instances, aromatic hydrocarbon resins derived from petroleum and coal tar can be used as extending resins. They are made from mixtures of monomers including α-methylstyrene (**10**), vinyltoluenes (**11**), indene (**12**), *etc.*, *via* carbocationic polymerisation using, for example, a BF_3 initiating system (*cf.* Chapter 6). They are available as liquids or solids with varying softening points and can be chemically modified to enhance compatibility with other resin types. For example, the incorporation of phenol into the initial monomer blend introduces hydroxyl groups into the resin to improve compatibility with epoxies.

Vinyltoluenes

(**10**)   (**11**)   (**12**)

The cost of hydrocarbon resins can be as little as 35% of an epoxy resin and they can be incorporated into epoxy paints at levels of up to 40% by weight on binder without any detrimental effects on performance. In fact, in some situations, the inclusion of a hydrocarbon resin will reduce costs and actually improve performance by increasing the flexibility of the paint film. Generally, the levels of modification are kept below 20%.

11.6.2   Economic Use of Pigments

Colouring pigments, particularly organics, are generally expensive and their use significantly affects the cost of paint. It is essential, therefore, to produce the maximum colour from the minimum amount of material. To maximise colour development, pigments must be extremely well dispersed, and once in this state they must be stabilised to prevent flocculation and subsequent loss of colour strength. As described in Chapter 10, wetting and dispersing agents are very effective in achieving stable dispersions of pigments. By determining the correct type and optimum level of wetting and dispersing agent, it is possible to minimise the amount of pigment needed to achieve the required colour strength and opacity, and thereby produce the most cost-effective formulations.

TiO$_2$ is the main opacifying pigment and it has a primary particle size of approximately 0.2–0.3 μm. At this particle size, opacity increases up to 20–30% PVC. Further increasing PVC gives no improvement unless the particle size of the TiO$_2$ also increases. This is because maximum scattering efficiency occurs when the particle centres are spaced at a distance of one wavelength of the incident light, enabling them to reflect each light wave at the same point in the waveform. At high PVCs, TiO$_2$ particles with diameters of 0.2–0.3 μm pack too closely for optimum scattering. The correct spacing can be achieved by incorporating a fine particle size extender into the formulation. The extender particles lie between the TiO$_2$ particles and spaces them at the required distance for maximum opacity. This utilises the TiO$_2$ more efficiently, and has the consequent benefit of reducing the cost.

Ultrafine particle size (typically 0.2 μm) hydrous kaolin, or china clay, is used as a 'spacer' pigment in paint formulations. Kaolin is hydrated aluminium silicate with a laminar structure and an aspect ratio of approximately 25 : 1. It can be used to replace 8–12% of the TiO$_2$ without compromising performance of gloss paints.

Specially developed calcined kaolins are used in decorative paints formulated above the CPVC. These particles have sealed voids containing air and are highly effective in scattering light and producing opacity. This effect is a form of 'dry hiding' and up to 20% replacement of TiO$_2$ is possible.

Hollow polymer and glass spheres containing air voids also give increased opacity, but they are more expensive than mineral extenders and find limited use in cost optimisation exercises.

11.6.3   Economic Use of Solvents

Solvent blend optimisation can also produce relatively small but useful savings. When replacing one solvent blend by another, it is important to maintain the same solvency and evaporation rate. This is facilitated by the use of computer programs that take into account factors such as solubility parameters, fractional polarity, hydrogen bonding index, viscosity, density and evaporation rates and use the data to calculate lower cost solvent blends for specific uses.

11.6.4   Economic Production of Coloured Paints and Inventory Problems

The number of coloured paints that can be prepared is enormous, and is further increased if these colours are duplicated in all types of paint, such as alkyd, epoxy, polyurethane, *etc.* For a paint supplier, the stock-keeping issues and working capital requirements become prohibitive, and clearly, stocking one tin of every possible colour is not a workable proposition. This is avoided by using tinting systems that provide a highly cost-effective method of producing coloured paints.

Apart from white and black, very few paints are produced from one pigment. Even if a colour can be made from a single pigment, it is difficult for the pigment manufacturer to produce a consistent colour, in every respect, from one batch to another. It is also difficult for paint manufacturers to consistently produce the same colour from one batch of paint to another since the degree of pigment dispersion will vary from batch to batch. To ensure that batches of paints have a consistent colour, adjustments are made using dispersions of coloured pigments, generally in resin, until the desired colour is achieved.

This principle of colour adjustment by addition of pigment has been extended and now almost all coloured paints are produced *via* mixing or tinting systems. In such schemes, a small range of so-called bases is prepared in a limited number of shades from light to dark. Some systems use only a light, medium and dark base. Highly concentrated pigment dispersions are then mixed into the base to produce the desired colour. For example, pale colours would be produced from the light base.

Colour measurement and adjustment is done by first preparing a paint film. The colour is measured using a reflectance spectrophotometer connected to a colour database. A light source illuminates the sample and the spectrum of reflected light measured at a predetermined angle. The result is compared with the colour standard in the database and adjustments are calculated. This is repeated until the desired colour is achieved.

Universal colour dispersions are capable of colouring white bases formulated from different polymers. White alkyd, urethane, and epoxy bases can all be tinted to shade using the same coloured dispersions. The technology depends upon the coloured pigments being dispersed in a medium compatible with all of the polymers used in the white bases. It is also possible to include water-borne white bases in the universal colouring systems.

Tinting or colouring systems can be used with any sized batch of coloured paint, be it one litre or a thousand litres. They can be used in factory-made colour paint processes or at point of sale processes such as the in store production of retail paints. In that way, rapid response can be given to requests for coloured paint in any reasonable quantity; stocks of finished coloured paints can be reduced to a minimum; and the number of manufacturing processes involving dispersion can be minimised, all of which ensures the most cost-effective production of coloured paint.

The cost of packaging can also be optimised. In a two-pack paint, each component is supplied in a container, whose cost forms part of the overall cost of the paint. By careful adjustment of the 'mix ratio' that determines the volume of each component, the most economic packaging cost can be achieved.

11.7 CONCLUSIONS

The role of the paint formulator is to exploit the properties of the many materials in his armoury for the best results available, and successful paint formulations are rightly regarded as proprietary information of the highest importance. The combinations frequently show synergistic effects whose discovery and exploitation represent a crucial aspect of the formulator's skill. There can also be antagonistic phenomena that lead to problems such as instability or reduced performance. The prediction of interactions is often not possible due to the complexity of the formulations and occasionally the novelty of the technology being investigated. Hence, the importance of comprehensive test protocols, and the continuing need for experienced and vigilant paint chemists.

11.8 BIBLIOGRAPHY

1. C.H. Hare, *Protective Coatings*, Technology Publishing Company, Pittsburgh, PA, 1994.
2. Various Authors, *Federation Series on Coatings Technology*, Federation of Societies for Coatings Technology, Philadelphia, PA.

CHAPTER 12

Application and Applications

A.R. MARRION

12.1   INTRODUCTION

Coatings are materials applied to all manner of surfaces to decorate them and protect them from the environment and other sources of harm. Sometimes they perform special functions such as providing an anti-fouling surface. All that from a film rarely more than a few millimetres thick, and usually only a few tens of microns. What is more, the coating must deliver its benefits with the minimum impact on the health of those exposed to it, or the wider environment.

And cheap too! The world demand for paints of all types in the year 2000 was of the order of 24 million tonnes, and worth approaching £40 billion. (The figures have risen steadily since.) Eighty percent of volume and value was accounted for by Western Europe, North America and the Asia-Pacific region, each taking a roughly equal share. On that basis, the average cost of a kilogram of paint, that might cover 10 m^2 of surface, was less than £2.00.

We have seen in Chapters 1–11 how much science and care is devoted to the design of coatings materials to ensure that they meet all the needs of their users. It remains to outline the last, and crucial, stage in the manufacturing of the coating itself, when the liquid (or potentially liquid) material is transformed into a solid film, with all the required properties, adhering tenaciously to the substrate.

The application process is usually carried out by the owner of the objects to be painted, or his agent, rather than the coatings manufacturer. The shift of responsibility must be handled with the greatest care to avoid damage to the delicate relationship between supplier and customer. For a satisfactory outcome, the supplier must have an intimate understanding of the industry he serves, and have taken his customer's needs into full account when designing the coating material. Equally, his client must be prepared to

347

ensure that the supplier's advice about his product is carefully followed. For that reason, paint manufacturers usually invest heavily in technical support staff who spend much of their time in the customer's premises observing the application process, and resolving the minor problems that arise, before they become major ones.

12.2   SUBSTRATES AND THEIR PREPARATION

Almost every substance from which objects can be made has been painted at some stage, and the nature of the substrate is the first consideration in designing a coating. Preparation of the substrate is, equally, the first consideration in applying a coating.

The familiar materials of construction are steel, wood, aluminium, plastics and concrete. However, it is probably true to say that the most widely coated substrate is paint itself since most successful schemes are multi-layer, often incorporating a primer whose main function is to adhere to the workpiece. The next coat contributes most of the thickness, fills minor imperfections, and provides a smooth base for the topcoat. The topcoat exists at the interface with the environment and contributes most of the cosmetic features such as colour and gloss, as well as being the first line of defence against irradiation, oxidation, hydrolysis and the action of aggressive fluids.

Substrate preparation is mainly about thorough cleaning to remove dust, rust, grease, moisture and loosely adhering paint residues. Abrasive paper, chemical paint strippers and detergents are usually all too familiar to the painter and decorator (and to most householders), but are very necessary if a satisfactory outcome is to be achieved.

Large steel structures are prepared by grit or shot blasting, when abrasive particles are fired at them by compressed air or pressurised water. The surface is not only cleaned, but also to some extent reshaped to provide the asperities that enable a strong mechanical bond to be formed with the coating. It is important to apply a coating as soon as possible after the blasting to prevent re-oxidation of the surface. In shipyards, the blocks from which the ships are to be assembled are blasted under cover and coated with a silicate 'shop primer' before being taken outside (*cf.* Chapter 9).

Smaller metal objects, amenable to production line treatment, are usually subjected to a 'conversion' process. Car bodies, for example, are usually 'phosphated'. Immersion in phosphoric acid produces a layer of iron phosphate, whilst acidic zinc phosphates produce a dense, uniform layer of mixed zinc and iron phosphate crystals adhering strongly to the surface. They have some protective effect in their own right and provide an ideal substrate for paint adhesion.

Aluminium has traditionally been 'chromated'. The oxide present naturally at the surface of aluminium is stripped off in an alkali bath. An oxide layer of controlled structure is then grown by exposing the workpiece to an acidic solution of chromium oxides. However, the use of chromates presents a health hazard that has led to increasing disfavour, and there is much interest, in for example the aircraft industry, in finding a suitable replacement.

Zinc has also been chromated, but the preferred treatment for exterior galvanised structures is to allow them to weather for 12–15 years before painting. It is an approach with obvious economic attractions.

Etch or 'wash' primers were developed to pre-treat steel, aluminium or zinc substrates as part of the priming process. They typically contained phosphoric acid and a reactive pigment, zinc tetroxychromate. Unfortunately, the health implications of hexavalent chromium have once again limited their acceptability.

Plastics substrates are very diverse and present significant difficulties to the coater. Some, such as polyethylene and polypropylene are difficult to wet. It is necessary to oxidise their surfaces chemically, or by flame-treatment, if a coating is to have any chance of adhering. Others are often contaminated by mould release agents that have to be removed by detergent or solvent, whilst avoiding the danger of solvent crazing that affects some plastics. Yet others may be significantly soluble in the paint solvents. Many are very flexible and require coatings of excellent flexibility, such as elastomeric polyurethanes. The impact resistance of components such as car bumpers can be compromised by the application of coatings. Not only is solvent crazing a possibility, but also a brittle surface layer may support the growth of cracks that develop sufficient velocity to continue into the substrate. Stoving coatings, of course, have to be designed with due regard to the thermal susceptibilities of the plastic substrate.

The curiously inverted relationship of glass reinforced polyesters and their gelcoats was described in Chapter 7. The gelcoat is applied to the inside of the mould, and the structural resin and glass 'layed-up' behind it. It is one example of a plastic substrate and coating that adhere extremely well to each other.

Coatings themselves often present difficult surfaces for coating. Epoxies, for example, often have a narrow 'overcoating window' during which the next layer can be expected to adhere well. If aged beyond the determined interval, it may be necessary to prepare the surface by sanding. Crosslinked coatings that have not aged for long enough may swell under the influence of the solvents in the next coat and produce a severely wrinkled surface. (Thermoplastic systems have a strong advantage in that regard, *cf.* Chapter 6.) Finally, a degree of compatibility is required if the second coat is to wet the first satisfactorily. Where that is not possible, resort is sometimes had to elaborate schemes involving tie-coats.

12.3 THE APPLICATION OF COATINGS

The transfer of liquid paint from its container to the substrate in the required thin layer is a problem of physics and engineering, much of which is beyond the scope of this book, though the rheological implications were discussed in Chapter 3. A summary of application techniques that have found favour in various markets is given in Figure 12.1.

12.3.1 Brushes

Early man in his painting experiments must have used his fingers in a daubing action. Later a stick may have worked better, especially if the end was frayed. So may have come the familiar paintbrush that can hold substantial amounts of paint between its bristles without dripping. Pigs, on

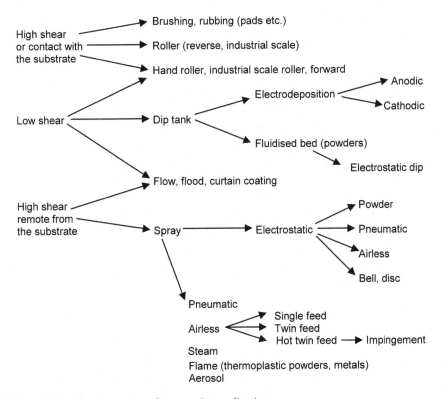

Figure 12.1 *Various approaches to paint application*

top of all their other gifts to mankind, traditionally provided the most suitable hairs for paintbrushes, as they had tapered shafts and were fibrillated at the tip. ('Split-ends' were desirable as they made for a smooth finish!) However, nylon and other synthetic materials can now be used to make equally effective bristles.

12.3.2 Rollers

Brushes are very versatile and effective but rather slow, and large, flat surfaces can be coated much more rapidly by hand-roller. Large rollers are also used for applying coatings in a continuous process, for example, to steel coil, and in various printing processes. Reverse rollers operate against the travel of the coil, so their action has more in common with a 'wiping' process.

Most of the 'daubing' techniques expose the paint to high shear at the point of application, so can benefit from the right kind of deviation from Newtonian viscosity. Forward rollers are an exception, *cf.* Chapter 3.

12.3.3 Dipping

The development of dipping processes followed the need to paint complex objects on a large scale. It initially seemed the ideal way to coat hidden areas of car-bodies, and other recessed areas, if holes were provided in enclosed sections to admit and drain the paint. However, it proved to be much less effective than expected, as the refluxing of solvents in such areas tended to wash the newly applied paint away.

Variants on the dipping process involve pumping liquid paint over suitable objects (flow or flood coating) or passing the workpieces through a sheet of falling paint (curtain coating). All are low shear processes. Powder coatings can similarly be applied by immersion of the heated substrates in a fluidised bed. The process is analogous to liquid dipping, but the initial heating is necessary to ensure that the powder particles do not fall off before being fully cured. Alternatively in electrostatic dipping, the particles are charged by corona points and the workpiece is passed through the cloud above the fluidised powder.

12.3.4 Electrodeposition

A better solution for the car industry came with the development of electrodeposition in the 1960s. The chemistry of electrodeposition was described in Chapter 8. In essence, a water-based coating is composed of an acid or amine functional polymer, 'peptised' with a low molecular weight amine or acid (anodic or cathodic systems, respectively). It is deposited on the workpiece under the influence of an applied potential of up to 200 V. The electrolysis of water produces acid or basic species in the vicinity of the workpiece to destabilise the aqueous paint.

Electrodeposition is highly amenable to production line applications, but requires a very large investment in plant and control. It would be normal for the paint supplier to sample the tank on a daily basis and prescribe adjustments to solids and acid or base content. Once established, the bath is fed with paint deficient in the neutralising counter-ion to offset its tendency to accumulate as polymer is removed. Acid or base is also removed by ultrafiltration.

The 'throwing power' of electropaint (*cf.* Chapter 8) means that all hidden areas can receive sufficient coating to resist corrosion, particularly if the coating requirements are 'factored in' to the design of the substrate. Car doors, for example, are nowadays designed with large holes in their inner skins to permit efficient electrocoating. By way of a reward, the corrosion-free lifetimes of cars have been substantially increased since the 1960s, first by anodic and then by cathodic electropaint.

12.3.5 Autophoresis

In a related process that has gained far less acceptance, both hydrofluoric acid and an oxidant are present in a bath containing a pigmented latex. The iron in the substrate is dissolved to produce ferrous ions, the oxidant preventing too much hydrogen evolution in the process. The ferrous ions cause coagulation of the latex so that it precipitates on the workpiece. When all areas have been coated, the reaction ceases and the coating can be taken out of the bath and stoved.

High performance coatings free of any throwing power restrictions can be achieved, though they contain large amounts of iron and are dark coloured. However, the use of hydrofluoric acid is a strong disincentive to their widespread adoption.

12.3.6 Pneumatic Spray

The earliest spray technique used a jet of compressed air at 2–3 atm to 'atomise' the paint. Small particles settled on the workpiece and coalesced to form very smooth, uniform films (when applied by an expert).

Pneumatic sprays are used, for example, in car-refinish markets and by the rather more adventurous DIY painter, being simple to use and robust. Their rate of coverage is rather limited however (0.25–0.5 L min^{-1}) and they are most effective with paints of low viscosity (0.05–0.1 Pa s).

12.3.7 Airless Spray

The efficient painting of large substrates such as ships, bridges and, to some extent, buildings requires a higher rate of delivery of paint than pneumatic spray permits. Airless spraying answers the need, being able to deliver paint of 0.5 Pa s at a rate of 2 L min^{-1}. Paint is compressed at up to 350 atm (using compressed air to drive a pressure 'intensifier') and forced through narrow nozzles, where expansion provides the atomisation and propulsion towards the substrate.

In large-scale operations the paint is drawn from barrels through hoses, often over a considerable distance. To facilitate the application and pumping, the paint can be heated, often as high as 75 °C, and the lines are 'traced' with electrical heating tape.

Where two-pack paints are employed, the components can be mixed prior to spraying. Such mixtures usually have a finite lifetime (the 'pot life') before they become too viscous to handle and would eventually gel with disastrous implications for the equipment. It is therefore essential to adopt a procedure that ensures use of all the mixed material in a predetermined time.

An alternative solution is plural component spraying, where the two components are pumped separately in the correct proportions to a mixing chamber close to the gun. Very fast reacting materials, those with a pot-life of a few seconds, cannot be handled with this arrangement as mixing even a few metres away from the gun could result in gelation in the line. They are mixed either in a specially designed chamber in the gun itself ('impingement mixing') or just after the components leave the spray tip. The latter technique is typically used to inject catalysts or initiators into the spray fan and is only useful in systems where incomplete mixing is not too disadvantageous. A modification of the approach allows finely chopped fibres of glass or other materials to be directed into the fan to produce reinforced coatings.

12.3.8 Electrostatic Spraying

Both pneumatic and airless spray application always suffer from 'overspray'. A proportion of the atomised paint particles miss the substrate, especially in windy conditions. Overspray not only represents a waste of material, but also can be a considerable nuisance since it is sometimes carried for miles. Gratuitous repainting is not always welcomed by the owners of cars parked in the vicinity.

Electrostatic painting was introduced in the 1940s to minimise the problem of overspray. In essence, the workpieces are earthed and attract the charged paint particles, ensuring much more efficient paint transfer. It is even possible to achieve a degree of 'wrap-around' where surfaces facing away from the sprayer can be coated.

Some inductive charging of sprayed particles is inevitable as they pass through the earth's natural electrostatic field, but is ensured by establishing a potential of up to 100 kV at the gun tip. Alternatively a corona discharge is achieved by applying similar voltages to a point at the end of the gun.

Electrostatic conditions can be used with pneumatic or airless spray, but are even more important in the case of powder coating where a charge is required to enable the particles to adhere to the substrate until fused in the oven. Powder particles are charged by corona discharge or by tribo-charging in their passage through a gun lined, for example, with polytetrafluoro-ethylene. The composition of the powder particle surface is critical to the efficiency of tribo-charging and additives are often incorporated. Hindered amines can be used in the original formulation, or alumina can be post-blended into the powder.

In large-scale painting lines, such as those in car plants, the atomisation of liquid paints has, since the 1950s, often been achieved using centrifugal forces rather than air. The paint is fed to a 'bell' or disc up to 600 mm in diameter, spinning at speeds up to 60,000 rpm. Overspray can be collected and recycled, though an air jet is sometimes employed to direct the spray towards the car bodies. An electrostatic potential is invariably used and rapid coverage is achieved. A similar arrangement can be used for spraying powders under production line conditions.

12.4 MARKETS FOR COATINGS

Some examples of different coatings businesses are discussed below. The list is confined to a selection of conventional paint markets since the intention is only to provide a flavour of the commercial constraints acting on paint chemistry. No consideration is given to the extensive commerce which exists in coatings for several types of vehicle, wooden and steel furniture, cloth, paper, leather, electronic circuit boards, roads and airstrips, fibre optics, photographic 'film', the human body, and so on. Indeed, printing inks and cosmetics are just two examples of huge industries, parallel to the coatings industry in terms of much of their technology. Printing inks have been based on alkyds and oleo-resinous media, with growing contributions from water-based and radiation cured systems, whilst a (small) proportion of the world's 5×10^6 or so square metres of human fingernail is coated with nitrocellulose lacquer. However, each has its own unique ethos that precludes informed comment in a book on coatings.

The most marked distinction between the various coatings markets is made by the possibility of using stoving processes or otherwise. Small metal objects (as large as car bodies) which are produced continuously on a production line lend themselves to rapid curing at elevated temperatures.

The coatings so produced are superior to those that have hardened at ambient temperature and indeed only such a rapid process is compatible with automated manufacture.

Many objects, on the other hand, are too large or too sensitive for stoving, and coatings drying at room temperature are applied. Ships and lips might serve as examples. Inspection of the following cases may make the distinctions clearer.

12.4.1 Buildings

Buildings of one sort or another account for about half the coatings materials sold worldwide. They include concrete, plaster, brick, wood, steel, galvanised steel and plastics substrates, some of which are exposed to the weather, and some of which are protected.

Concrete and plaster present porous, alkaline substrates. Consideration has to be given to the movement of water and salts and the tendency of sensitive organic groups to be hydrolysed. Polyester paints, for example, are not used on alkaline substrates in moist locations. Wood is also porous and inclined to expand and contract with variations in humidity. Brittle paints, or those that embrittle with time such as the traditional alkyds, are likely to have a rather short lifetime on exterior woodwork unless particular care has been taken in the selection and pretreatment of the timber. More flexible alkyds have been developed over recent years and show superior performance, but a more effective treatment may be the application of a biocidal preservative without real film-forming character.

Alkyd-based paints have traditionally been used on most parts of buildings, but they are increasingly being replaced by latex paints that dry by loss of water (the 'emulsion' paints familiar for many years on interior walls and ceilings). The coatings are applied by brushing or roller except on large industrial buildings where airless spray allows rapid coverage.

12.4.2 Steel Structures

Large land-based steel structures such as bridges, pipelines, tanks and other chemical plant are most frequently coated with epoxies pigmented with micaceous iron oxide, for example, to give maximum corrosion resistance. They are in some cases overcoated with alkyds or two pack urethanes to provide enhanced weatherability and the required colour. (A tank may be painted white to protect its contents from undue solar heating.)

Again, water-based systems are making significant inroads, subject to difficulties in getting them to form satisfactory films under humid conditions. Bare steel, before fabrication, is often thinly coated with etch primers or zinc silicate primers and the emission of harmful fumes during welding operations is a significant consideration.

Steel structures for the most part require of their coatings, good resistance to weather and a few bold colours. They are not much subject to mechanical abuse such as friction. When galvanised, as in the case of electricity pylons, they are allowed to weather for many years before painting. Large amounts of coating may be consumed in one project and the emphasis is very much on cost effectiveness. The coatings are usually applied by airless spray.

12.4.3   Ships

Ships and static marine structures, such as oil drilling platforms, have the same general requirements as their land-based counterparts, though the potential for vastly increased corrosion by seawater is a significant factor. Some parts are also subject to severe abrasion due to anchor cables, contact with jetties, contact with ice in polar waters, and the effect of air-borne ice particles. The interiors of cargo holds likewise experience heavy abrasion from cargoes such as coal, iron ore and pig iron. Ballast tanks have to contend with aggressive mixtures of fuel oil and seawater, alternating with periods of high humidity and occasionally temperatures as high as 100 °C. Cargo tanks must be able to withstand an unpredictable succession of fluids ranging from chemicals to foodstuffs, often held at elevated temperatures to facilitate unloading.

Tank interiors are coated with densely crosslinked epoxies. Other surfaces are coated with materials often superficially similar to those used on land-based structures, but formulated to withstand the special conditions they encounter. They are applied by airless spraying, often at considerable thickness.

12.4.4   Yachts

Yacht coating offers a striking contrast to that of large ships since yachts are objects of luxury, often painstakingly coated by their owners. Economy is given a lower priority than appearance and weather resistance. Timber surfaces are often varnished to retain and enhance the beauty of their grain. Glass-fibre composite hulls are provided with a coloured 'gel-coat' in manufacture, but are often painted subsequently to change their colour or improve their gloss. Alkyds of high quality and two-pack polyurethanes are most often used and are applied by brush or conventional (pneumatic) spray.

12.4.5   Aircraft

Aircraft exterior coatings are amongst the most severely tested in service. Flight at high altitude subjects them to intense ultraviolet irradiation, salty air is liable to initiate corrosion which may be promoted by mechanical stress, rapid temperature changes may cause cracking, and (as if that was not enough) leaks of fuel and phosphate ester hydraulic fluid have to be reckoned with.

The former practice was to employ an epoxy primer, heavily pigmented with a leachable chromate to passivate any metal exposed by minor damage. However, chromates are, as noted above, no longer acceptable and a completely satisfactory replacement has yet to be found. Water-borne epoxies are increasingly popular.

The topcoat is an isocyanate cured polyester for maximum weatherability and chemical resistance. A stoichiometric excess of isocyanate is often used to introduce an element of moisture cure. The resulting urea linkages bestow increased chemical resistance on the composition. Fluoropolymers have recently been introduced to aircraft topcoats to provide further enhancements in performance.

It is necessary to strip the coatings from aircraft at intervals to inspect the airframe. It must be done very rapidly to minimise time out of service, and with agents which do not attack the metal. Efforts have been made to develop coatings which, paradoxically, possess all the above attributes whilst being easily strippable.

12.4.6 Motor Cars

Car plants are nearly always very large-scale operations – the archetypal production lines. Bodies are manufactured continuously and undergo coating processes before being built up into the finished vehicles. Car owners demand a very high standard of appearance and resistance to corrosion, and most manufacturers use cathodic electropaint as a primer. It is covered by a relatively thick surfacer coat to hide imperfections in the metal, and to provide a smooth substrate. The topcoat may be a melamine cured polyester or acrylic, or less frequently a thermoplastic acrylic, colour matched to the current fashion with the greatest care.

Metallic finishes found their first use in the automotive industry where they contributed materially to body design. They have given rise to the widespread adoption of 'clear over base' systems where a thin pigmented thermoplastic layer is overcoated with a clear, often in a 'wet on wet' process.

Automotive topcoats are electrostatically sprayed, often by means of the 'high speed bell' described above. They are stoved at about 130 °C. The earlier coats are designed to be stoved at somewhat higher temperatures so that they are not too severely affected by later heat treatment. They are sometimes also under-stoved so that the later application of heat completes the process.

12.4.7   Car Refinish

Car refinishers operate much smaller concerns than original equipment coaters. In some cases they are one-man operations using hand spray guns, and buying just enough material to coat one car at a time, though the trend over recent years has been towards larger operations. The supplier's ability to match any of the enormous number of colours on the road is of overriding importance, and it is necessary to be able to obtain high gloss under conditions that may be less than ideal. It is impossible to stove the car, which is now equipped with all sorts of heat-sensitive items, from rubber tyres to the plastic sunglasses in the glovebox. However, it is often possible to provide a 'low-bake' or 'forced cure' below about 100 °C.

Traditional refinish paints were based on nitro-cellulose and dried by evaporation of solvent. Being rather soft they lent themselves to polishing to remove defects. More recently though, two-pack systems based on hydroxyacrylics cured with isocyanates have gained popularity since they can achieve a high gloss 'straight from the gun'.

12.4.8   Food and Beverage Cans

Tinplate cans for food or drink need to be lined with a continuous film of highly resistant material if they are not to be attacked by their contents with resultant metallic taints or even leaks. It is also necessary for the coatings to withstand considerable deformation without failure, to be free of any extractable components that might themselves taint the contents, and to be recognised as safe for food contact. Similar constraints even extend to the inks used on can exteriors.

Epoxy compositions such as epoxy-phenolics are used in most cases, and stoved at temperatures in the 250–300 °C range for a few seconds, consistent with the very high production rates of modern can lines. As in other markets, water-based can lacquers are rapidly gaining in popularity, and are typically based on high molecular weight epoxies modified with acrylics.

Traditional three-piece cans were made from tinned steel strip, roller coated in the flat state then formed into cylinders with soldered seams before tops and bottoms were fitted. More recently, the need for economy of materials (and the high performance of coatings) has led to the development of two-piece cans with bodies extruded from single tinplate or aluminium coupons. The walls of such cans are much thinner than in three-piece cans. Whilst aluminium cans do not require internal coating, steel cans have much thinner tin layers than the three-piece equivalents, and they are probably discontinuous, so still greater reliance is placed on the organic lining. The need to coat the inside of a can body uniformly and with an extremely low incidence of 'holidays' (holes) in a fraction of a second, places enormous demands on application technology. It is normally achieved by an automated spray arrangement, though other systems such as electrocoating have received active consideration.

12.4.9 Coil

Coil coating is the application of polymer film to steel (or aluminium) strip prior to fabrication. Application is by roller, and the strip is fed at high speeds through high temperature ovens. The process lends itself to automation and is usually carried out by the manufacturers of the metal stock.

The coating must come unscathed through the deformation involved in subsequent fabrication, and arrangements must be made to minimise the effects of corrosion at cut edges. Coated coil is ideally suited to the manufacture of rather flat objects such as domestic appliance panels, garage doors and cladding for buildings. For exterior use, extreme ductility has to be reconciled with high weather resistance.

Poly(vinyl chloride) plastisols enjoy wide popularity (*cf.* Chapter 7), but for highest endurance under sunny conditions, fluoropolymers are unrivalled and are gaining ground.

12.5   CONCLUSION

Our brief glance at the multitude of different applications for coatings and the various ingenious methods for applying them concludes this review of coatings chemistry and physics.

We have seen something of the benefits that coatings can bestow, the legislative lengths that are gone to ensure that they do no harm, and the enormous effort expended to enable them to deliver the service required.

The coatings industry continues to be a vital exemplar of the benefits to be had when scientific investigation and technological creativity are harnessed to a strong commercial vision.

Glossary

Abhesion	The condition of having minimal adhesion
Acid rain	Acidification of rainfalls caused by certain pollutants, such as sulfur dioxide, which may damage vegetation and the aquatic life in fresh water
Accelerated weathering	A test, often using a proprietary "QUV-A" or other Weatherometer whereby the effect of UV radiation and moisture on a coating can be assessed.
Adhesion	A measure of the strength of interaction between two bodies in intimate contact.
Adipic acid	Hexanedioic acid
Alkyd	A coating polyester bearing pendent fatty acid residues
Alkylidenemalonate	1,1-dicarboalkoxyalkene
Amine Bloom	Deposits of amine bicarbonate/carbamate that appear on certain amine-epoxy coatings on exposure to the atmosphere
Anthropogenic releases	Emission of pollutants resulting from human activities, as distinct from natural emissions, *e.g.* from trees
Anti-thixotropic behaviour	A slow reduction over time, of the consistency imparted to a sample by shearing
Apparent viscosity	The shear stress divided by the shear rate. It is not a constant coefficient
Arrhenius' Law	An equation describing the temperature dependence of the rate constant for elementary chemical reactions
Associative Thickener	A polymeric material used for thickening waterborne coatings. It creates microstructural domains by associating with itself and colloidal material.
Autooxidation (auto-oxidation, autooxidation)	A process in which organic materials react with atmospheric oxygen to produce free radicals. Consequent cure or degradation (or both) result.

360

Azeotrope	A liquid mixture of two or more substances where the vapour has the same composition as the liquid
Aziridine	Ethyleneimine
BAT	Best Available Technique
β-Transition	A secondary motion in amorphous polymer materials derived from rotational degrees of freedom in the backbone or in side-chains. It is sometimes invoked to explain the unexpectedly high toughness of some polymer systems.
Benzophenone	Diphenylmethanone
BET	Brunauer, Emmett, and Teller. An equation, and a method for determining the surface area of a material.
Bisphenol A	2,2(*bis*-4-hydroxyphenyl) propane
Bisphenol F	*bis*-(4-hydroxyphenyl) methane
Coefficient of Viscosity	The shear stress divided by the shear rate. A constant, defining a Newtonian fluid.
Caprolactam, (ε-caprolactam)	Hexahydro-2H-azepin-2-one
Cardura E10®	Glycidyl "versatate"®, glycidyl neodecanoate
Cathodic Protection	A system wherein a steel structure is protected against corrosion by the application of an electrical potential and current. The potential required to prevent corrosion is -850 mv (measured against a silver/silver chloride electrode) and is generated by the use of a sacrificial anode such as zinc, or by the application of an external potential.
CEPE	European Council of Paint, Printing Ink and Artists' Colours Industry
CFCs	Chlorofluorocarbons or chlorinated/fluorinated hydrocarbons
Chain-growth	A polymerisation process which proceeds through an initiation step to produce an active centre, followed by fast propagation reactions to build up of the polymer chain length from the very beginning of the reaction, and finally termination of the active centre. Molecular

weight may be controlled by the concentration of initiating species or by the addition of chain terminating agents.

Chalking — The production of a friable layer at the surface of a paint film on exterior exposure caused by degradation of the binder.

Chlorendic anhydride — 1,4,5,6,7,7-Hexachloro-5-norbornene-2,3-dicarboxylic anhydride

Chlor-rubber (chloro-rubber) — Chlorinated rubber usually containing approximately 65% by weight of chlorine. Used as a film-forming resin in solvent-borne paints

Coalescence — The fusion on a substrate, of droplets of sprayed paint, or polymer particles in a water-borne paint, to produce a continuous film.

Colour drift — The change in colour of a paint film over time, with or without external exposure.

Consistency — A general term for the property of a material by which it resists permanent change of shape

Continuous phase — The medium or matrix in which a discrete second phase is enclosed.

Controlled radical polymerisation — A radical polymerisation with features resembling those of living polymerisation

Conversion — The extent of reaction of a functional group expressed as the ratio of the amount reacted to the original amount.

Convertible coatings — Coatings that harden by chemical reaction

Craters, cratering — The formation of small bowl-shaped depressions in a paint film. Associated with air entrapment during spray application.

Creep compliance — A method of measuring viscoelasticity.

Crosslinker — A multifunctional molecule, usually but not necessarily of low molecular weight, primarily designed to harden a second material or film-former by reaction

Crosslinking — The process by which a low molecular weight functional polymer is converted to an infusible, insoluble, three dimensional, polymer network by the chemical reaction of multifunctional species

Cyclodextrins — Cyclic oligosaccharides of D-glucose having relatively rigid cone shaped doughnut structures

Degree of polymerisation	The number of repeat units in a polymer molecule
Diacetoneacrylamide	1,1-dimethyl-3-ketobutyl propenoamide
Dicyandiamide	Cyanoguanidine
Dilatency	An increase in volume caused by shear. (Often confused with shear thickening.)
Dimethylolpropionic acid	2,2-*bis*-hydroxymethylpropionic acid
Dipentaerythritol	2,2'-[Oxy-*bis*(methylene)]*bis*[2-(hydroxymethyl)-1,3-propandiol]
Disperse phase	A separate phase, such as a latex or a pigment, dispersed in a continuous medium
Dodecyl-mercaptan	Dodecanethiol
Dry hiding	The use of voids or pores in particles to scatter light thereby producing white paint films.
Dryer	A siccative, a metal salt or soap or other compound that promotes the drying of auto-oxidisable systems
Drying oil	A triglyceride whose component fatty acids can be auto-oxidised to yield a dry, rubbery material
Elastically effective network junctions	Points in a polymer network at which more than two chains emanating from similar points in the infinite network converge, and are bonded. The chains connecting these junctions are load bearing and contribute to the elastic modulus of the network under conditions where $T > T_g$
Elasticity	The full or partial recovery of a deformation created by an imposed stress on its release.
Elastomer	A lightly crosslinked polymer in the rubbery state
Electrical double layer	A diffuse aggregation of positive and negative electric charges surrounding a suspended colloidal particle.
Enamel	A glossy paint that dries by chemical reaction. (Vitreous enamel is an exception.)
End-functional	A polymer or oligomer whose reactive functions tend to be at its extremities
Entanglement molecular weight	The weight average molecular weight above which a polymer molecule can become physically entangled with its neighbours.

	The viscosity of the polymer increases much more sharply with increases in molecular weight above this value
Etch primer	A chromate/phosphate conversion coating applied to metal surfaces. Also called 'wash primers'.
Ethyl acrylate	Ethyl propenoate
Ethylene	Ethene
Ethylene oxide	Oxirane
Eutrophication	An increase of nitrogen content in the soil due mainly to ammonia emissions from agricultural sources
Film formation	The process by which an applied coating is transformed from a liquid to a solid film
Flooding and Floating	Preferential migration of pigments in mixed pigment coatings leading to colour differences on the surface and a mottled appearance.
Fluidity	The inverse of viscosity
Forced oscillation	(Strain oscillation, stress oscillation.) A method of measuring viscoelasticity.
Formaldehyde	Methanal
Fracture	The point at which an increasing strain in an object causes catastrophic breaking as the material can no longer support the load.
Fracture energy	The energy required to generate unit area of fracture surface when a crack propagates through a material
Fracture toughness	A measure of a material's resistance to crack propagation
Free radical	An atom or molecule that contains one or more unpaired electrons
Free volume	The 'empty' space in an amorphous polymer. The difference between the total volume of the sample and the volume occupied by the polymer segments
Functionality	The number of functional groups on a molecule
Galvanised steel	Steel coated with zinc to provide sacrificial corrosion protection
Gelation	A singular event during all network formation processes, in which once a (predictable) chemical

conversion has been achieved, a single molecule spans the whole reaction vessel and the sample weight average molecular weight (and viscosity) tends to an infinite value

Glass transition temperature, T_g	The temperature (actually a range) over which the backbone segments of an amorphous polymer achieve sufficient mobility to transform from a vitrified solid (glass) to a liquid (uncrosslinked polymers), or to a rubber (crosslinked polymers). Sometimes known as the α-transition (see β-transition)
Glycidyl methacrylate	2,3-epoxypropyl 2-methylpropenoate
Glyoxal	Ethanedial
Greenhouse effect	Global warming caused by certain atmospheric gases, such as carbon dioxide
Hooke's Law	The stress is proportional to the strain, and is reversible.
IPPC	EU Directive on Integrated Pollution Prevention and Control
Imidazolidine	Dihydroimidazoline; 1,3-diazacyclopentane
Isophthalic acid	Benzene-1,3-dicarboxylic acid
Immersion testing	A test involving the immersion of a painted panel in a liquid such as sea-water or a chemical such as methanol and assessing the resistance of the coating over a period of time.
In-can settlement	The gravity driven settlement of pigment in a tin of paint. The sediment can be quite soft or extremely hard, depending on the paint.
Internal stress	Stresses generated in a coating, most typically tensile stresses as a result of constrained shrinkage during film formation in the glassy state, after vitrification has occurred.
Isophoronediamine	1-amino-3-aminomethyl-3,5,5-trimethylcyclohexane
Isoprene	2-methyl-1,3-butadiene
Itaconic acid	Prop-1-ene-2,3-dicarboxylic acid
Junction functionality	The number of chains extending from a cross-link site in a polymer network, which are bonded to similar sites elsewhere in the network

Krafft point	The point where a transition occurs between the dissolution of a surfactant to form ions and its dissolution to form micelles
Lacquer	A coating that dries only by loss of solvent. (Chinese lacquer is an exception.)
Lacquer drying	Hardening by purely physical processes such as solvent loss
Ladder experiments	A series of experiments where one variable at a time is increased, whilst keeping the rest constant.
Latent cure	A curing reaction delayed by an additional agency so that, for example, a highly reactive system can display an extended pot-life
Le Chatelier's Principle	A postulate that in any system at equilibrium, which is subsequently perturbed, the system will react so as to minimise the effect of the perturbation.
Ligand	A molecule or anion bonded to the central metal atom in a coordination compound
Loss modulus	The rheological response to a viscous process, a fluid property.
Living polymerisation	A chain growth polymerisation in which there are no transfer or termination reactions
Linear viscoelasticity	A viscoelastic response in which the stresses and strains are linearly related, *i.e.* proportional. This usually requires small deformations.
Maleic anhydride	Butenedioic anhydride
Melamine	2,4,6-*tris*-amino-1,3,5-triazine
Methacrylate	2-methylpropenoate
Methylethyl ketone	Butan-2-one
Michlers ketone	*Bis*[4-(dimethylamino)phenyl]methanone
Microemulsion	A transparent thermodynamically stable emulsion in which the interfacial energy approaches zero
Micro-structure	Small scale structure in fluids created by colloidal materials present.
Mix ratio	In a two component paint, the volume of each component required to achieve the desired stoichiometry for the reactive groups.
Molecular weight between crosslinks	The total molecular mass that connects two multifunctional junction points in a polymer

network. The molecular weight between cross-links is a very important parameter in controlling the mechanical properties of polymer networks.

Naphthenic acid A mixture of various carboxylic acids derived from petroleum and containing cyclopentane residues. Their general formula is $C_nH_{(2n-2)}O_2$, where $n = 8 - 12$.

National emission ceilings Levels of pollutant emissions not to be exceeded by Member States of the EU at a fixed deadline (currently 2010)

n-butyl acrylate Butyl propenoate

Newton's Law Viscosity is invariant with either shear rate or shear stress

Newtonian fluid A fluid for which the shear stress is proportional to the shear rate.

Non-Newtonian fluid A fluid for which the proportionality between shear stress and shear rate is not constant with shear rate

Novolac A phenol (form)aldehyde condensate terminated in hydroxyphenyl groups

Number average molecular weight The total mass of a polymer sample divided by the number of molecules in the sample. It is also given as the sum of the products of the mole fraction of each individual component and its individual molecular weight

Oleic acid: *Cis*-octadec-9-enoic acid

Oleoresinous media Coating materials made by heating triglyceride oils with natural resins such as rosin (NB, in botanical terminology, "oleoresin" indicates a mixture of solid and liquid terpenes)

Orange Peel A surface defect comprising regular undulations on a surface coating resembling the surface of an orange.

o-phthalic acid Benzene-1,2-dicarboxylic acid (usually available as phthalic anhydride)

Ostwald ripening The growth of large droplets in a dispersion at the expense of the smaller ones

Overspray Atomised paint produced during spraying that does not reach the substrate to be painted.

Oxazolidine	Dihydrooxazoline; 1-oxa-3-azacyclopentane
Pascal (Pa)	The unit of stress
p-chlorobenzotrifluoride	4-chloro-α,α,α-trifluorotoluene
Pendent-functional	A polymer or oligomer having reactive functions spaced along, and extending from, its backbone
Pentaerythritol	2,2-*bis*(hydroxymethyl)-1,3-propanediol
Peptisation	Neutralisation of groups on a polymer to induce water-dispersibility
Permeability studies	Experiments designed to measure the rate of transport of oxygen, water vapour or ions through a paint film.
Photo-oxidation	The degradation of a polymer sample due to hydrogen abstraction reactions with peroxy radicals which are generated photochemically by reaction of adventitious radicals with oxygen
Pinholes	Tiny deep holes in a coating, often exposing the substrate. Associated with poor quality atomisation during the spraying process.
Plastic	A material which flows when a yield stress is exceeded
Plasticisation	The depression of polymer T_g by means of external additives or by chemically attaching flexibilising groups to the polymer
POG	Product Orientated Group. A subgroup formed under the CEPE organization
Polyacrylate	A polymer made from acrylate monomers having pendent carboalkoxy or related groups
Polycondensation	A polymer-forming process in which a low molecular weight species, such as water, is eliminated to form linkages
Polydispersity (molecular weight)	The ratio of weight average to number average molecular weight in a polymer
Polystyrene	Poly(ethenylbenzene)
Polytelechelic	A radiate or star molecule with functional groups on the ends of its arms
Powder coating	A coating applied as a dry powder which is fused to a coherent film by heating
Product stewardship	An initiative agreed by Chemical Manufacturers and Coatings Associations throughout the

world. It requires an organisation to take care and responsibility for reducing the adverse health and environmental effects of products throughout every step of their life cycle.

Propylene oxide — Methyloxirane

Pseudo-plastic — Shear thinning. Often used in a context where the shear stress is linear with shear rate at high shear rates, but no yield stress can be detected

PVB — Polyvinylbutyral. A vinyl resin made by the condensation of polyvinyl alcohol with butyraldehyde

QUV-A Weatherometer — A proprietary testing cabinet in which a painted panel is exposed to UV-A radiation and (usually) water spray.

Radiation cured coating — A coating cured with UV light, electron beam, or other radiation. Often solvent-free or containing a low amount of VOC

REACH — 'Registration, Evaluation and Authorisation of Chemicals'. New EU legislation (under preparation) intended to control use of chemicals in products and articles.

Reactive diluent — An additive which behaves like a solvent in reducing viscosity, but combines with coating polymers as they cure

Relative humidity — The actual amount of water vapour in the air compared with the amount in the same volume of air if it was saturated with water vapour, expressed as a percentage.

Relaxation spectrum — A rheological response as a function of frequency from forced oscillation. Derived from the storage and loss moduli of a material.

Relaxation — Structural reorganisation in an amorphous polymer in the non-equilibrium glassy state

Resin — A material used as the binder in a coating. Often a polymer or oligomer, but non polymeric materials (such as rosin) also qualify. Plant resins (such as rosin) are properly terpene materials as distinct from gums (water soluble polysaccharide materials).

Resole	A phenol (form)aldehyde condensate with hydroxymethyl or alkoxymethyl termini
Rheology	The science of material flow
Rheopexy	A time dependent increase in a rheological property, leading to gelation, created by the gentle oscillation motion applied to a sample.
Rosin	Colophony, a natural resin derived from pine trees
Rubber elasticity	An entropy based theory describing the equilibrium mechanical properties of lightly crosslinked flexible polymer chains. The theory can be applied to more densely crosslinked systems in the rubbery state.
Sagging	A gravity driven process causing paint to flow down a vertical surface.
Salt spray resistance	A test where a painted panel is exposed to an atmosphere of saturated salt spray. Used to check the anti-corrosive properties of paint.
Schiff's base	Araldimine
SED	'Solvent Emission Directive' 1999/13/EC, the European Directive on the limitation of emissions of volatile organic compounds due to the use of organic solvents in certain activities and installations
Semi-drying oil	A triglyceride with auto-oxidisable character, but insufficient functionality to form a network
Service life	The time over which a particular component or object is expected to carry out its function before properties degrade to a level at which it can no longer perform adequately.
Shear	The change of angle in a deformed body
Shear rate	The change of shear per unit time.
Shear stress	The force per unit area parallel to the area.
Shear thickening	The increase in viscosity with increasing shear rate.
Shear thinning	The decrease in viscosity with increasing shear rate.
Site average functionality	A much better measure of the functionality of a polymer or mixture than the average number of functional groups per molecule, the site average functionality controls the molecular weight

	build and gelation behaviour of a polymer in a step growth polymerisation/crosslinking reaction.
Solvent-borne coatings	Coatings, whose viscosity is adjusted by the use of organic solvent
Spreading rate	The area covered by a specified volume of paint at a particular film thickness *e.g.* 1 litre of paint will cover 5 m^2 at a wet film thickness of 200 microns.
Stearic acid	Octadecanoic acid
Step-growth	A polymerisation process which proceeds by a series of independent steps, and is characterised by a slow build up of polymer chain length at the beginning of the process and more rapid increase in chain length over the last few percent of the reaction. Molecular weight is easily controlled in copolymerisations by means of the stoichiometry of the reaction. In homopolymerisations, molecular weight is controlled by the addition of chain terminating species. Many important crosslinking reactions are of the step growth type.
Storage modulus	The rheological response to an elastic process, a solid property.
Strain	A dimensionless measure of the change in length of a material when a force is acting to distort it. The strain is the ratio of the change in length to the original length.
Styrene	Ethenylbenzene
Supercritical carbon dioxide	Highly compressed, liquified carbon dioxide. Supercritical fluids combine properties of both gases and liquids.
Surfactant	A surface active agent. Synonymous with emulsifier in this book.
Telechelic	A linear polymer, virtually all of whose molecules have functional groups on both ends, as distinct from other end-functional polymers
Terephthalic acid	Benzene-1,4-dicarboxylic acid
Thermoplastic	Polymers that have not undergone a chemical crosslinking process and can be softened by heating or by contact with suitable solvents

Thermosetting	Polymers that can harden by chemical reaction – not necessarily at elevated temperature
Thixotropy	A slow recovery of the consistency lost by shearing, when a sample is at rest
Throwing power	The ability of an electropaint to coat recessed or remote areas
Time dependency	The change of a rheological property with time on the application of a constant rate of deformation, a constant stress, or a gentle oscillation.
Tinplate	Thin steel sheet coated with tin
Trimellitic acid	1,2,4-benzenetricarboxylic acid
Troposphere	The level of the atmosphere close to the Earth
VeoVa 10®	Vinyl "versatate"®, vinyl neodecanoate
Vinyl acetate	Acetoxyethene
Vinyl chloride	Chloroethene
Vinyl fluoride	Fluoroethene
Vinyl *neo*-nonanoate, deconate, undeconate	Ethenyl esters of trisubstituted acetic acids, having 9, 10, 11 total carbons respectively. (Proprietary VeoVa 9, 10, 11)
Vinyl, vinyl resin	Generally, a group of polymers made up of vinyl chloride, vinyl acetate, vinyl acetal, vinyl alcohol and vinyl pyrrolidone.
Vinylidene fluoride	1,1-difluoroethene
Vinyltoluene	Methylstyrene, vinylmethylbenzene
Viscoelasticity	The property of a material exhibiting both viscous and elastic responses at the same time, to an imposed deformation.
Viscosity	The resistance of a substance to flow. Viscosity is given by the ratio of the stress generated in a sample by a shearing motion to the strain rate
Vitrification	The transformation of an amorphous polymer in the liquid or rubbery states to a glassy solid, by passing through the glass transition.
VOC	'Volatile Organic Compounds'. Any organic compound having a high volatility (exceeding a certain threshold expressed as vapor pressure or boiling point measured at a standard pressure) so that it is emitted to the atmosphere under the normal conditions of use

Volume solids	The ratio of the volume of the non-volatile material in a paint to its total wet volume, expressed as a percentage.
Water-borne coatings	Coatings, whose viscosity is adjusted by the use of water
White spirit	An essentially paraffinic hydrocarbon solvent with a boiling range of 150–190 °C
Weight average molecular weight (M_w)	The weight average molecular weight is a better indicator of where the bulk of the mass of the polymer resides than M_n. It is given by the sum of the products of the weight fractions of each individual component and its individual molecular weight
Yield	The point at which an increasing strain in an object causes an irreversible deformation in a material, sometimes known as plastic flow
Yield stress	The stress which must be exceeded for a non-recoverable (viscous) deformation to result.
Zwitterion	An ion that contains both positive and negative charges

Subject Index